Lecture Notes
in Business Information Processing   255

Series Editors

Wil van der Aalst
  Eindhoven Technical University, Eindhoven, The Netherlands
John Mylopoulos
  University of Trento, Povo, Italy
Michael Rosemann
  Queensland University of Technology, Brisbane, QLD, Australia
Michael J. Shaw
  University of Illinois, Urbana-Champaign, IL, USA
Clemens Szyperski
  Microsoft Research, Redmond, WA, USA

More information about this series at http://www.springer.com/series/7911

Witold Abramowicz · Rainer Alt
Bogdan Franczyk (Eds.)

Business
Information Systems

19th International Conference, BIS 2016
Leipzig, Germany, July, 6–8, 2016
Proceedings

 Springer

Editors
Witold Abramowicz
Department of Information Systems
Poznań University of Economics and
  Business
Poznan
Poland

Bogdan Franczyk
Information System Institute
University of Leipzig
Leipzig, Sachsen
Germany

Rainer Alt
Information System Institute
University of Leipzig
Leipzig, Sachsen
Germany

ISSN 1865-1348   ISSN 1865-1356 (electronic)
Lecture Notes in Business Information Processing
ISBN 978-3-319-39425-1   ISBN 978-3-319-39426-8 (eBook)
DOI 10.1007/978-3-319-39426-8

Library of Congress Control Number: 2016939365

Printed on acid-free paper

This Springer imprint is published by Springer Nature
The registered company is Springer International Publishing AG Switzerland

Preface

The 19th edition of the International Conference on Business Information Systems was held in Leipzig, Germany. Since its first edition in 1997, the BIS conference has become a well-respected event and has gathered a wide and active community of researchers and business practitioners.

The BIS conference follows trends in academic and business research. Since there is much interest in Big and Smart Data research, we decided to further discuss this topic. Thus the theme of the BIS 2016 conference was "Smart Business Ecosystems." This recognizes that no business is an island and competition is increasingly taking place between business networks and no longer between individual companies. Information technology has been an important enabler with systems for enterprise resource planning, business intelligence, or business process management. Driven by new Internet-based technologies, such as cloud, mobile, social, and ubiquitous computing as well as analytics and big data, so-called smart infrastructures are emerging. From the business side, they are conducive to new, more dynamic relationships between companies, suppliers, and customers, which are referred to as ecosystems. Resources are crowd-sourced as services via platforms that yield new business knowledge and business processes. A variety of aspects are relevant for designing and understanding smart business ecosystems. They span from new business models, value chains, and processes to all aspects of analytical, social, and enterprise applications and platforms as well as cyber-physical infrastructures.

The first part of the BIS 2016 proceedings is dedicated to ecosystems research. It is closely related to the next two chapters: "Smart Infrastructure" and "Big/Smart Data." However, during the conference other research directions were also discussed, including process management, business and enterprise modeling, service science, social media as well as applications of the newest research trends in various domains.

The Program Committee consisted of 110 members who carefully evaluated all the submitted papers. Based on their extensive reviews, a set of 34 papers were selected.

We would like to thank everyone who has helped build an active community around BIS. First of all, the reviewers for their time, effort, and insightful comments. We wish to thank all the keynote speakers, who delivered enlightening and interesting speeches. Last but not least, we would like to thank all the authors who submitted their papers.

July 2016

Witold Abramowicz
Rainer Alt
Bogdan Franczyk

Organization

Conference Organization

BIS 2016 was co-organized by Poznań University of Economics and Business, Department of Information Systems, and Leipzig University, Information Systems Institute.

Program Committee

General Co-chairs

Witold Abramowicz	Poznań University of Economics and Business, Poland
Rainer Alt	Leipzig University, Germany
Bogdan Franczyk	Leipzig University, Germany

PC Members

Frederik Ahlemann	University of Duisburg-Essen, Germany
Stephan Aier	University of St. Gallen, Switzerland
Antonia Albani	University of St. Gallen, Switzerland
Dimitris Apostolou	University of Piraeus, Greece
Timothy Arndt	Cleveland State University, USA
Maurizio Atzori	University of Cagliari, Italy
David Aveiro	University of Madeira, Portugal
Eduard Babkin	Higher School of Economics, Russia
Alexander Benlian	Darmstadt University of Technology, Germany
Morad Benyoucef	University of Ottawa, Canada
Maria Bielikova	Slovak University of Technology, Slovakia
Tiziana Catarci	Sapienza University of Rome, Italy
Wojciech Cellary	Poznań University of Economics and Business, Poland
Francois Charoy	Université de Lorraine, France
Tony Clark	Middlesex University, UK
Rafael Corchuelo	University of Seville, Spain
Andrea de Lucia	University of Salerno, Italy
Zhihong Deng	Peking University, China
Barbara Dinter	Chemnitz University of Technology, Germany
Josep Domingo-Ferrer	Universitat Rovira i Virgili, Spain
Suzanne Embury	University of Manchester, UK
Vadim Ermolayev	Zaporozhye National University, Ukraine
Werner Esswein	Dresden University of Technology, Germany

Gianvito Pio	Università degli Studi di Bari Aldo Moro, Italy
Geert Poels	Ghent University, Belgium
Jaroslav Pokorný	Charles University, Czech Republic
Birgit Pröll	Johannes Kepler Universität Linz, Austria
Elke Pulvermueller	University of Osnabrück, Germany
Ulrich Remus	University of Innsbruck, Austria
Fenghui Ren	University of Wollongong, Australia
Stefanie Rinderle-Ma	University of Vienna, Austria
Antonio Rito Silva	Instituto Superior Técnico, Portugal
Stefan Sackmann	Martin Luther University Halle-Wittenberg, Germany
Virgilijus Sakalauskas	Vilnius University, Lithuania
Sherif Sakr	University of New South Wales, Australia
Demetrios Sampson	University of Piraeus, Greece
Adamo Santana	Federal University of Pará, Brazil
Jürgen Sauer	University of Oldenburg, Germany
Stefan Schulte	Vienna University of Technology, Austria
Ulf Seigerroth	Jönköping University, Sweden
Elmar J. Sinz	University of Bamberg, Germany
Alexander Smirnov	Russian Academy of Sciences, Russia
Stefan Smolnik	University of Hagen, Germany
Henk Sol	University of Groningen, The Netherlands
Andreas Speck	Kiel University, Germany
Athena Stassopoulou	University of Nicosia, Cyprus
Darijus Strasunskas	Norwegian University of Science and Technology, Norway
York Sure-Vetter	Karlsruhe Institute of Technology, Germany
Jerzy Surma	Warsaw School of Economics, Poland
Bernhard Thalheim	Kiel University, Germany
Robert Tolksdorf	Free University Berlin, Germany
Genny Tortora	University of Salerno, Italy
Nils Urbach	University of Bayreuth, Germany
Herve Verjus	University of Savoie, France
Herna Viktor	University of Ottawa, Canada
Mathias Weske	Hasso Plattner Institute for IT Systems Engineering, Germany
Krzysztof Węcel	Poznań University of Economics and Business, Poland
Anna Wingkvist	Linnaeus University, Sweden
Axel Winkelmann	University of Würzburg, Germany
Guido Wirtz	University of Bamberg, Germany
Qi Yu	Rochester Institute of Technology, USA
Slawomir Zadrozny	Polish Academy of Sciences, Poland
John Zeleznikow	Victoria University, Australia
Jozef Zurada	University of Louisville, USA

Organizing Committee

Karen Heyden (Co-chair)	Leipzig University, Germany
Elżbieta Lewańska (Co-chair)	Poznań University of Economics and Business, Poland
Wilfried Röder (Co-chair)	Leipzig University, Germany
Christian Franck	Leipzig University, Germany
Barbara Gołęiewska	Poznań University of Economics and Business, Poland
Włodzimierz Lewoniewski	Poznań University of Economics and Business, Poland
Bartosz Perkowski	Poznań University of Economics and Business, Poland
Milena Stróżyna	Poznań University of Economics and Business, Poland

Additional Reviewers

Bartoszewicz, Grzegorz	Haag, Steffi	Osmani, Dritan
Beese, Jannis	Harrer, Simon	Palomba, Fabio
Benedict, Martin	Heim, David	Paton, Norman
Betke, Hans	Hoffmann, David	Rehse, Jana-Rebecca
Blaschke, Michael	Hornung, David Julian	Risi, Michele
Boillat, Thomas	Hornung, Olivia	Rößler, Richard
Böhmer, Kristof	Hummel, Dennis	Salas, Julián
Braun, Richard	Kaes, Georg	Sarigianni, Christina
Brenig, Christian	Kaiser, Marcus	Schacht, Silvia
Clees, Kevin	Klima, Christoph	Seyffarth, Tobias
Di Nucci, Dario	Kolb, Stefan	Shekhovtsov, Vladimir A.
Dittes, Sven	Kühnel, Stephan	Sorgenfrei, Christian
Drechsler, Andreas	Laifa, Meriem	Steinberger, Claudia
Dumont, Tobias	Lamek, Anna	Stolarski, Piotr
Faber, Anne	Landthaler, Jörg	Tarasov, Vladimir
Filipiak, Dominik	Lübbecke, Patrick	Thakurta, Rahul
Fischer, Marcus	Malyzhenkov, Pavel	Turi, Abeba Nigussie
Gall, Manuel	Marrella, Andrea	Waizenegger, Lena
Geiger, Manfred	Michel, Felix	Wiesneth, Katharina
Gouveia, Duarte	Nadj, Mario	Zahoransky, Richard M.

Contents

Smart Infrastructures

Process Management

Business and Enterprise Modeling

Service Science

Social Media

Applications

Ecosystems

High-Frequency Trading, Computational Speed and Profitability: Insights from an Ecological Modelling

Alexandru Stan[✉]

Business Information Systems Department, Babeş-Bolyai University,
Cluj-Napoca, Romania
alexandru.stan@econ.ubbcluj.ro

Abstract. High-frequency traders (HFTs) account for a considerable component of equity trading but we know little about the source of their trading profits and how those are affected by such attributes as ultra-low latency or news processing power. Given a fairly modest amount of empirical evidence on the subject, we study the relation between the computational speed and HFTs' profits through an experimental artificial agent-based equity market. Our approach relies on an ecological modelling inspired from the r/K selection theory, and is designed to assess the relative financial performance of two classes of aggressive HFT agents endowed with dissimilar computational capabilities. We use a discrete-event news simulation system to capture the information processing disparity and order transfer delay, and simulate the dynamics of the market at a millisecond level. Through Monte Carlo simulation we obtain in our empirical setting robust estimates of the expected outcome.

Keywords: Artificial markets · Ecological modelling · High-frequency trading · Profitability

1 Introduction

Nowadays, financial markets are defined by a continuous enhancement of the information processing power and the trading infrastructure [1,2] needed to exploit the processed information. Absolute computational speed becomes of paramount importance to traders due to the innate volatility of financial quotations. Relative processing speed, i.e. reacting faster than the rest of the market participants, is also fundamental as it enables fast traders to take advantage of any profit opportunity by rapidly reacting to news and market activity.

In this race for speed, a large fraction of automated trades comes from short-term investors known as high-frequency traders (HFTs). HFTs use state-of-the-art computational resources and trading infrastructures to analyse news and market activity and execute large number of orders at extremely high speeds. In a increasingly competitive context, HFTs profits are far from being secured [3] as markets also accommodate ultra high-frequency traders (UHFTs) which

© Springer International Publishing Switzerland 2016
W. Abramowicz et al. (Eds.): BIS 2016, LNBIP 255, pp. 3–14, 2016.
DOI: 10.1007/978-3-319-39426-8_1

pay for access to specific exchanges showing price quotes prior to the others. Thus, the HFT niche has become dense [4] with high-speed trading algorithms competing against each other and further squeezing profits.

The "Concept Release on Equity Market Structure" of the Security and Exchange Commission estimates that High Frequency Trading (HFT) surpassed in recent years 50 % of total volume in U.S.-listed equities, and concludes that it stands for a "dominant component of the current market structure and likely to affect nearly all aspects of its performance". Although they play a central part in current equity markets, we genuinely do not know much about the source of their profits and how these are impacted by ultra-low latency trading and processing speed when HFTs do not act as passive market makers but as aggressive liquidity consumers [5,6] pursuing fundamental value changes.

Given a fairly modest amount of reliable empirical evidence on the subject, we study the relation between the computational speed and HFTs' profits through an experimental artificial agent-based equity market. Our approach makes use of an ecological modelling in order to offer a better understanding of the extent to which ultra-low latency trading and asymmetrical information processing power affects HFTs' profit in news dominated markets. We examine the HFT profitability principally in relation with the shortness of the holding period. Through repeated computer simulations, we investigate essential aspects of HFT profitability levels, generally difficult to observe and accurately analyse in traditional analytical models.

In our study, we clearly segregate between passive HFTs, which solely place limit orders that are not instantly marketable and behave therefore like liquidity providing market makers, and aggressive HFTs, which originate exclusively market orders, and have to pay the execution costs of crossing the spread. In this study we focus only on aggressive HFTs' profits, since this type of HFT activity is problematic (passive HFT is benign providing only price and liquidity improvement to the market).

The remainder of the paper is organized as follows. Section 2 reviews a number of important results in the field. Section 3 details our trading model, Sect. 4 presents a condensed list of experimental results for a primary implementation of the model, and conclusion ends the paper.

2   Related Work

There is no general agreement in the academic world on the profitability level of HFT activities. [3] provide an analysis on the subject and argue that it should not perform particularly better than other types of lower frequency trading. [7] report the results of an extensive empirical study estimating the maximum possible profitability of the most aggressive HFT activities, and arrive at figures that are unexpectedly modest. For 2008, they indicate an upper bound of $21 billion for the entire aggregate of U.S. equities at the longest holding periods, and a bound of $21 million for the shortest holding periods. In addition, they consider these numbers "to be vast overestimates of the profits that could actually be achieved in the real world", and oppose these figures with the $50 trillion annual trading volume in the same markets. Other studies range HFT profits between $15–25 billion per year. Schack and Gawronski of Rosenblatt Securities claim that all of these figures are too high without specifying any numbers. [8] estimate the profits for the aggressive HFT alone and range them between $1.5–3 billion per year.

On the other hand, [9] find that HFT is highly profitable especially for liquidity-taking HTFs who can generate very high Sharp Ratios. They point out to three other interesting findings. HFTs get their profits from all other market participants. The HFT profits are persistent and concentrated around a core of dominant participants. New HFT industry entrants are more likely to under-perform and exit, while the fastest actors (in absolute and in relative terms) constitute the upper tail of performance. [9] also describes several stylized facts about the profitability of the HFT industry and the distribution of profits across firms. [10] analyse the optimal trading strategy of informed HFTs when they can trade aggressively ahead of incoming news and find that computational speed is very important for HFT profitability as high-frequency trades are correlated with short-run price changes. [10] also predict higher liquidity levels for stocks with more informative news in spite of intensive liquidity-consuming HFT activities.

A series of other studies result in less determined findings. For instance, a theoretical study of latency costs done by [11] concludes that low latency profit impact is considerable only in times of high volatility and small bid-ask spread. [12] analyses and exposes the type of information HFTs are best at processing and exploiting. [13] provides limited empirical evidence regarding HFT financial performance, trading costs, and effects on market efficiency using a sample of NASDAQ trades and quotes that directly identifies HFT participation. [5] proposes a strategy for identifying the fractions of HFT profits stemming from passive market making and aggressive speculation. [14,15] report that about 30 percent of HFT trades are of aggressive nature. [16] propose a low-latency activity metric as a proxy for HFT trading which can be used to estimate overall HFT profits.

Contrary to most of the aforementioned investigations, which make use of proprietary data, on which they apply estimates of the actual profits, we propose a flexible agent based model (ABM) inspired from ecology for assessing the impact of computational speed on HFT profits. We focus explicitly on profits stemming from aggressive HFT strategies, as this type of market activity possesses the greatest potential for harm.

3   The Agent Based Market Model

The model we put forward is inspired from the r/K selection theory [17] and is designed to assess the financial performance of two classes of HFT agents endowed with asymmetric computational capabilities. The agents interact within a real-time artificial financial marketplace. We use a discrete-event news simulation system to capture the information processing asymmetry and order transfer delay, and simulate the interactions between agents at a millisecond level.

3.1   A Brief Motivation for the Modelling

In ecology, the r/K selection theory describes two environmental extremes [17], and the strategies a population exhibits in order to exploit either condition. One extreme is an r-selective environment, characterized by the presence of freely available resources. It frequently arises when predators hold a population much lower in size than the maximum number of individuals the habitat can support (the carrying capacity of the environment). The r-selection species (r-strategists) are commonly smaller organisms with short life spans, living in transitory or unstable environments, and producing many offspring at low cost per individual. The other extreme is represented by a K-selective environment where a population lives at densities close to the carrying capacity of the environment. K-selected species (K-strategists) produce a low number of offspring highly adapted to their environment. They live in roughly stable environments and have longer life spans.

The r/K selection theory states the natural tendency of species to adopt behaviours which most efficiently help them exploit their environment. Using a similar argument, we isolate two contrasting trading behaviours which favour either trade speed and quantity, or trades' quality. In our modelling, high-frequency traders can be assimilated to r-strategists, since they focus on the traded quantity by placing at low costs large amounts of small size orders during very short periods of time. HFTs enter and close short-term positions at high speeds aiming to capture sometimes a few basis points on every trade. HTFs typically compete against each other [18,19], rather than against the buy-and-hold traders (BHTs). They dominate the volatile and unpredictable news market environments due to their capacity to adjust trading volumes to the brutal variations of highly dynamic trading contexts. High-frequency automata allow HTFs to trade plenty of times on the same signal. Their ability of flooding the market with orders constitute an adaptation to significant variability in the "charge capacity" of the market. Our model assumes that the charge capacity of the market randomly evolves driven by the emergence of trading signals.

Conversely, the K-strategists are traders employing buy-and-hold strategies, and concentrating on trades' quality. They dominate stable market environments effectively engaging in fewer an larger size trades. Their larger volume orders will have longer "life spans" as they take more time to be completed. Their main feature is a low value of the r parameter.

In our model r and K parameters are accounting for the order latency coefficients. We choose the latency parameters to be constant with $r \gg K$. High values for the parameters indicate high order volume placed at ultra-low latencies. The r and K are measures of the instantaneous rate of change in the volumes of orders; they are expressed in numbers per unit time per order. During simulations, we fluctuate the market conditions from one dominant mode to the other thereby supporting r or K selected strategies.

3.2 Model Generalities

Our model considers a market with one risky asset and four types of trading agents: one competitive market maker, buy-and-hold traders (K-strategists) and two categories of HFTs (r-strategists). The HFTs receive news signals about the asset and trade them against the rest of the market participants. Each signal is either reliable or not with some given probability. At the arrival of each signal, the HFTs cannot determine if it is true or false, before processing it. Therefore, they can trade on the signals before or after processing them. According to their processing capacities we segregate the r-strategists in two categories: the standard HFTs and the ultra-rapid HFTs (UHFTs). The UHFTs are HFTs characterized by superior signal processing power and ultra-low latency trading. Given their information processing limitation, the fast traders can adopt one of the following four strategies [10,20]: (a) ignore the signal and stay out of the market, (b) attempt to exploit it without processing it, (c) attempt to exploit it only after learning its value of truth (d) if the signal is false, apply a price reversal strategy.

Both classes of HFT agents are associated with a latency factor r and an information processing time distribution P, which model the speed the agents can respond to market changes. Lower latency and appropriate responses should naturally lead to better financial performances. The latency factor r represents the intensity of a homogeneous Poisson process at which the agents can have access to the market. P is a specific low mean/low variance log-normal processing time distribution.

The stylized BHTs of the model are assumed to be high-latency manual traders if the order is of medium size. For the execution of larger orders they apply volume-weighted average price (VWAP) algorithms which are known to have certain comparative advantages over manual execution.

The trading behaviour of the agents is calibrated in order to obtain aggregate order volumes miming r/k selection ecosystems. The equations in the system below describe the expected order volumes evolutions for each class of market participants in our ecosystem.

$$
\begin{cases}
\frac{dV_{bh}}{dt} = r_{bh} V_{bh} \left(1 - \frac{V_{bh}}{K_{bh}^t}\right) \\
\frac{dV_{hf}^u}{dt} = r_{hf}^u V_{hf}^u \left(1 - \frac{\alpha V_{bh}}{K_{hf}^t}\right) \\
\frac{dV_{hf}}{dt} = r_{hf} V_{hf} \left(1 - \frac{\beta V_{hf}^u + \alpha V_{bh}}{K_{hf}^t}\right)
\end{cases}
\tag{1}
$$

where

- V_{hf}^u, V_{hf} and V_{bh} are the order volumes stemming from three categories of market participants (UHFTs, HFTs and BHTs);
- r_{hf}^u, r_{hf} and r_{bf} are the corresponding computational speed coefficients. We choose the speed coefficients to be constant with $r_{hf}^u > r_{hf} \gg r_{bf}$. High values for the coefficients indicate high order volume placed at ultra-low latencies;
- α indicates how BH orders impact the overall HFT order volumes;

- β is a measure of increased HFT orders cancellation rate due to UHFT incoming orders;
- K_{hf}^t stands for the carrying capacity of all HFTs. It will be approximated to a fraction of the aggregate maximum stock level of their inventories. This parameter plays an important role as sizeable quantities of the risky asset are purchased by HFTs before being repurchased by more permanent holders;
- K_{bh}^t stands for the carrying capacity of the rest of the market participants. It is a fraction of the aggregate maximum stock level tolerated by the risk neutral competitive market maker.

The first equation in our modelling indicates that the volume dynamics of slow frequency traders follow a logistic differential equation. The related carrying capacity K_{bh}^t is modelled to be a positive first-order stationary process. The second and third equations describe the overall impact that the slow-frequency volume has on the fast volume (through the α parameter) and the inter-specific competition between the two kinds of HFTs (through the β parameter).

While the three computational speed parameters r_{hf}^u, r_{hf} and r_{bf} are known and built into the algorithmic behaviour of each trading agent, the α and β coefficients are to be estimated at the end of the simulation phase in order to understand the compound interaction between the three classes of agents and the weight each class has on the other two.

3.3   Empirical Strategy

We implemented our model within an adapted derivation of the java-based artificial financial framework named ATOM. ATOM is a general environment for agent-based simulations of stock markets developed within the Lille University. This simulator is able to closely reproduce the behaviour of most order-driven markets. ATOM generates order flows for intra and extra-day sessions and can be used to design experiments where fast trading agents respond to market changes almost instantaneously.

We observe a single case study designed to understand the manner in which the financial performance of HFTs is impacted by ultra-low latency trading and processing speed when HFTs do not act as passive market makers but as aggressive speculators pursuing fundamental value changes. To this end, we populate the marketplace with four types of participants with different behaviours and functions:

- Buy-and-Hold traders which contribute through their buying and selling activities to the price creation. At any single moment t, the BH trading population give forth to balanced streams of buy and sell orders of roughly homogeneous Poisson intensity. The decision to place new orders and their size is drawn at random from the distributions of the agent class;
- One competitive market maker which stands ready to buy/sell asset at its posted bid and ask prices at any time;

– HFTs which try to exploit buying or selling news signals as soon as they emerge into the market. Through the relative magnitude of their market activities and the correlation of their behaviour, they may induce high imbalances between supply and demand volumes resulting in significant price reversals. They generate Poissonian streams of buy and sell orders at high intensity. The market mechanism allows these agents to make decisions, place, execute or cancel orders on an extremely short time scale;
– Performance enhanced UHFTs with even faster information processing abilities and ultra-low latency trading.

The design of our discrete-event news simulation takes into account the fundamental tension of aggressive HFT, which exists between the length of holding period and the costs of trading [7], i.e. the positions must be held open long enough so that advantageous price movements can compensate the trading costs. The buy or sale by aggressive HFTs cause transaction costs for each order of at least the sum of the spreads at the onset and liquidation of the position. Since aggressive trading will typically widen the spreads, in order for the trades to be profitable, change in price must be high enough to cover larger than usual spread-based transaction costs. Therefore, we generate news signals indicating corrections in the fundamental value of the security at least five times higher than the mean value of the bid-ask spread.

The model was implemented in the java language. Before the simulation, we had to calibrate it in order to reproduce [21] a number of real stylized facts in stock markets. Amongst those, we focus on the three features that are the most ubiquitous for financial price series: the fat-tailed asset return distribution, the volatility clustering, and the unit-root property.

Each simulation comprises 20 consecutive trading sessions during which we alternate the intensity of news trading signals. We start off the experiments with a trading panel encompassing the BHTs and the market maker. As news trading signals come in, the HFTs invest the market and try to exploit them. We change the signal processing speed and the latency for the two populations of HFTs, and store the price and traders' wealth time series. We compare their dynamics and final outcome. We repeat the experimentation through Monte Carlo simulation by varying the size of the trading populations. For each set of parameters in our empirical setting the experiment is repeated 100 times so as to obtain robust estimates of the expected outcome.

At the beginning of the simulation (t = 0), each BHT is endowed with an amount of cash and a number of risky assets. The distributions of cash and risky assets both follow power laws. HFTs and UHFTs posses the same amount of aggregate wealth at the beginning of the simulation. All traders stop trading when they run out of either cash or stock. We assume that the inventories of the HFTs are much smaller than the inventories of the other market participants.

In the experimental setting, we limit the population size for each class of traders to a few hundreds agents as the simulations are computationally intensive. All traders, aside from the MM, are initiating order flows at Poisson intensities in order to comply with the restrictions previously stated in the section.

4 Experimental Results

In our experiments we used a mean order delay time for UHFTs equal to 5 ms and considered holding periods up to one second. Panel (a) of Fig. 1 exhibits a simple but interesting finding. For each of the holding periods, it plots the percentage of the total profits realized in our experiments by the enhanced UHFTs within that time range. We can see that most of the profit of the UHFTs is generated by positions held for extremely short periods. We notice that UHFTs speed advantage is not lost to slower HFTs, which access the exchange slower. Consequently, a sizeable part of the HFTs' profitability is captured right after the trades execution by quickly liquidating the positions. Profitable short holding periods are due to relatively sharp price movements and higher volatility levels. Panel (b) of Fig. 1 presents a different picture for the HFTs. We can interpret these values as the expected time horizon profitability for HFTs in presence of computationally faster UHFTs. We can see that profitability is substantially lower for holding time shorter than 25 ms. UHFTs combine aggressive order placement with very short holding periods to amputate the profitability of the HFTs. The profits realized over a longer horizon are comparable to those obtained by UHFTs. In a highly competitive environment, the cause of the decline in the HFTs profits for short holding periods is a reduced number of trading opportunities due to the presence of UHFTs. Panels (a) and (b) of Fig. 2 show the mean profit per trade executed by UHFTs and HFTs. For instance, the ratio of profitability per transaction at a less than 25 ms holding time is 0.0267. Thus the UHFT profitability is strongly correlated with the number of trading opportunities. Panel (b) displays HFTs' profit decline for short holding periods consistent with the previous results. The correlation is less strong for them, the ratio of profitability for the same holding time range is only 0.0036. We conclude that regular HFTs are fully dominated by the UHFTs on the very short horizon position segment: their profits decrease fast, and the explanation is that UHFTs' high profits are obtained at the expense of all the other types of market participants including the HFTs. Over this segment the aggregate volume of profitable HFT originated trades and returns is relatively low.

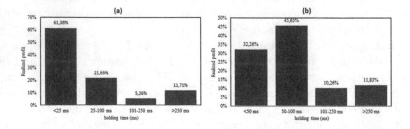

Fig. 1. The distribution of the aggregate profits of (a) UHFTs and (b) HFTs as a function of the holding period.

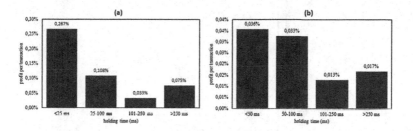

Fig. 2. The mean profit per trade executed by (a) UHFTs and (b) HFTs.

Figure 3 illustrates the financial performance of the HFTs as a function of latency. The performance is captured through the Sharpe Ratio as a measure for calculating risk-adjusted return. For very short performance differentials between HFTs and UHFTs, the level and consistency of their profits as well as their risk-return trade-off demonstrate surprisingly strong performance. We observe that the computational speed benefits go sharply down as the performance differential between the two classes of traders deviates from zero. Figure 3 shows an almost exponential decline of HFTs profitability in both relative latency and relative information processing speed. We also analysed the persistence of the trading profits by assessing how one trading session's profits impact the next session's profits through an autoregressive AR(1) model $P_t = c + \varphi P_{t-1} + \varepsilon_t$. For φ we obtained higher positive values closer to 1 parameter in the case of the UHFTs. It resulted that all HFTs' profits are persistent, with increased persistence concentrated around the dominant class of fast traders. It results that increased computational speed enhances not only the magnitude but also the consistency of the financial performance.

We observed fundamentally different profitability profiles when exploiting reliable and false news for the two classes of fast traders. While regular HFTs are incurring losses when trying to exploit this type of events the UHFTs perform extremely well. The false signals followed by signal correction are a source

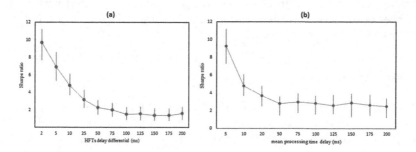

Fig. 3. An illustration of HFTs' profit decline as a function of performance differential between them and UHFTs. The performance differential is displayed for both (a) order transfer delay and (b) information processing time.

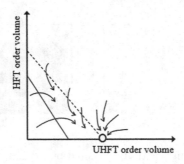

Fig. 4. Extinction of the HFTs' trading activity in a market environment strongly dominated by UHFTs

of negative autocorrelation for the HFTs' order imbalances. Another observation relates to the fact that the profitability profiles are not related to the capitalization of the agents but rather normally distributed within the two classes.

We estimated the values of the α and β coefficients in order to understand the compound interaction between the three classes of agents and the weight each class has on the other two. We obtained close to zero values for the *alpha* coefficient and statistically significant positive values for β. The HTFs' order volume (V_{hf}) is mainly impacted by the UHFT incoming orders (V_{hf}^u) through the β parameter. This observation is in line with the findings of [18,19] that HTFs typically compete against each other, rather than against the BHTs. We identified a market condition which is frequently associated with the loss of profitability for the under-performing HFTs. We empirically observed that negative financial performances in HFTs' market activities occur whenever the following condition holds true for a longer period of time:

$$\beta \overline{V}_{hf}^{u} > \overline{V}_{hf} \tag{2}$$

This inequality corresponds to a differential of competitiveness between the two classes of fast traders high enough to exclude a long-term coexistence of the two classes of competitors in the same market, and is associated with frequent occurrences of unreliable trading signals and substantial gaps in computational speed (Fig. 4).

5   Conclusions

High-frequency traders (HFTs) account for a considerable component of equity trading but we know little about how their trading profits are affected by ultra-low latency and information processing power. Given a fairly modest amount of empirical evidence on the subject, we study the relation between the computational speed and HFTs' profits through an experimental artificial agent-based equity market. Our approach relies on an ecological modelling inspired from the r/K selection theory, and is designed to assess the relative financial performance of two classes of aggressive HFT agents endowed with dissimilar computational capabilities. Our model considers four types of trading agents: one competitive market maker, buy-and-hold traders (K-strategists) and two categories of HFTs (r-strategists), the standard HFTs and the ultra-rapid HFTs. The UHFTs are HFTs characterized by superior signal processing power and ultra-low latency trading.

In our study, we clearly segregate between passive and aggressive HFTs, and focus only on analysing aggressive HFTs' profits. All aggressive HFTs trade in the direction of fundamental value changes and in the opposite direction of transient price deviations. They adversely select through marketable orders liquidity supplying non-marketable orders. The short-lived informational advantage given by the computational speed is large enough to cross the bid-ask spread and generate positive trading revenues. We use a discrete-event news simulation system to capture the information processing disparity and order transfer delay, and simulate the dynamics of the market at a millisecond level.

Our study presents three findings that provide insight into the prime movers of HFT profitability. First, relative computational speed is of paramount importance in securing outstandingly high levels of profitability in HFT activities. We observed an exponential decline of HFTs profitability in both relative latency and relative information processing speed. This implies that the HFT arms race for speed should continue or even strengthen in the computerized financial trading. The aggregate order volume originated by the regular HFTs, and thereby their gain opportunities, is primarily impacted by the activity of faster UHFTs and much less by the activity of the low-frequency actors in the market. Trading revenues without fees are zero sum in the aggregate and the high profits of the UHFTs are done at the expense of all the other types of market participants: market maker, BHTs, and HFTs.

Second, we analyse the persistence of HFT profits by assessing how one trading session's profits impact the next session's profits. We obtain that all HFT profits are persistent, with increased persistence concentrated around the UHFTs class of dominant market traders. It results that increased computational speed enhances not only the magnitude but also the consistency of the financial performance.

Third, we identified a market condition which is frequently associated with the loss of profitability for under-performing HFTs. The condition corresponds to a differential of competitiveness between the two classes of fast traders high enough to exclude a long-term coexistence of the two classes of competitors in the same market, and is correlated with frequent occurrences of unreliable trading signals and substantial discrepancy in computational speed.

Acknowledgement. This work was financed by UEFISCI, under project PN-II-PTPCCA-2013-4-1644.

References

1. Diaz-Rainey, I., Ibikunle, G., Mention, A.L.: The technological transformation of capital markets. Technol. Forecast. Soc. Chang. **99**, 277–284 (2015)
2. Kauffman, R.J., Liu, J., Ma, D.: Innovations in financial IS and technology ecosystems: high-frequency trading in the equity market. Technol. Forecast. Soc. Chang. **99**, 339–354 (2015)
3. Moosa, I., Ramiah, V.: The profitability of high-frequency trading: is it for real? In: Gregoriou, G.N. (ed.) The Handbook of High Frequency Trading, Chap. 2, pp. 25–45. Academic Press, San Diego (2015)
4. Lhabitant, F.S., Gregoriou, G.N.: High-frequency trading: past, present, and future. In: Gregoriou, G.N. (ed.) The Handbook of High Frequency Trading, Chap. 9, pp. 155–166. Academic Press, San Diego (2015)
5. Menkveld, A.J.: High frequency trading and the new market makers. J. Finan. Markets **16**(4), 712–740 (2013)
6. Bookstaber, R., Paddrik, M.E.: An agent-based model for crisis liquidity dynamics (2015)
7. Kearns, M., Kulesza, A., Nevmyvaka, Y.: Empirical limitations on high frequency trading profitability (2010)
8. Arnuk, S., Saluzzi, J.: Latency arbitrage: The real power behind predatory high frequency trading (2009)
9. Baron, M., Brogaard, J., Kirilenko, A.: The trading profits of high frequency traders (2012)
10. Foucault, T., Hombert, J., Rosu, I.: News trading and speed. J. Finan. (2015)
11. Moallemi, C.C., Saglam, M.: The cost of latency in high-frequency trading (2013)
12. Zhang, S.S.: Need for speed: an empirical analysis of hard and soft information in a high frequency world (2012)
13. Carrion, A.: Very fast money: high-frequency trading on the NASDAQ. J. Finan. Markets **16**(4), 680–711 (2013)
14. Jovanovic, B., Menkveld, A.J.: Middlemen in limit order markets (2015)
15. Hagstrmer, B., Norden, L.L.: The diversity of high-frequency traders (2013)
16. Hasbrouck, J., Saar, G.: Low-latency trading. J. Finan. Markets **16**(4), 646–679 (2013)
17. Roff, D.: Evolution of Life Histories: Theory and Analysis. Springer, New York (1993)
18. Easley, D., López de Prado, M.M., O'Hara, M.: The Microstructure of the 'Flash Crash': Flow Toxicity, Liquidity Crashes and the Probability of Informed Trading. Social Science Research Network Working Paper Series, October 2010
19. Vuorenmaa, T.A., Wang, L.: An agent-based model of the flash crash of may 6, 2010, with policy implications (2014)
20. Dugast, J., Foucault, T.: False news, informational effciency, and price reversals (2014)
21. Lebaron, B.: Agent-based financial markets: matching stylized facts with style. In: Colander, D. (ed.) Post Walrasian Macroeconomics, pp. 221–236. Cambridge University Press, Cambridge Books Online (2006)

A Methodology for Quality-Based Selection of Internet Data Sources in Maritime Domain

Milena Stróżyna[1(✉)], Gerd Eiden[2], Dominik Filipiak[1],
Jacek Małyszko[1], and Krzysztof Węcel[1]

[1] Poznań University of Economics and Business, Poznań, Poland
{milena.strozyna,dominik.filipiak,jacek.malyszko,
krzysztof.wecel}@kie.ue.poznan.pl
[2] LuxSpace Sarl, Betzdorf, Luxembourg
eiden@luxspace.lu

Abstract. The paper presents a methodology for identification, assessment and selection of internet data sources that shall be used to supplement existing internal data in a continuous manner. Several criteria are specified to help in the selection process. The proposed method is described based on an example of the system for the maritime surveillance purposes, originally developed within the SIMMO research project. As a result, we also present a ranking of concrete data sources. The presented methodology is universal and can be applied to other domains, where internet sources can offer additional data.

Keywords: Internet data sources · Quality assessment · Selection methodology

1 Introduction

Each information system need to be supplied with data in order to fulfil its functions. The data can come from various sources, depending on the system, its purposes and operating context. In systems used by organisations, sources of data can be internal (e.g. transactional data) or external, coming from the outside of an organisation (e.g. sensors, external systems and databases, Internet). Irrespective of the type of data used, each potential data source for a system needs to be appropriately defined and assessed. This procedure is crucial, while designing and developing a system [1].

The goal of this paper is to present a methodology for a selection of open internet sources that can be treated as an data source for an information system. Data from the Internet is used then to enhance data from other sources, such as legacy systems, sensors, internal databases, etc.

The general scope of this paper encompasses a procedure for identification, assessment and selection of internet data sources. The process of designing the proposed methodology was driven by the standard approach to the data quality, which defines quality as *"the totality of features and characteristics of a product*

© Springer International Publishing Switzerland 2016
W. Abramowicz et al. (Eds.): BIS 2016, LNBIP 255, pp. 15–27, 2016.
DOI: 10.1007/978-3-319-39426-8_2

or service that bears its ability to satisfy stated or implied needs" [2]. In case
of information systems, these needs mean the functional and non-functional
requirements. Therefore, we assumed that each potential data source for the
system should be analysed and assessed taking into account two elements: (1)
system's requirements; (2) a selected set of quality criteria. In the first step we
select a set of quality measures, which are then used to assess data sources. Then,
for each measure a rating scale and a weight is assigned. Finally, a method for
calculating a quality grade and setting a selection threshold is specified.

The paper is built around the use case of an information system from the
maritime domain, shortly described in the next section. Section 3 presents the
related work in the area of data sources selection for the information systems.
Then, in Sect. 4 a proposal of applied research methodology for identification and
selection of internet data sources is presented. Section 5 describes the results
of the project's work, where the proposed approach for sources selection was
applied. In Sect. 6 we summarise the results.

2 System for Intelligent Maritime Monitoring

Nowadays, with growing importance of the maritime trade and maritime economy, one of the key priorities and critical challenges is to improve the maritime
security and safety by providing appropriate level of maritime surveillance. This,
in turn, can be realised by providing tools supporting maritime stakeholders in
analysis of the current situation at sea – creation of the so-called "Maritime
Domain Awareness (MDA)". MDA implies collection, fusion and dissemination
of huge amount of data, coming from many, often heterogeneous, sources. However, current capabilities to achieve this awareness are still improving and there
is a need for development of dedicated information systems and tools. This need
concerns especially systems, able to fuse in real-time data from various heterogeneous sources and sensors. To our best knowledge, currently there exists no
maritime surveillance system which would automatically acquire and fuse AIS
data with information available in internet sources. As a result, there is also no
standard methodology for selecting and assessing the quality of internet sources
to be used in such systems.

This challenge was addressed by the SIMMO project[1]. Within the project a
system has been developed aiming at improving the maritime security and safety
by providing high quality information about vessels and automatically detecting
potential threats (i.e. suspicious vessels). The concept of the SIMMO assumes
constant retrieval and fusion of data from two types of data sources, namely:

1. Satellite and terrestrial Automatic Identification System (AIS)[2], which provides inter alia information about location of ships and generic static information about them.

---

[1] SIMMO – System for Intelligent Maritime MOnitoring, the JIP-ICET 2 project
financed by the Contributing Members of the European Defence Agency.

[2] http://www.imo.org/en/OurWork/Safety/Navigation/Pages/AIS.aspx.

2. Open internet sources that provide additional information about ships, not included in AIS (e.g. flag, vessel type, owner).

In general, the SIMMO system integrates information from these two types of sources, what is essential for the better identification of vessels, and then detects suspicious vessels. For the efficient use of sources of the second type, a need to select appropriate sources emerged. The methodology for the quality assessment of potential internet sources had to be defined and adopted.

3 Related Work

3.1 Internet Sources Related to the Maritime Domain

The creation of the enhanced Maritime Domain Awareness and detection of suspicious vessels requires usage of different data sources. The sources that are applicable in the maritime surveillance domain can be divided into three categories.

The first and the most widely used are sensors. Sensors provide kinematic data for the observed objects in their coverage area and can be further divided into active (e.g. radar, sonar) and passive (which rely on data broadcasted intentionally by objects, e.g. AIS). A survey on sensors used in maritime surveillance can be found in [3].

The second category includes authorised databases, containing information about vessels, cargo, crew etc. However, most of them are classified and encompass inter alia port notifications sent by ships, HAZMAT reports, The West European Tanker Reporting System (WETREP), LRIT data centers, SafeSeaNet [4].

The first and the second category are basically accessible only to the maritime authorities, such as the coastguard. Therefore, they can be referred to as closed data sources. Moreover, most of them are not published in any form on the Internet.

The third category consists of data sources, which are publicly available via Web (hereinafter referred to as internet data sources). This data includes inter alia vessel traffic data, reports and news. More specifically, they can be divided as follows [5]:

- open data sources, in which data is freely accessible and reusable to the public (no authorisation required),
- open data sources with required authorisation and free registration,
- closed data sources with required authorisation and non-free access.

The term open data refers to the idea of making data freely available to use, reuse or redistribute without any restriction [6]. In the maritime context, there are organisations and communities that provide their maritime related data online and make it accessible for the public. Examples are ports, publishing vessel traffic data as well as blogs, forums and social networks, which share information about maritime events [5].

Although there are various categories of data sources, in the existing maritime information systems usually only data received from sensors are used. The research in this area focuses on collection of sensor data, such as SAR, AIS, IR, video and radar data [3,7] or fusion of sensor and non-sensor data (for example inclusion of expert knowledge) [8–10]. The research, which additionally focuses on usage of open data for the purpose of maritime surveillance, is presented in [5].

3.2 Data Quality Assessment

There is no uniform definition of data quality nor a standard or a commonly used approach for assessment of data quality. ISO9000:2015 defines data quality as the degree to which a set of characteristics of data fulfils requirements [11]. In the information systems literature, a lot of various data quality attributes can be found. The examples are: completeness, accuracy, timeliness, precision, reliability, currency and relevancy [12]. Other such as accessibility and interpretability are also used. Wang et al. [13] identified nearly 200 such quality attributes. Still, no general agreement exists either on which set of dimensions define the quality of data or on the exact meaning of each dimension. Batini et al. [14] present different definitions of popular quality attributes provided in the literature.

Taking into account the fact that there is little agreement on the nature, definition, measure and meaning of data quality attributes, the European Parliament decided to propose its own uniform standards for guaranteeing quality of results for the purposes of the public statistics, described in the ESS Quality Assurance Framework [15]. In this standard, seven quality criteria were defined [16]: (1) *relevance* (the degree to which data meets current and potential needs of the users); (2) *accuracy* (the closeness of estimates to the unknown true values); (3) *timeliness* (the period between the availability of the information and the event or phenomenon it describes); (4) *punctuality* (the delay between the date of the release of the data and the target date); (5) *accessibility and clarity* (the conditions and modalities by which users can obtain, use and interpret data); (6) *comparability* (the measurement of the impact of differences in applied measurement tools and procedures where data are compared between geographical areas, sectoral domains or over time); (7) *coherence* (the adequacy of the data to be reliably combined in different ways and for various uses).

When the quality attributes are defined, the next step is data quality assessment. Also in this matter, the literature provides a wide range of techniques to assess and improve the quality of data. In general, the assessment consists of several steps [14]: (1) *data analysis* (examination of data schemas, complete understanding of data and related architectural and management rules); (2) *data quality requirements analysis* (surveying the opinion of users and experts to identify quality issues and set quality targets); (3) *identification of critical areas* (selection of databases and data flows); (4) *process modelling* (a model of the processes producing or updating data); (5) *measurements of quality* (selection of quality attributes and definition of corresponding metrics). The measurement of quality can be based on quantitative metrics, or qualitative evaluations by data experts or users.

There exist a number of methodologies for quality assessment and quality measurement. Batini surveyed thirteen of them [14]. Nauman et al. [17] propose a quality driven source selection method using Data Envelopment Analysis technique. A data source is described by three qualities in this method: *understandability* (a subjective criterion), *extent* (an objective criterion), and *availability* (an objective criterion), whereas the *efficiency* of a given data source is the weighted sum of its quality scores. Weights are calculated using a linear programming. An important feature of this method is the fact that it focuses on each data source selectively. With regard to the step of quality measurement, it can be performed with different approaches, such as questionnaires, statistical analysis and involvement of the subject-matter experts (expert or heuristic techniques) [18].

4   Methodology

While designing and developing an information system, a key role plays the selection of data sources. These sources can be either internal or external (coming from outside the organisation). Irrespective of the type of data used, each potential data source needs to be identified and assessed. This procedure consists of several steps: (1) identification of potential data sources; (2) definition of quality criteria; (3) analysis of identified sources and assessment with regard to defined requirements (quality measurement); (4) selection of sources for a system. For selected sources a detailed design of data acquisition procedures takes place, including cooperation model (e.g. politeness policy). When data is obtained it has to be fused, i.e. a common data model meeting the initial system requirements has to be developed and used to organise new data in a homogeneous and integrated form. It entails semantic interoperability problems related to the interpretation of data coming from different sources. Although covered in the SIMMO project, the process of fusion data from various sources is a separate process from the source selection and as such is out of scope of this paper. In the following paragraphs, we describe in details the steps of the proposed method, using a use case from the SIMMO project. The method is presented in Fig. 1.

In order to identify, assess and select internet data sources for the SIMMO system and then to set up a cooperation model with the selected sources, a specific methodology has been followed. In the first step, potential data sources related to maritime surveillance were identified. To this end, conventional search engines (like Google) as well as meta search engines like Dogpile[3], Mamma[4], Webcrawler[5] were used. Apart from the search engines, also other data sources were analysed, including sources indicated in [19] and suggested by maritime experts, who were interviewed during requirements analysis for the SIMMO system. In this paper, we focused only on data sources which are used by the system regularly, meaning that they are constantly monitored for changes and

---

[3] http://www.dogpile.com/.

[4] https://mamma.com/.

[5] http://www.webcrawler.com/.

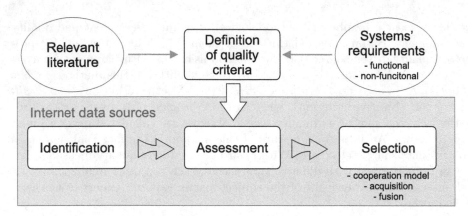

Fig. 1. Source selection method

data from them is retrieved at defined time intervals. The conducted analysis does not encompass the data sources which are to be used only once and will not be monitored for changes, e.g. internet source with a list of all ports worldwide.

In order to select sources of the highest quality and best suited to the SIMMO system, the identified data sources are assessed using specific quality criteria. For each identified data source, its features and characteristics are analysed and assessed taking into account the functional requirements defined for the SIMMO system and the selected set of quality criteria. Each potential source is assessed using the same set of quality measures.

In the proposed methodology, the data quality measures proposed by the European Statistical System (ESS) were adopted [15]. As a result, the following six quality measures are used: (1) Relevance (Usefulness); (2) Accuracy and Reliability (Completeness); (3) Timeliness and Punctuality; (4) Coherence and Comparability; (5) Accessibility (Availability); (6) Clarity (Transparency). These criteria can be adjusted to the specific of the developed system. For each measure a four-level rating scale is used (high $= 3$, medium $= 2$, low $= 1$, $N/A = 0$) and a weighting factor is assigned. After the assessment in each criteria, a final quality grade is calculated. The internet sources with the final mark above a defined threshold is then selected as a source for the system. In case of SIMMO, the threshold was set to 85 %. The weighting factors and quality threshold were assigned using the Delphi method [20].

In the final step, a cooperation model for each selected source is defined. The cooperation model should present how a cooperation with data provider (selected data source) will look like, including politeness policy and time intervals between data updates. For this end, each source had to be analysed with regard to existence of a defined politeness policy or terms of use.

5 Results

5.1 Identification of Internet Data Sources

As indicated in Sect. 4, the first step of the methodology for system's sources selection is the identification of potential sources. In the SIMMO case, potential data sources related to maritime surveillance and maritime domain were identified, using search engines, literature review and consultations with subject matter experts. As a result, over 60 different data sources available on the Web were found. The identified data sources are part of both the shallow and the deep Web. They provide information in a structured, semi-structured and unstructured manner. The list of identified internet data sources is presented in Table 2. From the point of view of data access, we divided them into four categories:

1. Open data sources (O) – websites that are freely available to Internet users,
2. Open data sources with registration and login required (OR) – websites that provide information to users only after registration and logging (e.g. Equasis),
3. Data sources with partially paid access (PPA) – websites that after paying a fee provide a wider scope of information (e.g. MarineTraffic),
4. Commercial (paid) data sources (PA) – websites with only paid access to the data (fee or subscription required).

From all the identified sources, for further analysis we selected only the open data sources. At this stage, we eliminated the commercial data sources and websites that provide access to data only after prior authorization. The elimination of commercial websites results from the fact that they provide only very general, marketing information about data they have and access to any data is available only after paying a fee or signing a contract. Moreover, an attempt to make a contact with these data providers in order to get access to a sample data failed. Therefore, we have not been able to verify the data model or scope of data provided by these sources. As a result, only sources with a public content (open data sources) were selected for further analysis.

Similarly, two other data sources (IALA, SafeSeaNet) were sorted out due to the fact that access to the data requires application of the long-lasting procedure for the needed data with no guarantee that the access will be granted.

5.2 Internet Data Sources' Assessment

As a result of initial selection, 43 sources were assessed in details. As described in Sect. 4, each source was analysed from the point of view of six quality measures. Definitions of these criteria were adjusted to the specific of the SIMMO project (see Table 1). For each measure a weight was assigned, denoting an importance of a given measure in the final quality grade.

Table 1. Quality measures used to assess Internet data sources

Name	Description	Weight
Accessibility	A possibility to retrieve data from a source; website structure and stability	0.3
Relevance	How well the data are fitted to the use-cases and system's requirements	0.3
Accuracy & Reliability	Data scope, Missing elements, Ship coverage	0.2
Clarity	Explanation of source's metadata model, Data provider	0.1
Timeliness & Punctuality	Data update, Time delay in publishing the data	0.05
Coherence & Comparability	Definition of a described phenomenon and units of measure	0.05

Source: Own work.

Each source was assessed taking into account the following measures:

1. **Accessibility (A)** – here it is assessed, whether it is possible to retrieve data from a source using a crawler. It takes into account a structure of a source, technologies used in its development, a form in which data is published as well as source stability (changes of a structure, errors, unavailability of a service). This includes also such aspects like terms of use, privacy policy, a requirement for login or registration, access to data (fees, subscriptions) etc.,
2. **Relevance (R)** – what kind of information is provided by a source and whether this information matches the user requirements of the SIMMO system,
3. **Accuracy & Reliability (AR)** – it is assessed, whether information provided is reliable (a source (owner) of information is trusted). It evaluates also data scope (how much information is available), ship's coverage (information about how many ships is provided) as well as data accuracy (number of missing information),
4. **Clarity (C)** – it is assessed, whether a source provides appropriate description and explanation of data model and source of data (who is a data provider),
5. **Timeliness & Punctuality (TP)** – it is evaluated, how often data is updated (time interval between data availability and an event, which it describes) as well as what is a time delay in publishing updated information,
6. **Coherence & Comparability (CC)** – it is compared, whether the data provided in a source describes the same phenomenon or has the same unit of measure like data available in other sources.

The assessment of the identified sources was conducted by the SIMMO project's team, being experts and having experience in data retrieval from various data sources, including structured and unstructured internet sources. For this step, the Delphi method was utilized. For the selected sources, the experts assigned a grade in each criterion, using a four-level rating scale: high (grade 3), medium (grade 2), vlow (grade 1), N/A (grade 0).

The rate N/A means that an information required for a particular criterion (e.g. update interval or ship coverage) is not specified by a source and as a result, it was not possible to assess a source in this matter. In case of Accessibility measure, the rate N/A means, that due to the terms of use or privacy policy, it is prohibited to automatically retrieve or use data published in a given source The results of quality assessment for each source is presented in Table 2.

5.3   Final Selection of Sources

After sources' assessment, the final selection took place. Firstly, all sources with Accessibility measure equals to *N/A* or *Low* were sorted out. This elimination results from the reasons indicated in the previous paragraph and the prohibition of using information from these sources.

With regard to data sources with *Low* Accessibility, this encompasses the sources with unstructured information (e.g. text in a natural language). We excluded them due to the fact that an automatic retrieval of this information would require a large amount of work and developing methods in the field of Natural Language Processing. As a result, we decided to include in the project only sources with structured or semi-structured information.

Secondly, sources with Relevance equal to *Low* were eliminated as well. It results from the fact that it's pointless to retrieve data that are not well-suited to the use-cases or requirements defined for the SIMMO system.

In the next step, a final quality grade for each source was calculated according to the formula:

$$X_s = \sum_{i=1}^{n} \frac{\frac{x_i}{3} w_i}{\sum_{j=1}^{n} w_j} * 100\,\%,$$

where s means number of the analysed sources, $n = (1, 6)$, x_i means the grade assigned by the experts to a given quality measure i, and w_i means the measure's weight.

Based on the calculated quality grade, a ranking of sources was created (see Table 2). From the ranked list of sources, only sources with a final grade above a defined threshold were selected for usage in the SIMMO system (the bold Source Names in Table 2). The quality threshold was defined by the experts at the level of 85 %.

5.4   Model of Cooperation with Data Owners

In the final step, of the applied methodology, a model of cooperation with external data providers was defined. By external data providers we understand the internet data sources, selected for the SIMMO system. For each selected source, a separate cooperation model was designed and described in the project's documentation. In defining the model, the following aspects were taken into account:

Table 2. List of assessed Internet data sources

Type of information	Source Name	Type	A	C	R	TP	CC	AR	Quality Grade	Selected
General vessel data	**Marine Traffic**	PPA	H	M	H	H	H	H	98,33%	Yes
	US Coast Guard	O	H	H	H	H	M	H	98,33%	Yes
	Maritime mobile Access and Retrieval System (ITU MARS)	O	H	H	H	H	M	M	91,67%	Yes
	Maritime-connector	O	H	H	H	N/A	H	H	90,00%	Yes
	ShipFinder	O	H	M	M	H	M	L	73,33%	No
	AIS HUB	O	H	M	M	M	M	L	70,00%	No
	Equasis	OR	N/A	H	H	H	M	H	68,33%	No
	IMO GISIS	OR	L	H	H	N/A	M	H	68,33%	No
	Vessel finder	O	N/A	L	H	H	H	H	66,67%	No
	FleetMon	OR	N/A	M	H	H	M	H	66,67%	No
	Lloyd's Register Ship in Class	OR	M	H	M	L	M	H	65,00%	No
	ShipSpotting	OR	N/A	H	H	H	H	L	56,67%	No
	World Shipping Register	PPA	N/A	M	M	M	H	H	55,00%	No
	IHS	PA	-	-	-	-	-	-	-	No
	Clarkson	PA	-	-	-	-	-	-	-	No
	Internet Ships Register	PA	-	-	-	-	-	-	-	No
	Grosstonage	PA	-	-	-	-	-	-	-	No
	Lloyd's List Intelligence	PA	-	-	-	-	-	-	-	No
	Vessel Tracker	PA	-	-	-	-	-	-	-	No
	International Association of Lighthouse Authorities (IALA)	PPA	-	-	-	-	-	-	-	No
	SafeSeaNet Vessel Traffic Monitoring and Information System	OR	-	-	-	-	-	-	-	No
Ship owners	InfoMare	O	H	L	M	N/A	M	N/A	55,00%	No
	Seaagent	PPA	N/A	L	H	N/A	L	N/A	33,33%	No
Weather	ICM Meteo	O	L	H	L	M	L	L	40,00%	No
	Meteooffice	O	L	H	L	M	L	L	40,00%	No
	Sailwx	O	N/A	L	L	M	L	M	33,33%	No
Classification of ships	**International Association of Classification Societies (IACS) Vessel in class**	O	H	H	H	H	H	H	100,00%	Yes
	American Bureau of Shipping (ABS)	O	H	H	H	M	M	M	88,33%	Yes
	International Association of Classification Societies (IACS) Transfer of Class	O	H	H	M	H	M	M	81,67%	No
	ClassNK	O	N/A	H	M	M	M	M	58,33%	No
	Leonardo Info	OR	N/A	H	H	N/A	M	H	58,33%	No
	Bureau Veritas Group	PA	-	-	-	-	-	-	-	No
	China Classification Societies	PA	-	-	-	-	-	-	-	No
	International Register of Shipping	PA	-	-	-	-	-	-	-	No
PSC / Banning / Detentions	**Thetis Company Performance**	O	H	H	H	H	L	H	96,67%	Yes
	Tokyo Mou	O	H	H	H	H	L	H	96,67%	Yes
	Mediterranean MoU	O	H	H	H	H	L	H	96,67%	Yes
	Black Sea MoU	O	H	H	H	H	L	H	96,67%	Yes
	Government of Canada - Port State Control	O	H	H	M	H	H	H	90,00%	Yes
	Indian Ocean MoU	O	M	H	H	H	L	H	86,67%	Yes
	Riyadh MoU	O	M	H	H	H	L	M	80,00%	No
	Latin America Mou	O	M	M	H	H	L	M	78,33%	No
	Paris MoU	O	N/A	H	H	H	L	H	66,67%	No
	Abuja MoU	O	N/A	H	H	M	L	H	63,33%	No
	Caribbean MoU	O	N/A	H	H	N/A	L	H	56,67%	No
Maritime crimes	ICC Commercial Crime Services	PPA	M	H	L	H	L	M	60,00%	No
	Maritime Safety Information	O	L	H	L	M	L	H	53,33%	No
Tankers	**Q88.com**	PPA	H	H	M	M	M	M	88,33%	Yes
	Auke Visser's International Supertankers	O	L	M	H	L	M	M	63,33%	No
	International Association of Independent Tanker Owners	PA	-	-	-	-	-	-	-	No
Container ships	Containership-info	O	H	M	L	M	M	H	73,33%	No
Fishing vessels	Commission for the Conservation of Antarctic Marine Leaving Resources (CCAMLR)	O	H	H	L	H	H	M	73,33%	No
	International Convention for the Conservation of Atlantic Tunas (ICCAT)	O	H	H	L	N/A	H	H	70,00%	No
	Indian Ocean Tuna Commission (IOTC)	O	H	H	L	N/A	H	H	70,00%	No
	Western & Central Facific Fisheries Commission (WCPFC)	O	H	H	L	N/A	H	M	63,33%	No
	FAO Vessel Record Management Framework (FVRMF)	O	M	H	L	L	M	H	61,67%	No
LNG vessels	Zeus Intelligent	PA	-	-	-	-	-	-	-	No
	LNG World	PA	-	-	-	-	-	-	-	No
Oil platforms	Oil and gas: offshore maps in UK	O	H	H	L	H	H	H	80,00%	No

Legend: A - Accessibility; C - Clarity; R - Relevance; TP - Timeliness & Punctuality; CC - Coherence & Comparability; AR - Accuracy & Reliability; H - High; M - Medium; L - Low; N/A - Not available; O - Open; OR - Open with registration; PPA - Partially paid access; P - Paid access

- scope of available information – what kind of information is available in a source,
- scope of retrieved information – which information pieces will be retrieved from the source by the SIMMO system,
- type of source – whether retrieved content is published in shallow or deep Web and in what form data is available, e.g. internal database, separate xls, pdf or csv files,
- update frequency – how often information in a source is updated; whether the whole content is updated or only new information appears,
- politeness policy – what kind of robot exclusion protocol the website administrators defined, if any, defining which parts of their Web servers should not be accessed by crawlers as well as indicating the number of seconds to delay between requests,
- re-visit approach – how often the SIMMO system will retrieve information from a source, i.e. the intervals between consecutive downloads from the source, taking into account the politeness policy, if defined.

To sum up, the application of the proposed sources' selection methodology in the SIMMO project allowed us to identify, assess and finally choose open internet data sources of the highest quality, which are about to supply the SIMMO system with the maritime data. However, it needs to be stressed that the whole assessment procedure did not focus on the quality of data available in a given source, rather than on the quality of the source itself. The aspects of data quality retrieved from the selected sources (e.g. data completeness, validity, consistency, ambiguity etc.) were dealt in the project at the later steps of the system development. Due to limited volume of this paper it is impossible to present them here.

6   Discussion and Conclusion

The goal of this paper was to propose a methodology for identification, assessment and selection of internet data sources, which are about to be a source of information for an information system. In a nutshell, the proposed methodology consists of 5 steps, starting from potential sources identification and ending with definition of cooperation model. It can be used in designing an information system in any domain which requires acquisition of data available in the Internet. In the paper, the method is described and evaluated based on the running example of the information system from the maritime domain.

The performed analysis gave us an overview on the scope of the data related to vessels and maritime domain which is available on the Web and can be freely used in the maritime information systems. Moreover, the conducted analysis revealed that there is plenty of data sources with valuable information that unfortunately cannot be used due to strict terms of use or policies regarding prohibition on the use of any techniques for automatic retrieval of data published by a given source. There are also sources, which require prior written authorization to use their data. However, an attempt to get such authorization failed (there was no response from the information provider regarding our request for access) or the whole procedure is long-lasting and requires engagement of public authorities.

At the moment, the proposed method has not been validated in other domains or industries. Nevertheless, we believe that it could be utilized for assessment of potential internet sources for traffic monitoring systems used in other transportation areas, such as railway, road or air. Analysis of possible exploitation in these domains may be the subject of the future work on the proposed methodology.

Moreover, the future work may encompass proposing additional analysis steps, which would focus more on the quality of the data itself (not only on the quality of the source). It would require specification of additional quality measures and development of method(s) for automatic assessment of the data quality as soon as the data is acquired. Also inclusion of additional attributes, e.g. domain-related, for assessing the source quality may also be considered.

Acknowledgements. This work was supported by a grant provided for the project *SIMMO: System for Intelligent Maritime MOnitoring* (contract no. A-1341-RT-GP), financed by the Contributing Members of the JIP-ICET 2 Programme and supervised by the European Defence Agency.

References

1. Robey, D., Markus, M.L.: Rituals in information system design. MIS Q. **8**, 5–15 (1984)
2. International Organization for Standardization: ISO 8402–1986 (GB/T6583-1992): Quality-Vocabulary, June 1986
3. Vespe, M., Sciotti, M., Battistello, G.: Multi-sensor autonomous tracking for maritime surveillance. In: International Conference on Radar, 2008, pp. 525–530. IEEE (2008)
4. European Commission: Integrated Maritime Policy for the EU. Working document III on Maritime Surveillance Systems (2008)
5. Kazemi, S., Abghari, S., Lavesson, N., Johnson, H., Ryman, P.: Open data for anomaly detection in maritime surveillance. Expert Syst. Appl. **40**(14), 5719–5729 (2013)
6. Alonso, J., Ambur, O., Amutio, M.A., Azañón, O., Bennett, D., Flagg, R., McAllister, D., Novak, K., Rush, S., Sheridan, J.: Improving access to government through better use of the web. World Wide Web Consortium (2009)
7. Rhodes, B.J., Bomberger, N.A., Seibert, M., Waxman, A.M.: Maritime situation monitoring and awareness using learning mechanisms. In: Military Communications Conference, MILCOM 2005, pp. 646–652. IEEE (2005)
8. Fooladvandi, F., Brax, C., Gustavsson, P., Fredin, M.: Signature-based activity detection based on Bayesian networks acquired from expert knowledge. In: 12th International Conference on Information Fusion, FUSION 2009, pp. 436–443. IEEE (2009)
9. Riveiro, M., Falkman, G., Ziemke, T.: Improving maritime anomaly detection and situation awareness through interactive visualization. In: 11th International Conference on Information Fusion, 2008, pp. 1–8. IEEE (2008)

10. Helldin, T., Riveiro, M.: Explanation methods for Bayesian networks: review and application to a maritime scenario. In: Proceedings of the 3rd Annual Skövde Workshop on Information Fusion Topics (SWIFT 2009), pp. 11–16 (2009)
11. Peter, B.: Data quality. The key to interoperability (2010)
12. Wang, R.Y., Reddy, M.P., Kon, H.B.: Toward quality data: an attribute-based approach. Decis. Support Syst. 13(3), 349–372 (1995)
13. Wang, R.Y., Strong, D.M.: Beyond accuracy: what data quality means to data consumers. J. Manage. Inf. Syst. 12, 5–33 (1996)
14. Batini, C., Cappiello, C., Francalanci, C., Maurino, A.: Methodologies for data quality assessment and improvement. ACM Comput. Surv. 41(3), 16:1–16:52 (2009)
15. European Statistical System: ESS handbook for quality reports. Eurostat (2014)
16. European Parliament: Regulation (EC) No 223/2009 of the European Parliament and the Council of 11 on European statistics and repealing Regulation (EC, Euratom). Official J. Eur. Union 52 (2009)
17. Naumann, F., Freytag, J.C., Spiliopoulou, M.: Quality-driven source selection using data envelopment analysis. In: Proceedings of the 3rd Conference on Information Quality (IQ), Cambridge, MA (1998)
18. Dorofeyuk, A., Pokrovskaya, I., Chernyavkii, A.: Expert methods to analyze and perfect management systems. Autom. Remote Control 65(10), 1675–1688 (2004)
19. Kazemi, S., Abghari, S., Lavesson, N., Johnson, H., Ryman, P.: Open data for anomaly detection in maritime surveillance. Expert Syst. Appl. 40(14), 5719–5729 (2013)
20. Brown, B.B.: Delphi process: a methodology used for the elicitation of opinions of experts. Technical report, DTIC Document (1968)

Towards Identifying the Business Value of Big Data in a Digital Business Ecosystem: A Case Study from the Financial Services Industry

Anke de Vries[1(✉)], Claudia-Melania Chituc[2], and Fons Pommeé[3]

[1] DAF, Information Technology Division, Eindhoven, The Netherlands
Anke.de.Vries@daftrucks.com
[2] Eindhoven University of Technology, Eindhoven, The Netherlands
C.M.Chituc@tue.nl
[3] Capgemini, I&D Risk, Regulations & Compliance, Utrecht, The Netherlands
Fons.Pommee@capgemini.com

Abstract. In today's increasingly digital business ecosystem, big data offers numerous opportunities. Although research on big data receives a lot of attention, research on the business value of big data is scarce. The research project presented in this article aims at advancing the research in this area, focusing on the identification of opportunities towards determining the business value of big data. The goal of the research project pursued is to develop a framework that supports decision makers to identify opportunities for attaining business value form big data in the financial services industry. The proposed framework was constructed based on information collected by performing an in-depth literature review and interviews with experts in the area of big data and financial services industry, and it was empirically validated via a questionnaire sent to experts. A comparative analysis was also performed, emphasizing the strengths of the proposed framework over existing approaches.

Keywords: Big data · Business value · Financial services industry · Case study · Digital business ecosystem

1 Introduction

The business world is rapidly digitalizing, creating new opportunities for companies [1]. Data and information are currently becoming primary assets for many organizations [2], which make extensive use of information and communication technologies (ICT) to collect, store and process digital data. This trend and the recent ICT developments determined the use of the term big data, which involves storing and analyzing data, and transforming data into knowledge (and information) that ultimately contribute to the value of an organization. Characterized by the enormous volume, variety, and velocity of data [3–6], big data brings numerous opportunities for businesses in today's increasingly digital business ecosystem. Examples of applications of big data in commerce and business, society/administration, and scientific research are illustrated in [7]. Analytical tools coming from big data can make complex insights easier to understand and make

© Springer International Publishing Switzerland 2016
W. Abramowicz et al. (Eds.): BIS 2016, LNBIP 255, pp. 28–40, 2016.
DOI: 10.1007/978-3-319-39426-8_3

this information immediately ready to use at every point in the organization and at every skill level [8]. The potential of big data to create value is also emphasized in a recent report by McKinsey Global Institute [9].

Although the potential of big data to generate value is recognized, the review of previous studies revealed that this topic is not extensively explored, and, in general insufficiently understood. Most studies are focusing on technological aspects of big data, neglecting business-related aspects. The business value of big data is of high relevance especially for the financial services industry. In fact credit card companies represent one of the main investors in big data [10]. Banks make use of big data analytics for customer segmentation and profiling, product cross selling based on the profiling, spending patterns of customers, channel usage, sentiment and feedback analysis and security and fraud management [11]. However, banks are struggling to profit from increasing volumes of data [12]. Additionally, a comprehensive approach to identify the potential value of big data that can support stakeholders or decision makers in the financial services industry is not yet available.

The research project presented in this article addresses this gap. The goal of the research project pursued is to develop a framework that supports decision makers to identify opportunities for attaining business value[1] out of big data in the financial services industry. The proposed framework was constructed based on information collected by performing an in-depth literature review and interviews with experts in the area of big data and financial services industry, and it was empirically validated via a questionnaire. A comparative analysis was also performed, emphasizing the strengths of the proposed framework.

This paper is structured as follows. Section 2 presents the methodology followed. Relevant concepts are referred next. The proposed framework is described in Sect. 4. The case study validation is presented next. A comparative analysis of the proposed framework with existing approaches is then discussed. The paper concludes with a section summarizing the results and addressing the needs for future research.

2   Methodology

The main research question that guided this research work is: *How can the business value of big data be determined in the area of financial services?*

The research approach developed to answer this research question is illustrated in Fig. 1, following a design science approach [13]. Firstly, the environment is established. In this research study, the need of the design of a framework illustrating the business value of big data in the financial services industry is posed. The knowledge base is used as a foundation to build the artefact, which is the actual conceptual framework. A document study (Step 1) and interviews (Step 2) are performed to build the first draft of the conceptual framework that indicates how business value can be created from investing

---

[1] Within the scope of this research project, the term business value refers to the financial gains (e.g., expressed in monetary units, such as increased profit), and non-financial gains (e.g., competitive advantage, productivity enhancement) of an organization.

in big data (Step 3). Interviewees were chosen based on their affiliation with business intelligence and big data. This process was repeated multiple times to refine the developed framework. The knowledge base is also used to provide methodologies for the evaluation step (Step 5). In the first three steps the elements for the framework are selected. The validity of the cause and effect relations is verified via a questionnaire. The survey, with the conceptual framework as input, is only distributed to persons with knowledge on business intelligence and big data. The results of the survey are processed in the final framework in consultation with experts. As a last step, the developed framework is empirically validated. Step 5 also served to gain practical insights with respect to the use of the proposed framework in practice. Altogether, the last part of the research project concerned the analysis of the results, final refinement of the framework and the elaboration of a set of recommendations for future research work.

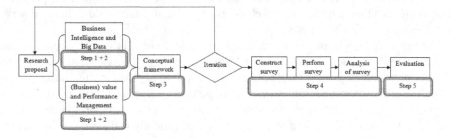

Fig. 1. Research methodology

3 Business Value and Big Data: An Overview

In basic terms, big data refers to datasets that are large in volume, diverse in data sources and types, and created quickly, resulting in bigger challenges in harvesting, managing and processing them via traditional systems and capabilities [14].

Companies must undergo substantial changes in order to make use of big data, which requires a significant investment in technology, and the formulation of an adequate strategy within the organization [15]. Numerous challenges are associated with big data, e.g., difficulties in data capture, data storage, data analysis and data visualization [7]. Designing, developing, and implementing data management systems that live up to the access and speed requirements is a highly complicated process [10]. Big data solutions become actually successful only when people, processes, and technology are integrated [10, 16–18].

Big data gets increasing attention in the financial services industry because accurate, consistent management of both financial data and customer information is essential to be successful in this industry [10]. The variety of data can have an impact on organization's risk measurements and its trading and investment performance. According to [5], using big data can translate into operational, financial, and business gains, and quicker access to cleaner, more relevant data to drive insights and optimize decision making. An additional benefit for this sector refers to the use of big data analytics for regulatory compliance [10]. The impact of big data analytics in the banking sector in India is discussed in [11], with emphasis on six aspects: customer segmentation and profiling, product cross selling based on the profiling, spending patterns of customers, channel usage, sentiment and feedback analysis and security and fraud management. However, a comprehensive approach to identify the business value of big data in this sector that can be used by decision makers is not provided.

As emphasized in [19], the ultimate role of data is to support decision making. Executives now want to run businesses on data-driven decisions, to understand optimal solutions based on complex business parameters or new information and they want to take actions quickly [8]. The utility of big data for data-intensive decision-making is emphasized in numerous works, e.g., [7, 20]. Researchers, policy and decision makers have to recognize the potential of harnessing big data that can generate growth in their fields, and the potential benefits it brings, including: increase in operational efficiency, identification and development of new products and services, identification of new customers and markets [7]. Companies can significantly improve their business performance simply by focusing on how operating data can inform daily decision making [21]. Top-performing organizations actually use analytics five times more than lower performers [8]. A reason for which companies do not make better use of data analytics might be that their management practices have not caught up with their technology platforms [22]. Besides the technological shift, the difficulties related to the cultural shift made by adopting an evidence-based decision making tool needs to be considered.

Research on economic and business-related aspects of big data is scarce. Although several approaches are available to measure value (e.g., value of IT, value of business intelligence, such as: [23–25]), scarce research focuses on the business value of big data. Banks are striving to profit from increasing volumes of data [4]. Benefits of big data are described, but no concrete process or framework is advanced that shows how business value can be created from investing in big data. However, the importance to understand the business value of big data is highly emphasized, e.g., [24, 26]. Preconditions that need to be met to create business value for a specific environment are discussed in [27]: technical development and project management, strategic alignment, process engineering, change management. Three main ways for performance assessment are identified in [28]: performance indicators, benchmarking methods and frameworks for performance assessment. In the context of big data, frameworks are recommended to be used to capture the (potential) business value of big data, representing the most qualitative method for performance assessment.

Several performance measurement frameworks were identified and examined considering the scope of this research work: the Balanced Scorecard of Kaplan and Norton [29, 30], Performance Pyramid of Lynch and Cross [31], Kanji's Business Scorecard [32], and the Performance Prism [33]. These performance frameworks all have predefined measurements that are used to compare the organization's results with predefined standards. The Benefit Logic Model developed by Capgemini [34] was also analyzed in this project. Traditionally, the Benefit Logic Model was used to define a cause-effect diagram that shows how solutions contribute to cash flow generation. The final solutions presented on the left side of the model are linked to the drivers presented in the middle of the model that represent the cause and effect relations to reach the final goal. Only the solutions and drivers that are mentioned more than once are used in the conceptual framework. This final goal is presented at the right side of the model, which is cash flow generation in the traditional diagram. Cash flow generation can be created by increasing revenue or decreasing costs. Breaking down these two sides into smaller parts will lead to improvement opportunities that can be reached by implementing the solutions [16].

Although the (potential) benefits of big data and analytics are generally recognized, a key adoption obstacle remains the lack of understanding of how to leverage analytics for business value. This research work covers this gap by advancing a value framework for investing in big data, following the Benefit Logic Model by Capgemini [34]. This model is chosen as a basis for the development of a value framework of big data for the financial services industry because of the clear overview, the focus on value, and the structured approach of the model. The Benefit Logic Model describes how value can be created without quantifying it, which makes it very usable for this study. Moreover, with the Benefit Logic Model it is possible to provide insight into the value of big data without other elements, e.g., infrastructure, actors.

4   Framework Design and Validation

4.1   Data Collection

The conceptual framework developed and presented in this article was elaborated after pursuing an in-depth literature review (using various retrieval systems e.g. ABI/Inform, ScienceDirect, and Emerald) and interviewing experts in the fields of financial services and big data[2]. The key design issue was capturing the value of investing in big data for the financial services industry. The information from the document study was used as input to define the questions for the semi-structured interviews. Specific information about factors that influence the value of big data was also considered, e.g., drivers leading to positive cash-flow, solutions that could be implemented right away. The structure of the interview and examples of questions in each category are provided in the *Appendix*.

4.2   Framework Elements

The collected information determined the identification of preconditions, drivers and solutions that need to be implemented to gain value from investing in big data in the financial services industry. These elements were included in the first draft of the conceptual framework that shows the cause and effect relationships between all the factors. Two iterations were performed to optimize the quality of the conceptual framework and to ensure the focus on big data and financial services industry.

---

[2] In total eight semi-structured interviews with experts in the field of financial services and big data were conducted. Semi-structured interviews were used because they give the companies' perspective on this topic and could confirm insights that come from the document study. The purpose of the interviews was to acquire information about their vision on big data, possible approaches to assess the business value of big data and the potential benefits an organization would gain by investing in big data.

Three preconditions that have to be met at all times in order to gain value from the investment were included in the framework: privacy, security, and compliance to regulations. The solutions that have to be implemented to gain value refer to: installing the hardware and software required by big data, performing analytics and applying visualization techniques, creating a project team to create culture, methods, and provide training, and attracting new employees who fit best with a big data environment. It is important that the hardware and novel software tools are used in combination with the existing IT landscape to be efficient and cost-effective, and the new processes derived from using big data techniques are synchronized with the current processes to ensure that inter- and intra-organizational business processes are executed without errors. Note to neglect is that employees and management need to change the mindset to a new way of working with big data and that they need specific trainings to use the new technology. When employees are trained to work with big data and master it, they will probably be more satisfied, leading to a decrease in employee costs. The relation between attracting new employees, employees' satisfaction, and decrease in employee costs is the only relation not shown in literature, being based only on information from the interviews. All the other relations are supported by the document study and interviews with experts.

The drivers for value were also identified. Only the solutions and drivers that were mentioned more than once were included in the framework. The driver 'event-driven marketing' was combined with the driver 'customized offering' because both aim at providing a personalized offer for a customer. The drivers 'real time information delivery', 'increase of agility/velocity', and 'discovering the total experience of customers' are linked to the driver 'process insight' because they all refer to the same factor. This driver includes (real-time) insights in the behavior of customers and processes and delivers the information as fast as possible. The solution 'change in mindset of employees' was combined with the solution 'investing in change management' since it is actually a part of change management.

4.3 Case Study Validation

To validate the developed framework, a case study was performed. The collected information allowed the refinement of the developed framework. The case study analysis was performed through a questionnaire. Every question included in the questionnaire corresponded actually to a line of the framework, indicating to what extent the respondent agrees with this linkage between two factors. All questions are presented with the key design issue of capturing the value of big data in the financial services industry. The questions were closed questions with the answering options based on a five point Likert scale going from 'strongly disagree' to 'strongly agree'. Additionally, personal information was requested to characterize the sample. At the end of the questionnaire it was possible for respondents to provide comments.

51 experts responded the questionnaire, from which 19 completely filled in the questionnaire. To check if the missing data of the respondents is random or systematic, the Little's MCAR $\chi2$ test was performed. The significance of the test is 1,000 (> 0,05), which means that the missing data is completely at random. Therefore it is only possible to use regression based methods to fill in the missing values [35]. The cases with more than 15 %

of missing values had to be deleted because completing them would cause too much bias [35]. The questionnaire contains 69 variables. Thus, the cases with more than 10 missing values were deleted. This resulted in a deletion of 24 cases, leaving 27 cases of which 8 were incomplete. These 8 cases were completed using the regression based method mean substitution to have a slightly bigger dataset. Accordingly, the blank spots were replaced by the mean of that variable based on answers from other respondents. Ultimately, a dataset of 27 complete cases was constructed and used for analysis.

4.4 Characterization of Respondents

The respondents were asked to provide information about certain demographics, such as: gender, age, nationality, education, current industry sector, current company, job status, and the number of years they worked at the current organization. From the 27 respondents, 21 are males (77,8 %), and 12 respondents belong to the category 21–35 years old (44,4 %), 11 belong to the category 36–50 (40,7 %), and 4 belong to the category 51–65 years old. This division relates to the range of the working population. 20 respondents were Dutch (74,1 %), 6 Indian, and 1 was Norwegian. For the 27 respondents, 3 completed high school, 9 achieved a bachelor degree, and 15 accomplished a master degree. This means that 88,9 % of respondents belong to the category higher education, which is not a representation of the working population. 20 respondents currently work in the financial services industry (74,1 %), 2 in retail, 1 in care, and 4 are consultants in the IT sector (14,8 %). For the 27 respondents, 2 are first level supervisor, 5 are in middle management, and 20 have non-managerial jobs (74,1 %), which means that top management is not represented in this dataset. For the 27 respondents, 1 works at the current organization for less than 1 year, 10 work for 1–2 years at the current organization (37 %), 4 work for 3–5 years at the current organization, 4 work for 6–10 years at the current organization, and 8 work for more than 10 years at the current organization.

4.5 Reliability and Validity

The reliability of the data was tested with the reliability coefficient Cronbach's Alpha [35, 36]: $\alpha = 0,983$ and remains around 0,983 when a construct would be deleted. All the relations, in both directions, are included, which results in 6o items that are used to calculate Cronbach's Alpha. The validity of data represents if a measure assesses the construct that it is intended to measure [35, 36]. Convergent validity and discriminant validity are commonly used. A correlation matrix was used to test them. A correlation matrix provides the strengths of a relationship between two variables without providing a direction of the relationship. Because the data set contains only 27 cases, it was not possible to assume that the data is normally distributed. Therefore normality tests have been performed to test the normality of the data and it was chosen to use Spearman's correlation coefficient to check the validity of the data, because this coefficient does not require data to be normally distributed [35]. Because the proposed relationships between variables are directional (the relations between cause and effect are tested one-directional), a one-tailed test was used [35]. By analyzing the correlation matrix, it was concluded that the variables that should have a strong relationship correlate highly and variables that should not have a relationship correlate on a low level. Though, three correlations are striking. Firstly, the processes payments, savings, financings, investments, and insurances correlate highly with each other for variables such as customized deals and synchronizing IT, often a correlation of 0,9 or higher.

This is not a problem for the validity; however it could be a signal that the processes should be combined into one variable. The second correlation that is noticeable is the relation between cross-selling and retention of the customer, which is quite low while it was expected to be high. The last correlation is the relation between the employee costs and costs in general, which is actually lower than expected. Although these last correlations indicate that the discriminant validity is affected, overall the correlations do not show many deviations. Therefore, the requirements of convergent and discriminant validity are met.

4.6 Value Framework for the Financial Services Industry

The results obtained determined the development of the final version of the framework: Big Data Value Framework for the Financial Services Industry portrayed in Fig. 2. This framework supports decision makers to identify opportunities for attaining business value out of big data in the financial services industry. The elements and arrows illustrate how value can be obtained from each solution. The links are independent of each other, indicating that reaching a goal is already possible when meeting one of the sub goals.

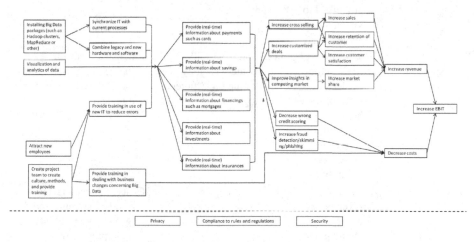

Fig. 2. Conceptual framework for identifying opportunities to attain business value out of big data [16]

The solutions that are presented in the framework and have to be implemented to gain value were grouped on the left side of the framework. The drivers that are representing the cause and effect relations to eventually get to the final benefit (increasing EBIT) are shown in the middle of the framework. The implementation of the solutions supports the access to accurate real-time information related to the main processes in the financial services industry: payments, savings, financings (credits, lending, and mortgages), investments, and insurances. It also provides support to decision makers, improving the accuracy of decisions. These more complete insights will lead to an increase in better targeted marketing (e.g., cross selling, customized deals), better insights in the competing market, and it also decreases wrong credit scoring and stimulates fraud detection. When customers are better targeted with marketing it is likely that the sales, the retention of customer, and the satisfaction of customer will increase because the customer gets an individual treatment. Moreover, better insights in the competing market can lead to an increase in the market share because an organization now has information on how to improve the organization to outperform the competition. All together this will lead to an increase in revenue.

Table 1. Comparison of Value Framework.

Framework/ Approach	Basic elements/ dimensions	Main strengths for big data value assessment	Main weaknesses for big data value assessment
Balanced score-card [20, 21]	Four perspectives (Customer, financials, internal business processes, learning and growth) which are linked to vision and strategy	-Customer, financial, learning and growth dimensions, and their relations with strategy and vision is relevant for big data; -Identification of different objectives and measures	Technology is not addressed
Performance prism [33]	Five perspectives: stakeholder satisfaction, stakeholder contribution, strategies, processes, capabilities	Focus on stakeholders to indicate what they need from the big data technology	Extensive focus on processes, which is not of (high) relevance for big data
Value Framework for the financial services industry	Costs and revenue, technology, learning and growth, internal business processes, customer relationship management	-Focus on big data	Costs and revenue, technology, learning and growth, internal business processes, customer relationship management

At the cost side, the decrease in wrong credit scoring will decrease the overall costs because credits are provided based on real insights in behavior and not on predetermined categories. Together with an increase in fraud detection and a decrease in employee costs the overall costs will decrease. The increase in revenue and decrease in costs will eventually lead to a positive cash flow generation.

The preconditions that have to be met at all times are privacy, compliance to rules and regulations, and security. These preconditions are independent of the solutions, drivers, and final goal. Personal information should only be used when rules are not trespassed and the person accepts that the organization uses the information. Moreover, an organization should be compliant to the rules and regulations that are defined. Especially in the financial services industry there are many different rules which change constantly. Last, the security of information should be optimal at all times to make sure it will not be used by non-authorized persons.

5 Comparative Analysis and Discussion

The proposed framework was compared to other two approaches for value assessment: the Balanced Scorecard [29, 30], and Performance Prism [33]. The main elements of each approach are summarized in Table 1, and the main strengths and weaknesses for big data value assessment are indicated. The main strengths of the framework advanced in this article compared to the other two approaches are: (i) it addresses technology-related aspects which are of high importance in the context of big data, (ii) it provides support in identifying value creation from an investment, (iii) supports a cause-effect analysis between the elements included in the framework, (iv) it was constructed explicitly to address the specificities of the financial sector, which is targeted in this research project.

6 Conclusions and Future Work

Most organizations nowadays are fundamentally dependent on their data and information handling services and tools. Big data services are evolving in an ecosystem in which diverse technologies, market needs, social actors and institutional contexts fuse and change over time [37]. Research on big data focuses mainly on technological aspects (e.g., data interoperability, algorithms, development of tools), and research on business-economic aspects is scarce. Although big data is associated with numerous benefits, the literature review performed showed that a value framework for big data, in general, or for the financial services industry, is not yet available. The Big Data Value Framework for the Financial Services Industry advanced in this article (illustrated in Fig. 2) represents a step towards overcoming this gap. It embeds elements gathered from an in-depth literature review and interviews with experts in the areas of big data and financial services industry, and it was validated by industry representatives. Accordingly, the solutions (e.g., installing the hardware and software, create project team) and the drivers (e.g., customized deals, cross selling, fraud detection) will lead to an increase in revenue and decrease in costs. However, the preconditions privacy, security, and compliance to rules and regulations have to be met at all times to generate business value.

Although the research work performed was accurately conducted and follows a well-defined methodology, some limitations of the present work can be identified. Only eight interviews are performed which provide insights in the opinions of eight experts. This also holds for the information collected via the questionnaire. With only 27 respondents (experts in the areas of big data and financial services industry) the framework cannot be generalized. However, it can be used as an indication of which variables have an influence on the business value of big data.

Future work will focus on the development of drivers at the cost side of the proposed framework. Mainly drivers for the revenue side were identified from the interviews and questionnaire, whereas the drivers of the cost side lagged behind, which might be caused by the respondents' focus on revenue instead of costs. It is also intended to develop specific metrics for the variables included in the framework to quantify the gains from investing in big data.

Appendix: Excerpt of Interview Questions

General

- For how long do you work at the current organization?

Business Intelligence

- What is your definition of business intelligence?
- What are the core elements of business intelligence?

Big Data

- What is your definition of big data?
- What are the core elements of big data?
- To what extent is it possible that big data provides new insights in information needs compared to business intelligence?

Value of Business Intelligence and Big Data

- Which measuring methods do you think are appropriate to measure the business value of big data?
- In which parts of the financial services industry does big data provide business value?

References

1. Weill, P., Woerner, S.L.: Thriving in an increasingly digital ecosystem. MIT Sloan Manag. Rev. **56**(4), 27–34 (2015)
2. Demirkan, H., Delen, D.: Leveraging the capabilities of service-oriented decision support systems: putting analytics and big data in cloud. Decis. Support Syst. **55**, 412–421 (2013)
3. IBM. Bringing Big Data to the Enterprise, 06 March 2013. http://www-01.ibm.com/software/data/bigdata/

4. Capgemini: Financial Services, 03 December 2012. http://www.capgemini.com/financial-services
5. Lopez, J.A.: Best Practices for turning Big Data into Big Insights. Bus. Intell. J. **17**(4), 17–21 (2012)
6. Won, T.: Big Data 3Vs, 7 June 2013. http://tedwon.com/display/dev/Apache+Hadoop#ApacheHadoop-BigData3Vs
7. Chen, C.L.P., Zhang, C.-Y.: Data-intensive applications, challenges, techniques and technologies: a survey on BIG Data. Inf. Sci. **275**, 314–347 (2014)
8. LaValle, S., Lesser, E., Shockley, R., Hopkins, M.S., Kruschwitz, N.: Big Data, analytics and the path from insights to value. MIT Sloan Manag. Rev. **52**(2), 21–31 (2011)
9. Manyika, J., Chui, M., Brown, B., Bughin, J., Dobbss, R., Roxburgh, C., Hung, B.A.: Big Data: the next frontier for innovation, competition, and productivity May 2011. http://www.mckinsey.com/insights/business_technology/big_data_the_next_frontier_for_innovation
10. Nasar, M., Bomers, J.V.: Data management and financial regulation: using a Big Data approach to regulatory compliance. Bus. Intell. J. **17**(2), 34–40 (2012)
11. Srivastava, U., Gopalkrishnan, S.: Impact of big data analytics on banking sector: learning from Indian banks. Procedia Comput. Sci. **50**, 643–652 (2015)
12. Capgemini, Big Data alchemy: how can banks maximize the value of their customer data? (2014). https://www.capgemini.com/resources/big-data-customer-analytics-in-banks. Accessed 24 Nov 2015
13. Hevner, A.R., March, S.T., Park, J., Ram, S.: Design science in information systems research. MIS Q. **38**(1), 75–105 (2004)
14. Bharadwaj, A., et al.: Digital business strategy: towards a next generation of insights. MIS Q. **37**(2), 471–482 (2013)
15. Davenport, T.H., Barth, P., Bean, R.: How 'Big Data' is different. MIT Sloan Manag. Rev. **54**(1), 22–24 (2012)
16. De Vries, H.A.: The Business value of big data: a framework proposal for the financial services industry. M.Sc. thesis, Eindhoven University of Technology, The Netherlands (2013)
17. Capgemini: big data: next-generation analytics with Dutch case studies. Capgemini. http://www.nl.capgemini.com/expertise/publicaties/big-data-nextgeneration-analytics-with-dutch-case-studies/. Accessed 17 Jan 2013
18. Capgemini: The deciding factor: Big Data and decision making. http://www.nl.capgemini.com/sites/default/files/resource/pdf/The_Deciding_Factor__Big_Data__Decision_Making.pdf. Accessed 23 Feb 2013
19. Regalado, A.: The power to decide. what's the point of all that data, anyway? It's to make decisions. MIT Technology Review (2014). https://www.technologyreview.com/s/523646/the-power-to-decide/
20. McAfee, A., Brynjolfsson, E.: Big data: the management revolution. Harvard Bus. Rev. **90**(10), 60–68 (2012)
21. Ross, J.W., Beath, C.M., Quaadgras, A.: You may not need big data after all. Harvard Bus. Rev. **91**(12), 90–99 (2013)
22. March, S.T., Hevner, A.R.: Integrated decision support systems: a data warehousing perspective. Decis. Support Syst. **43**, 1031–1043 (2007)
23. Lönnqvist, A., Pirttimäki, V.: The measurement of business intelligence. Inf. Syst. Manag. **23**(1), 32–40 (2006)
24. Hitt, L.M., Brynjolfsson, E.: Productivity, business profitability, and consumer surplus: three different measures of information technology value. MIS Q. **20**(2), 121–142 (1996)

25. Melville, N., Kraemer, K., Gurbaxani, V.: Information technology and organizational performance: an integrative model of IT business value. MIS Q. **28**(2), 283–322 (2004)
26. Sawka, K.: Are we valuable? Compet. Intell. Mag. **3**(2), 53–54 (2000)
27. Williams, S., Williams, N.: The business value of business intelligence. Bus. Intell. J. **8**(4), 30–39 (2003)
28. Chituc, C.-M., Nof, S.Y.: The Join/Leave/Remain (JLR) decision in collaborative networked organizations. Comput. Ind. Eng. **53**, 173–195 (2007)
29. Kaplan, R.S., Norton, D.P.: Translating Strategy into Action: The Balanced Scorecard. Harvard Business School Press, Boston, MA (1996)
30. Kaplan, R.S., Norton, D.P.: The Strategy Focused Organization: How Balanced Scorecard Companies Thrive in the New Business Environment. Harvard Business School Press, Boston, MA (2000)
31. Lynch, R.L., Cross, K.F.: Measure up! Yardsticks for Continuous Improvement. Blackwell Business, Cambridge, England (1995)
32. Kanji, G.K., e Sá, M.P.: Kanji's business scorecard. Total Qual. Manag. **13**(1), 13–27 (2002)
33. Neely, A., Adams, C., Kennerly, M.: The Performance Prism: The Scorecard for Measuring and Managing Business Success. Pearson Education, London, England (2002)
34. Wortmann A., Maree M.: De bedrijfseconomische benadering van ICT- investeringen; geen woorden maar waarde. Handboek Management Accounting (2001)
35. Field, A.: Discovering Statistics Using SPSS. SAGE Publications, London, England (2009)
36. Sekaran, U.: Research Methods for business: A Skill Building Approach. Wiley, Hoboken NJ (2003)
37. Chae, B.K.: Big data and IT-enabled services ecosystem and coevolution. IT Prof. IEEE **17**, 20–25 (2015)

Big/Smart Data

Flexible On-the-Fly Recommendations from Linked Open Data Repositories

Lisa Wenige[✉] and Johannes Ruhland

Chair of Business Information Systems,
Friedrich-Schiller-University, Jena, Germany
{lisa.wenige,johannes.ruhland}@uni-jena.de

Abstract. Recommender systems help consumers to find products online. But because many content-based systems work with insufficient data, recent research has focused on enhancing item feature information with data from the Linked Open Data cloud. Linked Data recommender systems are usually bound to a predefined set of item features and offer limited opportunities to tune the recommendation model to individual needs. The paper addresses this research gap by introducing the prototype SKOS Recommender (SKOSRec), which produces scalable on-the-fly recommendations through SPARQL-like queries from Linked Data repositories. The SKOSRec query language enables users to obtain constraint-based, aggregation-based and cross-domain recommendations, such that results can be adapted to specific business or customer requirements.

Keywords: Linked Data · Recommender systems · Query-based recommender systems

1 Introduction

Recommender systems (RS) are on the forefront of decision support systems within e-commerce applications [1,2]. Among the most known examples of recommender systems are the ones used by well-established e-commerce retailers such as Amazon[1] or Netflix[2]. Here, users receive personalized suggestions for items of the product catalog. But both content-based and collaborative filtering systems suffer from certain shortcomings, such as rating sparsity or limited content analysis [1]. To address the problem of data sparsity, the Linked Open Data (LOD) movement gave rise to the type of Linked Data recommender systems (LDRS). These systems tackle drawbacks of traditional approaches by enriching existing recommender systems with information from public data sources. But even though current LDRS show promising results, they do not yet take full advantage of the potential that LOD offers. The paper addresses this research gap through:

---

[1] http://www.amazon.de.
[2] http://www.netflix.com.

© Springer International Publishing Switzerland 2016
W. Abramowicz et al. (Eds.): BIS 2016, LNBIP 255, pp. 43–54, 2016.
DOI: 10.1007/978-3-319-39426-8_4

– Identification of central problems of different types of recommender systems (Sect. 2).
– Overview of current challenges for RS in e-commerce applications and how they can be addressed with the help of Linked Data technologies (Sect. 3).
– Technical description of the SKOSRec prototype, that implements these ideas (Sect. 4).
– Evaluation of the systems' key components (Sect. 5).
– Discussion of the main findings and of the practical implications of the approach (Sect. 6).

2 Related Work

2.1 Recommender Systems

Upon the presentation of the first systems in the 1990s, the area of recommender systems has been an established research field. The most common recommendation algorithms apply some form of collaborative filtering technique, where users are referred to preferred items of like-minded peers [3–5]. In contrast to that, content-based systems derive recommendations from feature information for items in the user profile [1,6,7,9,10]. Thus, similar items are detected. Content-based and collaborative filtering systems suffer from certain shortcomings. Even though both paradigms can be combined in hybrid systems [7], many of the problems still remain.

One of the main issues on the operational side is the data sparsity problem, where user preferences are rare. It mostly occurs, when new users or items are added to the system (cold start problem), but it can also arise when the amount of feedback information is simply not enough to derive meaningful recommendations. Especially in content-based systems, users can receive unfitting recommendations due to incomplete or ambiguous item feature information (limited content analysis) [1].

2.2 Linked Data Recommender Systems

Recently, researchers have started to utilize Linked Data information sources to address the problem of insufficient item feature information. The LOD cloud comprises data on almost any kind of subject and offers general purpose information (e.g. DBpedia[3]) as well as data from special domains [8]. LOD resources are usually identified through URIs. Thus, the LOD cloud provides less ambiguous information than text-based item descriptions [9]. Experiments on historic data showed that LDRS are at least competitive with classical systems and sometimes even outperform them in terms of accuracy [9–12,21].

But even though LDRS achieve considerable results, they do not yet take full advantage of the LOD cloud. Current approaches require a considerable amount

---

[3] http://wiki.dbpedia.org/.

of pre-processing, such as the selection and extraction of item features. Once a set of item features has been selected, the recommendation model is 'hard wired' into the system and can not be adapted to changing user or business demands.

2.3 Query-Based Recommender Systems

Due to the fact that most LDRS and non-LDRS are not capable of customizations, there have been efforts to enable systems to produce query-based recommendations. In the field of non-LDRS this is achieved through enhancing relational databases with recommendation functionalities [13]. For instance, the REQUEST system integrates the personalization process with OLAP-like queries, such that the selection of items/users can be based on certain conditions and aggregations. Thus, recommendation models are adaptable to different requirements at runtime [14,15]. But as information on user preferences is usually sparse, this information becomes even sparser when only certain items or users are selected. This often leads to unreliable recommendation results [15].

The issue of data sparsity could be addressed through content-based approaches that enhance item feature information with LOD resources. To date, there are only a few systems that consider user preferences in conjunction with Linked Data technologies, such as SPARQL queries [16–19]. With these systems, expressive user preferences can be formulated. But we argue that the potential of LOD is not yet fully exploited for recommendation tasks. Query-based LDRS either follow a fixed workflow of similarity detection and SPARQL graph pattern matchings [16,17] or face long execution times when processing a large number of triple statements [18].

Therefore, the main goal of the paper is the development of a system that flexibly integrates user preferences with SPARQL elements at reasonable computational cost and that provides novel and meaningful recommendations in update-intensive environments, where local databases do not provide sufficient data.

3   LOD for Flexible Recommendations in e-Commerce

The following aspects give an overview of current challenges of RS applications in the e-commerce sector and describe how they can be addressed with the help of Linked Open Data:

– **Comprehensiveness:** As stated in the previous sections, e-commerce RS have to deal with the issues of data sparsity and low profile overlap among users. Especially small sites do not have a customer base that is big enough to provide enough ratings [2]. That is why the integration of Linked Data sources into recommender systems could help to overcome existing limitations on the data side. The LOD cloud comprises billions of triple statements ranging from general purpose data to information sources from domains, such as media or geography. These datasets can be of value in multimedia retailing or online travel agencies [8].

- **Adaptability:** RS of e-commerce sites usually do not offer functionalities where customers can restrict recommendation results to specific criteria. But to achieve deeply personalized results, it would be desirable to apply pre- or postfiltering on the product catalog [2,13]. For instance, think of a customer of a media streaming site who, in spite of his/her purchasing history, wants to provide the information that he/she is strongly interested in European movies. Above that, in areas like tourism, user preferences depend on many factors, such as context, travel companions or travel destination preferences [20]. In addition to that, not only consumers could profit from customization functionalities, but also marketing professionals and administrators [2,15]. For instance, marketing campaigns could be fit to special holiday occassions of the year to promote long-tail items. These aspects require data-rich applications, that can be accessed with expressive queries.
- **On-the-fly recommendations:** Current RS usually rely on pre-computed recommendation results. But the aspect of adapatability is strongly tied to the aspect of just-in-time recommendations. As customer and business requirements can not be foreseen, recommendation models should be configurable at runtime, such that a user can select the right data when it is actually needed [21]. To enable flexible recommendations results from Linked Open Data repositories, efficient strategies for processing large numbers of triple statements have to be identified.

4 The SKOS Recommender

In this section we present SKOSRec, a system prototype that addresses the previously identified challenges of RS in e-commerce applications.

4.1 Scalable On-the-Fly Recommendations

Most LDRS identify similar items through their features. But considering the large amount of information, using all known features of a resource leads to poor scalability and long processing times. Thus, in the context of LDRS it was proposed to select certain item features (properties) for the recommendation model [9,10,22–24]. But due to the large amount of information in the LOD cloud, the selection process can be error-prone and time consuming. Thus we propose to perform similarity computation on URI annotations that are part of commonly used vocabularies, such as the Simple Knowledge Organization System (SKOS). SKOS vocabularies have become a de-facto standard for subject annotations, since a majority of Linked Data sets are annotated with SKOS concepts. We implemented a system, called SKOS Recommender that uses its own SPARQL-like query language (SKOSRec) to produce flexible on-the-fly recommendations. For identifying similar items the systems relies on SKOS concepts, but can be extended to other URI resources from the LOD cloud. The system uses Apache Jena[4] and can be applied on local as well remote SPARQL endpoints. The following section summarizes the general workflow of the SKOSRec system.

---

[4] https://jena.apache.org/.

1. **Parsing:** Parse the SKOSRec query.
2. **Compiling:** Decompose the query into the preferred input resource r (e.g. a movie) and a SPARQL graph pattern P.
3. **Resource retrieval:** Retrieve *relevant resources* from *SKOS annotations* of r in conjunction with P.
4. **Similar resources:** Score and rank the resources according to their *conditional similarity* with the input resource p.
5. **Recommendation:** Output the final recommendation results.

In the following, we will now rigorously define keywords in italics by using the notation for SPARQL semantics that was introduced by [25].

Definition 1 (SKOS annotations). *Let AG be the annotation graph of an RDF dataset D (AG $\subset$ D), where resources are directly linked to concepts c of a SKOS system via a predefined property (e.g. dct:subject). All nodes of the AG are IRIs and the annotations of an input resource r are defined as follows:*

$$annot(r) = \{c \in AG \mid \exists < r, subject, c >\} \tag{1}$$

Upon retrieval of input resource annotations, similarity calculation does not have to be performed on the whole item space.

Definition 2 (Relevant resources). *The mapping Ω of relevant resources and their annotations is obtained by retrieving all resources P_r that share at least one SKOS concept with resource r. In addition to that, relevant resources are joined with a SPARQL graph pattern P, so that resources are excluded when certain user requirements are defined.*

$$P_r = (r, subject, ?c) \ AND \ (?x, subject, ?c) \tag{2}$$

$$\Omega = \{\mu(?x)|\mu \in [\![P]\!]_D\} \bowtie [\![P_r]\!]_{AG} \tag{3}$$

After querying all relevant resources and their annotations, similarity values can be calculated. They are based on the Information Content (IC) of the shared features of two resources. This idea was introduced by [23], but is expanded to the case when the item space is restricted to match a user defined graph pattern.

Definition 3 (Conditional similarity). *Let annot(r) be the set of SKOS features of resource r and annot(q) the set of SKOS features of resource q and $q \in \{\mu(?x)|\mu \in \Omega\}$, then their similarity can be derived from the IC of their shared concepts $C = \{annot(r) \cap annot(q)\}$*

$$sim(r, q) = IC_{cond}(C) \tag{4}$$

Definition 4 (Conditional Information Content). *The IC of a set of SKOS annotations is defined through the sum of the IC of each concept $c \in$*

$\{\mu(?c)|\mu \in \Omega\}$, where $freq(c)$ is the frequency of c among all relevant resources and n is the maximum frequency among these resources.

$$IC_{cond}(C) = -\sum_{c\in C} log\left(\frac{freq(c)}{n}\right) \tag{5}$$

The retrieval of relevant resources and concept annotations can lead to long processing times, especially in cases when concepts are frequently used in a dataset. Hence, the number of records from SPARQL endpoints should be reduced. By knowing the length of the top-n recommendation list, it can be calculated which resources can be omitted without influencing the final ranking. This is the case when the maximum potential score for a certain number of shared features is smaller than the minimum potential score for a higher number of shared features. By this means, it is determined how many annotations have to be shared at least with an input resource (cut value) (see Eqs. 6 and 7).

$$\Omega_{cut} = \{\mu_{cut}(?x)|\mu_{cut} \in F_{count(?c)}(\Omega) > cut\} \tag{6}$$

$$\Omega_{reduced} = \Omega \bowtie \Omega_{cut} \tag{7}$$

4.2   Expressive SPARQL Integration

In the course of this paper, we are only able to give a short overview of the SKOSRec query language. Central to the idea of customizable on-the-fly recommendations is that both item similarity computation and querying of LOD resources can be flexibly integrated in a single query language. Even though there already exist some language extensions that combine SPARQL with imprecise parts [16,17], they do not take full advantage of the expressiveness of the RDF data model. Hence, we propose the SKOSRec query language that extends elements of the SPARQL 1.1 syntax (see underlined clauses in Listing 1) [26]. It enables flexible and powerful combinations of graph pattern matchings and subquerying with recommendation results. The 'RECOMMEND' operator issues the process of similarity calculation based on the input resource and potential user defined graph patterns, whereas the 'AGG' construct ensures that certain resources are exluded from the result set.

Listing 1. Grammar of SKOSRec

```
RecQuery        ::= Prologue SelectPart? SimProjection
                    Aggregation? ItemPart*
SelectPart      ::= SelectQuery
SimProjection   ::= RECOMMEND Var ItemLimit
Aggregation     ::= AGG IRIref
ItemLimit       ::= TOP INTEGER
ItemPart        ::= PREF (DECIMAL)? (VarPart | IRIPart)
IRIPart         ::= IRIref  (ConceptSim)?  (WherePart)?
VarPart         ::= Var (ConceptSim)? (WherePart)?
ConceptSim      ::= C-SIM Relation DECIMAL
Relation        ::= ( < | > | <= | >= | = )
WherePart       ::= BASED ON WhereClause()
```

The central contributions of the new language are summarized below.

- **Recommendations for an input profile:** Whereas recommendations can be generated from both user and item data [16], we argue that the integration of local customer information and LOD resources is not feasible. An e-commerce retailer might avoid such a solution because of privacy concerns and additional costs and would rather prefer to obtain recommendations from outsourced repositories.
- **Graph pattern matching for preference information:** The SKOSRec language allows expression of preferences for variables that are contained in graph patterns. Thus, users can formulate vague preferences.
- **Subquerying with recommendation results:** In some areas it might be helpful to reuse recommendation results as a SPARQL-like subquery. Thus, triple stores can be powerfully navigated.

5   Experiments

We conducted several experiments to evaluate the viability of our approach. The goal of the evaluation was to find out, whether it is possible to get meaningful recommendation results with highly expressive queries from existing LOD repositories at reasonable computational cost. For this purpose, we issued SKOSRec queries from different target domains (movies, music, books and travel) (Table 1) to a local virtuoso server containing the DBpedia 3.9 dataset.

5.1   Scalable On-the-Fly Recommendations

The effectiveness of the optimization approach presented in Sect. 4.1 (Eqs. 6 and 7) was examined on 4 different datasets, where each dataset comprised 100 randomly selected DBpedia resources from the target domains. The performance test was carried out on an Intel Core i5 2500, clocked at 3.30 GHz with 8 GB of RAM. Evaluation results showed that, even though our approach imposes overhead that leads to slightly increased computational cost for smaller datasets (e. g. Fig. 4), it considerably reduces processing times for a growing number of triple statements for bigger datasets (Table 2, Figs. 1, 2 and 3).

Table 1. Overview of the target domains

	Movie	Book	Music	Travel
rdf:type	dbo:Film	dbo:Book	Schema:MusicGroup	dbo:Place
# items	90,063	31,172	86,013	725,546
# annotions p. item (∅)	7.207	4.410	6.060	2.422

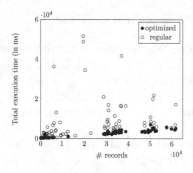

Fig. 1. Movie domain

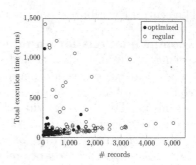

Fig. 2. Book domain

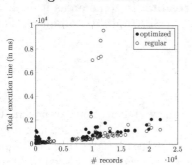

Fig. 3. Music domain

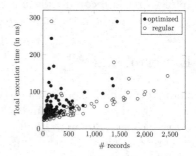

Fig. 4. Travel domain

Table 2. Results of the performance test

Domain	Exec. time in ms (∅)		# records (∅)	
	Regular	Optimized	Regular	Optimized
Movie	9,961	1,921	23,542	20,535
Book	303	92	1,206	837
Music	4,145	501	7,662	501
Place	218	73	621	134

5.2   Expressive SPARQL Integration

As former research on LDRS has already shown that the application of SKOS annotations leads to good precision and recall values in comparison to standard RS algorithms [9,10], we followed an explorative approach. We investigated, whether it is possible to formulate highly expressive SKOSRec queries that produce meaningful recommendation results from the DBpedia dataset. We issued advanced queries to showcase the viability of our language in several usage scenarios of the target domains.

Conditional Recommendations. This query template generates highly individual or business relevant recommendations. In the example depicted below, a marketer wants to obtain query results that are personalized and promote Christmas movies at the same time.

```
PREFIX dct: <http://purl.org/dc/terms/>
PREFIX skos: <http://www.w3.org/2004/02/skos/core#>
PREFIX dbc: <http://dbpedia.org/resource/Category:>

RECOMMEND ?movie TOP 3
PREF <r1> <r2>
WHERE {
 ?movie dct:subject ?c .
 ?c skos:broader* dbc:Christmas_films .
}
```

Input (r1, r2)	Output
The Devil Wears Prada	Love Actually
Bridget Jones's Diary	The Family Stone
	Scrooge
The Terminator	Ben-Hur
Raiders of the Lost Ark	Die Hard
	Trancers

Aggregation-Based Recommendations: Roll Up. When user preferences can be derived from their sublevel entities, the roll-up template might improve recommendations. Think of a travel agency intending to suggest city trips. Two customers would receive similar trip recommendations once they have been to the same cities even though they have visited different points of interest (POI). The example shows that it can be reasonable to instantiate the process of similarity calculation on sublevel entities to better fit recommendations to customer needs.

```
PREFIX rdf: <http://www.w3.org/1999/02/22-rdf-syntax-ns#>
PREFIX yago: <http://dbpedia.org/class/yago/>
PREFIX dbr: <http:dbpedia.org/resource>

SELECT DISTINCT ?place (count(?place) as ?count)
WHERE {
 ?sight ?locatedIn ?place .
 ?place rdf:type yago:City108524735 .
}
GROUP BY ?place
ORDER BY DESC(COUNT(?place))
LIMIT 5
RECOMMEND ?sight TOP 1000
AGG <http://dbpedia.org/resource/Berlin>
PREF <r1> ... <r5>
```

Input (r1 ... r5)	Output
Checkpoint Charlie	Moskau
East Side Gallery	East Berlin
Berlin Wall	Hamburg
DDR Museum	Trieste
Stasi Museum	Warschau
Re:publica	Vancouver
Berghain	London
Friedrichshain	Amsterdam
E-Werk	Paris
Bauhaus Archive	Montreal

Aggregation-Based Recommendations: Drill Down. Sometimes customers find it hard to concretize their preferences. They might have a vague understanding of what they like, but could not tell why. In this example a user knows that somehow he/she likes movies directed by Quentin Tarantino. A drill-down query would find the most similar films to those that were directed by him and aggregate the results, such that related directors and their movies would be recommended.

```
PREFIX dbo: <http://dbpedia.org/ontology/>
PREFIX dbr: <http://dbpedia.org/resource/>

SELECT DISTINCT ?director ?movie WHERE {
 ?movie dbo:director ?director .
}
RECOMMEND ?movie TOP 3
AGG dbr:Quentin_Tarantino
PREF ?prefMovie BASED ON {
 ?prefMovie dbo:director dbr:Quentin_Tarantino
}
```

?director	?movie
Robert Rodriguez	From Dusk till Dawn
Frank Miller	Sin City
Robert Rodriguez	Sin City
Tony Scott	True Romance

Cross-Domain Recommendations. Even though, standard collaborative filtering algorithms sometimes generate recommendations that are from a different domain than the items in a user profile, marketers cannot directly control the outputs. In contrast to that, the SKOSRec language enables explicit cross-domain querying. The following example shows that suggestions for novels (e. g. Beat novels) can be obtained by examining the user preference for a music group (e. g. The Beatles).

```
PREFIX dct: <http://purl.org/dc/terms/>
PREFIX skos: <http://www.w3.org/2004/02/skos/core#>
PREFIX dbc: <http://dbpedia.org/resource/Category:>
PREFIX rdf: <http://www.w3.org/1999/02/22-rdf-syntax-ns#>
PREFIX dbo: <http://dbpedia.org/ontology/>
PREFIX dbr: <http://dbpedia.org/resource/>

SELECT ?book (COUNT(?book) as ?count) WHERE {
 ?book dct:subject ?c .
 ?c skos:broader{,2} dbc:Novels . {
  SELECT ?book  WHERE {
   ?book ?p1 ?o . ?o ?p2 ?band . ?book rdf:type dbo:Book . }}}
GROUP BY ?book
ORDER BY DESC(COUNT(?book))
RECOMMEND TOP 10
PREF dbr:The_Beatles
```

Output (?book)
On the road
One Flew Over the Cuckoo's Nest
Sometimes a Great Notion

6 Conclusion

This paper demonstrated how Linked Open Data technologies can be utilized for highly flexible on-the-fly recommendations in e-commerce applications. Former LDRS calculated user preference predictions offline and thus prevented customizations and frequent updates of data sources as well as recommendation models. Although there have been efforts to enable user restrictions at runtime in query-based recommender systems, most of them either do not scale to large item spaces or do not handle data sparsity issues that arise when restricting the set of potential products to certain criteria. Above that, existing query-based LDRS do not yet take full advantage of the expressiveness that RDF and SPARQL graph patterns offer.

The SKOSRec prototype addresses this research gap by offering a powerful combination of similar resource retrieval and SPARQL graph pattern matchings from Linked Open Data repositories. Thus, individual and/or business preferences can be flexibly integrated. With the SKOSRec query language at hand, e-commerce retailers could define various recommendation workflows that can be adapted to specific usage contexts. For instance, the marketing department could use campaign templates and end users could enter their preferences through a user interface.

References

1. Adomavicius, G., Tuzhilin, A.: Toward the next generation of recommender systems: a survey of the state-of-the-art and possible extensions. IEEE Trans. Knowl. Data Eng. **17**(6), 734–749 (2005)
2. Schafer, J., Konstan, J., Riedl, J.: E-commerce recommendation applications. Appl. Data Min. Electron. Commer. **5**(1), 115–153 (2001)
3. Konstan, J., Miller, B., Maltz, D., Herlocker, J., Gordon, L., Riedl, J.: GroupLens - applying collaborative filtering to Usenet news. Commun. ACM **40**(3), 77–87 (1997)
4. Terveen, L., Hill, W., Amento, B., McDonald, D., Creter, J.: PHOAKS - a system for sharing recommendations. Commun. ACM **40**(3), 59–62 (1997)
5. Shardanand, U., Maes, P.: Social information filtering: algorithms for automating word of mouth. In: Proceedings of the SIGCHI Conference on Human Factors in Computing Systems, pp. 210–217 (1995)
6. Balabanovic, M., Shoham, Y.: Fab - content-based, collaborative recommendation. Commun. ACM **40**(3), 66–72 (1997)
7. Mobasher, B., Jin, X., Zhou, Y.: Semantically enhanced collaborative filtering on the web. In: Berendt, B., Hotho, A., Mladenič, D., van Someren, M., Spiliopoulou, M., Stumme, G. (eds.) EWMF 2003. LNCS (LNAI), vol. 3209, pp. 57–76. Springer, Heidelberg (2004)
8. Schmachterberg, M., Bizer, C., Paulheim, H.: State of the LOD Cloud 2014. http://linkeddatacatalog.dws.informatik.uni-mannheim.de/state/
9. Di Noia, T., Mirizzi, R., Ostuni, V., Romito, D.: Exploiting the web of data in model-based recommender systems. In: 6th ACM Conference on Recommender Systems, pp. 253–256 (2012)

10. Di Noia, T., Mirizzi, R., Ostuni, V., Romito, D., Zanker, M.: Linked open data to support content-based recommender systems. In: 8th International Conference on Semantic Systems, pp. 1–8 (2012)
11. Peska, L., Vojtas, P.: Enhancing recommender system with linked open data. In: Larsen, H.L., Martin-Bautista, M.J., Vila, M.A., Andreasen, T., Christiansen, H. (eds.) FQAS 2013. LNCS, vol. 8132, pp. 483–494. Springer, Heidelberg (2013)
12. Harispe, S., Ranwez, S., Janaqi, S., Montmain, J.: Semantic measures based on RDF projections: application to content-based recommendation systems. In: Meersman, R., Panetto, H., Dillon, T., Eder, J., Bellahsene, Z., Ritter, N., De Leenheer, P., Dou, D. (eds.) ODBASE 2013. LNCS, vol. 8185, pp. 606–615. Springer, Heidelberg (2013)
13. Koutrika, G., Bercovitz, B., Garcia-Molina, H.: FlexRecs - expressing and combining flexible recommendations. In: ACM SIGMOD International Conference on Management of Data, pp. 745–758 (2009)
14. Adomavicius, G., Tuzhilin, A.: Multidimensional recommender systems: a data warehousing approach. In: Fiege, L., Mühl, G., Wilhelm, U.G. (eds.) WELCOM 2001. LNCS, vol. 2232, pp. 180–192. Springer, Heidelberg (2001)
15. Adomavicius, G., Tuzhilin, A., Zheng, R.: REQUEST: a query language for customizing recommendations. Inf. Syst. Res. **22**(1), 99–117 (2011)
16. Ayala, A., Przyjaciel-Zablocki, M., Hornung, T., Schätzle, A., Lausen, G.: Extending SPARQL for recommendations. In: Semantic Web Information Management, pp. 1–8 (2014)
17. Kiefer, C., Bernstein, A., Stocker, M.: The fundamentals of iSPARQL: a virtual triple approach for similarity-based semantic web tasks. In: Aberer, K., et al. (eds.) ASWC 2007 and ISWC 2007. LNCS, vol. 4825, pp. 295–309. Springer, Heidelberg (2007)
18. Rosati, J., Di Noia, T., Lukasiewicz, T., Leone, R., Maurino, A.: Preference queries with Ceteris Paribus semantics for linked data. In: Debruyne, C., Panetto, H., Meersman, R., Dillon, T., Weichhart, G., An, Y., Ardagna, C.A. (eds.) OTM 2015. LNCS, vol. 9415, pp. 423–442. Springer, Sierre (2015)
19. Siberski, W., Pan, J.Z., Thaden, U.: Querying the semantic web with preferences. In: Cruz, I., Decker, S., Allemang, D., Preist, C., Schwabe, D., Mika, P., Uschold, M., Aroyo, L.M. (eds.) ISWC 2006. LNCS, vol. 4273, pp. 612–624. Springer, Heidelberg (2006)
20. Felfernig, A., Gordea, S., Jannach, D., Teppan, E., Zanker, M.: A short survey of recommendation technologies in travel and tourism. OEGAI J. **25**(7), 17–22 (2007)
21. Marie, N., Gandon, F., Ribiere, M., Rodio, F.: Discovery hub: on-the-fly linked data exploratory search. In: 9th International Conference on Semantic Systems, pp. 17–24 (2013)
22. Khrouf, H., Troncy, R.: Hybrid event recommendation using linked data and user diversity. In: 7th ACM Conference on Recommender Systems, pp. 185–192 (2013)
23. Meymandpour, R., Davis, J.: Recommendations using linked data. In: 5th Ph.D. Workshop on Information and Knowledge, pp. 75–82 (2012)
24. Ostuni, V., Di Noia, T., Di Sciascio, E., Mirizzi, R.: Top-n recommendations from implicit feedback leveraging linked open data. In: 7th ACM Conference on Recommender systems, pp. 85–92 (2013)
25. Pérez, J., Arenas, M., Gutierrez, C.: Semantics and complexity of SPARQL. ACM Trans. Database Syst. TODS **34**(3), 16 (2009)
26. SPARQL 1.1 query language. https://www.w3.org/TR/sparql11-query/

Effective Visualizations of Energy Consumption in a Feedback System – A Conjoint Measurement Study

Tobias Weiss[1]([⊠]), Madlen Diesing[1], Marco Krause[2],
Kai Heinrich[1], and Andreas Hilbert[1]

[1] Lehrstuhl für Wirtschaftsinformatik – Business Intelligence Research,
Technische Universität Dresden, Münchner Platz 3, 01187 Dresden, Germany
tobias.weiss@tu-dresden.de
[2] T-Systems Multimedia Solutions GmbH,
Riesaer Strasse 5, 01129 Dresden, Germany
marco.krause@t-systems.com

Abstract. Sustainable use of energy is one of the guiding principles of today's society. But there is a lack of comprehensive analysis solutions for the energy consumption of private households to provide real insights. In order to provide useful information, feedback systems may be the answer. Numerous studies about feedback systems have been conducted so far and each individual component of such a system has been tested. The combination of these components leads to a dashboard for decision support of private households. Within this study the individual components were combined in several configurations and implemented as a prototype dashboard. A Conjoint measurement is used for evaluation and observation of user preferences collected in over 1,000 questionnaires. The result, an evaluated dashboard, combines several effective feedback elements based on user preferences and helps to save energy based on decision support and transparency.

Keywords: Feedback system · Smart metering · Energy intelligence · Decision support · Conjoint

1 Introduction

Sustainability is one of the core concepts in economics and social research. A big focus within these fields of research relates to energy suppliers. These companies are responsible for shaping the future in terms of replacement of fossil fuels and nuclear power sources by renewable energy sources. Besides supporting the shift to renewable sources in the future, they are also able to create incentives towards responsible boundaries for energy consumption in private households [1]. A problem that often occurs in that context is the lack of information provided by the supplier companies. Information are often sparse and not suitable for detecting inefficient energy consumption patterns [2]. This mainly concerns the fact that billing is only provided on an annual level, at least, in Germany. That leads to unspecific, time delayed reactions on the consumer side. With the comprehensive rollout of smart meter technology a new

© Springer International Publishing Switzerland 2016
W. Abramowicz et al. (Eds.): BIS 2016, LNBIP 255, pp. 55–66, 2016.
DOI: 10.1007/978-3-319-39426-8_5

data source is available that yields high potential when it comes to optimization on the consumer side as well as the development of new business models on the supplier side [3, 4].

In order to supply consumers with the required information, energy feedback systems are designed. These systems can help to reduce energy consumption up to 20 % [4]. However, the research field of feedback systems in the area of energy consumption is comprehensive. Therefore we conducted a systematic review [5]. We found that while most of the systems suggest visualization concepts in order to give feedback to the consumer, they were limited to a certain part of feedback rather than looking at the whole system. The latter is important to optimize energy savings and present optimal visualization depending on the data and customer preferences. For example, a study examined the consumption reduction over a test period of 100 days using only an in-home display [5]. In this preliminary study of feedback systems all described elements were characterized and a systematic overview of possible components for feedback and goal setting systems was presented (Fig. 1):

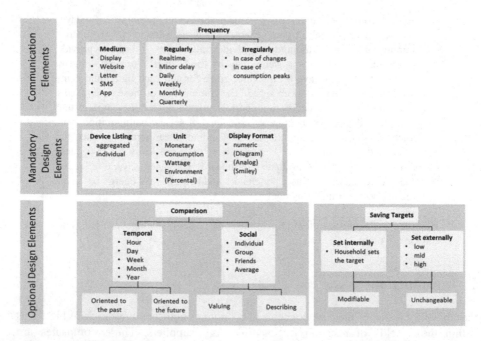

Fig. 1. Systematic overview over feedback system elements [5]

The problem of single components not being integrated into a feedback system which is able to visualize information for the consumer poses a challenge and acts as a basis for this paper.

According to [6], feedback systems are a special case of decision support systems and should employ rules of visualization in order to generate added value on the consumer side. Concepts of user friendly software and usability have to be applied to create an advantage for the consumer using these systems. This is an important factor considering the technology acceptance level of the consumer and the consumer motivation to use feedback systems.

2 Research Design and Methodology

Considering the information in [5], our goal can be stated as the evaluation of visualization components and their combination in order to design an energy feedback system for private households.

To evaluate user preferences for certain combinations of components we conduct a choice-based conjoint measurement analysis [7]. The conjoint measurement enables us to access a user's preferences for each component in the context of the whole feedback system [8]. We will refer to the feedback system as a product and the components as attributes of that product, since this terminology is more common in conjoint analysis.

2.1 Choosing the Attributes

The first step of developing conjoint measurements is to select product attributes which the user is required to access according to his preferences. First, the attributes selected for conjoint measurement have to be checked on the fulfilment of the general requirements of the conjoint analysis [8]. Table 1 shows the attributes according to [5] and their applicability according to the requirements of the study.

Table 1. Alignment of feedback elements with the conjoint measurement analysis [5]

Attribute	Description
Media and frequency	These attributes are highly correlated and cannot be separated. According to [5], electronic media has the highest potential for energy savings and will therefore be used in the study. As a representative media we choose "Website"
Device listing	Details on every electronic device are somewhat difficult from a technological perspective, so we only use aggregated figures over all devices
Units	Mainly monetary units are preferred by the users, as well as consumption- and environmental-focused units are used. To evaluate these findings the monetary unit "Euro" and consumption unit "kWh" are chosen
Visualization	Visualization concepts depend on the statement that should be visualized and therefore vary varied using a conjoint measurement
Social comparisons	User preferences regarding comparisons between households are very different. Therefore the attribute "Social Comparison" is included in the study with the values "Yes" or "No". We use comparisons between the target household and an average consumption of households with the same number of persons

(Continued)

Table 1. (*Continued*)

Attribute	Description
Comparisons over different time periods	Using information in order to make decisions regarding energy consumption in past periods can decrease energy consumption. Therefore this attribute acts as a basic element of every feedback system. While it is uncommon to vary this during the conjoint measurement we will later implement different visualizations for hourly, monthly and annual comparisons. Since there are no studies concerning consumption prognosis, we will vary future prognosis between annual and monthly comparisons. These time intervals were chosen because the month is a basic billing interval for monthly bills like rent, phone bill or salary. Summarizing the above, we used the attribute "Time Comparison" with the values of "Yes" and "No"
Goals	We only conduct the analysis with personal goals rather than pre-set goals from external sources. Since users do have different preferences concerning goals we choose to include this attribute in the analysis as the attribute "Goals" with the values "no goals", "detailed description", "rough description"

2.2 Visualization of the Components

We combined the attributes described above to products which had to be evaluated. In order to yield realistic visualizations we used standard consumption profiles from real households [9]. Figure 2 shows combinations of the attributes "Time Comparison" with monthly and annual comparisons.

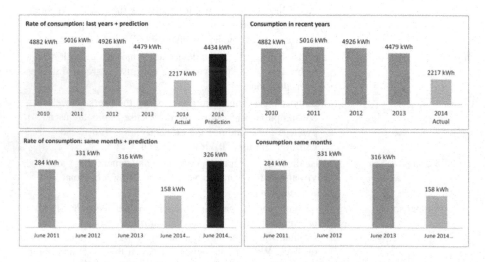

Fig. 2. "Time Comparison" with monthly and annual comparisons

2.3 Empirical Survey

An empirical survey for conjoint measurements consists of choice situations which themselves are orchestrated from certain stimuli. The subject chooses the product variation that earns the highest utility. In order to keep up motivation during the experiment only two stimuli are presented [10]. We used SPSS to generate the choice situations. We have included four attributes with two or three values and two stimuli per choice situation which results in a total of 16 choice situations. Since these are too many choice situations for one subject, we split the situations in two blocks of eight. You can find all stimuli in the appendix [14].

2.4 Quality of the Measurement

In order to measure the quality of the empirical design we use D-efficiency criteria [10]. In order to yield the design matrix with orthogonal coding we transform the original matrix, which is important to evaluate the balance and accuracy. The results are shown in Table 2 respectively for the first four stimuli. Now we calculate the D-efficiency of a $N_D \times p$ design matrix X as [10]:

$$D - efficiency = 100 \frac{1}{N_D \left| X'X^{-1} \right|^{1/p}} \tag{1}$$

Our design results in an efficiency value of 98.5 which is fairly close to the optimal value of 100 that represents perfect orthogonality of the design matrix. We can therefore proceed to parameter estimation of our conjoint model in order to create the stimuli for the conjoint analysis (please mind: this is a snapshot).

Table 2. Snapshot of the design matrix with binary coding (0: element variation not present, 1: variation present)

Comparisons over different time periods	Social comparison	Goals		Units	Selection situation – stimulus (block, please check [14] for a visual impression of the stimuli)
1	1	1	0	0	1-1 (1)
1	0	1	0	0	1-2 (1)
1	0	0	1	0	2-1 (1)
0	0	0	1	1	2-2 (1)

2.5 Model Specification and Estimation

Preference Model: In order to explain a subject's preferences we use the part-worth-model. This entails in comparison to the vector model or the ideal-point model to apply the variation onto qualitative attributes as well. The part–worth model only describes the utility of an attribute. The utilities are later aggregated to yield the overall utility model, so that we can calculate the utility of a stimulus.

Choice Model: In addition to the preference model, we need to define a choice model that describes how subjects based on the preference model will select certain products. We will employ the most common model at this, which is the multinomial logit choice model [10].

Estimation: We use maximum likelihood estimation to yield probabilities for the choice of a certain stimulus [10].

3   Executing the Survey

3.1   Survey Questionnaire

Introduction questions serve the purpose of familiarizing the subject with the topic and ease the participant into the questionnaire. For this purpose, we defined six questions about personal energy consumption that were personal, topic-based and easy to answer.

After introducing that the questionnaire is about energy consumption, we offer the subjects the possibility to monitor their energy consumption in an online dashboard (that serves as the feedback system). At this point the choice situations are presented and the choice based utility is measured in regard to the stimuli of every choice situation. The subjects are confronted with a detailed description of each choice situation that represents the main part of the questionnaire. Since we split the subjects up in two groups each subject was given eight choice situations. We choose the two blocks and choice situations in each block at random in order to avoid order effects. Following those choice situations a subject is confronted with five questions regarding comparison of energy consumption over time (comparison to historical data). The last part of the questionnaire consists of demographic items in order to match the answers of the preceding questions with a profile.

3.2   Pretest and Sample

We conducted a pretest according to [11] before conducting the study in an online survey tool. The pretest was conducted with 13 participants and led to marginal alterations of the questionnaire. Since the medium is an electronical feedback system implemented via website, we expect the main users to be digital natives who prefer graphical visualizations instead of plain text [12]. Therefore we focus our survey on that target group. We choose our subjects so that they make up a representative sample regarding the target groups of digital natives which in this case is our population with focus on motivated students [13]. We conducted the survey within 3 months collecting 1,207 questionnaires from which 1,072 were completed and contained no missing values.

4   Results

4.1   Demographics

The most important information about the subjects are given below:

- Gender: 40.5 % female, 57.6 % male
- Age: 84.1 % 20–29 years old
- Education: 49.6 % High school, 26.7 % College Degree
- Size of household: 21.6 % 1 person, 39.6 % 2 persons, 21.2 % 3 persons
- Type of household: 89.3 % rented apartment
- Monthly net income: 73.4 % below 1,000 Euro

4.2   Results of Introduction Questions

Only 13 % of the subjects have a clear understanding or knowledge of their annual power consumption and only 19 % of all subjects know what their annual energy bill states in terms of energy costs. This is supported by the facts that only 18 % look at their meters more frequently than once a year. However, more than 70 % of the subjects want to reduce their energy consumption. Also, 70 % of the subjects plan to reduce consumption due to environmental awareness.

4.3   Preliminary Rating of the Features

The feature rating uses a scale from 1 (I do not like it) till 5 (I like it very much). The result of the rating was shown in Table 3. This table shows the arithmetic mean and the standard deviation of the separate features in three dimensions. The first one gives the average-score for the appearance, the second one for the content and the third calculates the average for both categories (in case of mean).

Table 3. Results of the separate feature-rating

Feature	Arithmetic mean			Standard deviation	
	Appearance	Content	Both	Appearance	Content
Rate of consumption: last years	3.43	3.73	3.58	0.97	1.06
Rate of consumption: same months	3.43	3.74	3.58	0.98	1.03
Additional prediction	3.36	3.77	3.57	1.08	1.18
Comparison households with same size	2.99	3.46	3.22	1.17	1.24
Saving target: bar chart	3.16	3.39	3.28	1.07	1.16
Saving target: speedometer	3.49	3.39	3.44	1.20	1.18

With exception of the social comparison all features reach a score bigger than 3 in the dimensions appearance and content. The highest score of the appearance with a score average of 3.49 shows the saving target shown in a speedometer, followed by the rate of consumption of the last years and the rate of consumption of the same months about the last years with a score of 3.43.

In the dimension of content, the feature of prediction reached with 3.77 the highest score, even followed by the rate of consumption of the last years and the rate of consumption of the same months about the last years.

Weighted both categories with 0.5 the retrospective comparisons get with 3.58 the highest scores, followed by the prediction with 3.57. Whereas the social comparisons get the lowest score with 3.22. But this score is bigger than the average evaluation point 3 and shows in this way a positive trend as well.

The rating results of the feedback features show a mainly positive judgement. But especially the dimension of appearance offers potential for improvements as this category shows the worst results in comparison to the other dimension.

4.4   Results of the Conjoint Measurement

Table 4 shows the results regarding the attributes in product combinations.

Table 4. Part-worth utility of all attributes

Attribute j	Value m	Part-worth utility b_{jm}
Time comparison	1 Yes	$b_{11} = 0.285$
	2 No	$b_{12} = -0.285$
Social comparison	1 Yes	$b_{21} = 0.160$
	2 No	$b_{22} = -0.160$
Goals	1 No goals	$b_{31} = -0.343$
	2 Detailed goals	$b_{32} = 0.012$
	3 Rough goals	$b_{33} = 0.331$
Measurement unit	1 Euro	$b_{41} = -0.002$
	2 kWh	$b_{42} = 0.002$
None-option		$b_5 = -1.386$

We yield an increased utility for prognosis on a monthly or annual basis. The same result is true for the social comparison. Highest utility values within the goal category were achieved when visualizing "rough goals". The measurement unit "kWh" is preferred. We can now calculate overall utility values from the part-worth values. The highest utility (0.54) is given by stimulus 2 (see Fig. 3) in choice situation 3 in block two, using "kWh", "social comparison" and "prognosis". This is followed by stimulus 1, which is described exactly as above but using "Euro" as the unit of measurement.

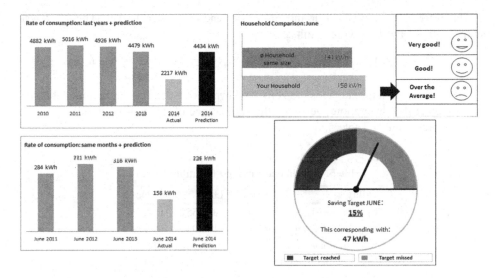

Fig. 3. Stimulus 2 in choice situation 3 with highest utility (Color figure online)

To yield attribute weights from the choice based analysis we use the range of the part-worth utility counts in order to derive relative importance of each attribute. The relative importance is shown in Fig. 4.

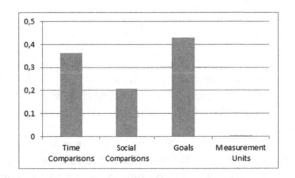

Fig. 4. Relative importance of attributes

From Fig. 4 we can state that, while the goals are the most important attribute, the measurement unit importance is only marginal (0.003).

4.5 Evaluation of Results

We employ a likelihood ratio test in order evaluate the goodness of fit regarding the model.

Therefore, we compare our model to the random choice model. The LLR test statistic yields the following results:

$$LLR = -2 \cdot (LL_0 - LL) = -2 \cdot (-9,430.488 + 7,993.058) = 2,839.558 \qquad (2)$$

This results in a p-value close to zero. The hypothesis that the random model holds can be rejected. Furthermore we check the significance of the part-worth utility counts that are presented in Table 5.

Table 5. Likelihood ratio test for utility values

Utility	LLR_j	p-value
b_{11} (Prognosis "yes")	353.552	0.000 %
B_{21} (Social comparison "no")	102.037	0.000 %
b_{31} (No goals)	256.950	0.000 %
b_{32} (Detailed goals)	0.229	0.640 %
b_{41} (Euro)	0.023	0.882 %
b_5 (None-option)	2298.717	0.000 %

While the utility values of prognosis of the attributes "yes", social comparison "no" and "no goals" are highly significant, the attributes "detailed goals" as well as measurement unit "euro" are not significant.

5   Conclusion

5.1   Summary

We conducted a conjoint measurement analysis among n = 1,072 students that naturally were assumed to be digital natives. A feedback system was designed according to principals of visualization in IS research. The subjects assessed the components with above average ratings.

The assessment of the visualization elements mostly revealed positive feedback, yet there is potential for improvement. Based on the gathered data we can suggest using saving goals, social comparisons and consumption prognosis as components for a feedback system. The relative importance of the attributes goals and social comparisons contradict the rejection of those components within the meta-study that was conducted as a preliminary research project [5]. A feedback system designed for digital natives should therefore contain the feedback elements prognosis, social comparison, saving goal which is visualized by a speedometer, history based comparisons of energy consumption visualized by bar charts as well as daily consumption visualized by line plots. The measurement units should be kWh as well as Euro.

5.2 Limitation

While conducting the analysis it was apparent that assessment of the content was more positive than assessment of the visualization. For future research we suggest improving visualization in feedback systems. Using final comments of the subjects, they often complained about the colors. This observation leads to the conclusion, that the standards of the visualization of information in business [6] are not fully applicable to private households, although they both follow similar goals. Furthermore, consumers have a low readiness to pay for a feedback system. On one hand, this can correlate with the opinion of the subjects that such system has to be provided for free as an additional benefit by the supplier companies. On the other hand, the small income of the subjects in this study can explain the low willingness to pay.

Another issue that the subjects mentioned in connection with feedback system was the worry of a potential lack of security of the private data by data transmission through a website. To be able to give answers on how data security can be guaranteed further research needs to be undertaken.

In general, to increase the acceptance for such systems by the consumer the benefits of using them need to be clarified and communicated in an appropriate way. Furthermore the developed feedback system didn't have only investigated about the user preferences but also regarding about the effect on the energy consumption of the households.

This study uses the theoretical concept of digital natives although the social science criticizes the quality and empirical evidence of this concept. Nevertheless this study has preferred a prototypical way to generate data and the technical affinity of the target group which is connected to this theoretical concept. This analysis can be used as a first step for the construction of an evaluation of a feedback system for energy consumption of private households. Further research should transfer this concept to other groups of consumer, besides the target group of this study.

Moreover, the possibilities and contribution of such a feedback system to the sustainability in energy consumption have to be explored. The focus for this research should lay on the effect of a feedback system for a long-term change of energy consumption by consumers.

References

1. DEA Deutsche Energie-Agentur: Die Energiewende - das neue System gestalten (2013)
2. Houwelingen, J.H., Raaij, W.F.: The effect of goal-setting and daily electronic feedback on in-home energy use. J. Consumer Res. **16**(1), S.98–S.105 (1989)
3. Davis, A., Krishnamurti, T., Fischhoff, B., Bruine de Bruin, W.: Setting a standard for electricity pilot studies. Energy Policy **62**, S.401–S.409 (2013)
4. Gans, W., Alberini, A., Longo, A.: Smart meter devices and the effect of feedback on residential electricity consumption: evidence from a natural experiment in Northern Ireland. Energy Econ. **36**, S.729–S.743 (2013)

5. Weiß, T., Diesing, M., Krause, M.: Die Wirkung von Feedback und Goal Setting auf den Energieverbrauch von Privathaushalten – Eine Meta-Analytische Untersuchung. In: Cunningham, D., Hofstedt, P., Meer, K., Schmitt, I. (Hrsg.) INFORMATIK 2015, Lecture Notes in Informatics (LNI). Gesellschaft für Informatik, Bonn (2015)

6. Kohlhammer, J., Proff, D.U., Wiener, A.: Visual Business Analytics. Effektiver Zugang zu Daten und Informationen. 1., neue Ausg. Dpunkt, Heidelberg (2013)

7. Herrmann, A.: Marktforschung. Methoden, Anwendungen, Praxisbeispiele. 2., aktualisierte Aufl. Gabler, Wiesbaden (2000)

8. Baier, D. (Hg.): Conjointanalyse. Methoden, Anwendungen, Praxisbeispiele. Springer, Berlin, Heidelberg (2009)

9. E.ON: Synthetisches verfahren (2014). http://www.eon-westfalenweser.com/pages/ewa_de/Netz/Strom/Netznutzung/Synthetisches_Verfahren/index.htm (zuletzt aktualisiert am 28.08. 2014, zuletzt geprüft am 28.08.2014)

10. Backhaus, K., Erichson, B., Weiber, R.: Fortgeschrittene multivariate Analysemethoden. Eine anwendungsorientierte Einführung. Springer Gabler, Berlin (2013)

11. Porst, R.: Fragebogen. Ein Arbeitsbuch. 4., erweiterte Aufl. 2014, Korr. Nachdruck 2013. Springer VS, Wiesbaden (2014)

12. Prensky, M.: Digital natives. digital immigrants part 1. On the Horizon 9(5), S.1–S.6 (2001)

13. Berekoven, L., Eckert, W., Ellenrieder, P.: Marktforschung. Methodische Grundlagen und praktische Anwendung. 9., überarb. Aufl. Gabler, Wiesbaden (2001)

14. Appendix: Online-Appendix for this publication (2016). https://www.dropbox.com/s/uklneq39rvz3bln/Appendix.zip

Specification and Implementation of a Data Generator to Simulate Fraudulent User Behavior

Galina Baader[(⊠)], Robert Meyer, Christoph Wagner, and Helmut Krcmar

TU Munich, Information Systems,
Boltzmannstraße 3, 85748 Munich, Bavaria, Germany
{galina.baader,robert.meyer,
christoph.wagner,helmut.krcmar}@in.tum.de

Abstract. Fraud is a widespread international problem for enterprises. Organizations increasingly use self-learning classifiers to detect fraud. Such classifiers need training data to successfully distinguish normal from fraudulent behavior. However, data containing authentic fraud scenarios is often not available for researchers. Therefore, we have implemented a data generation tool, which simulates fraudulent and non-fraudulent user behavior within the purchase-to-pay business process of an ERP system. We identified fraud scenarios from literature and implemented them as automated routines using SAP's programming language ABAP. The data generated can be used to train fraud detection classifiers as well as to benchmark existing ones.

Keywords: Data generation · Fraud scenarios · User simulation · SAP ERP · Purchase-to-pay process · ABAP · BAPI · BDC

1 Introduction

Fraudulent behavior is a prevalent issue with tremendous effects on the world economy, consuming an estimated 5 % of a typical organization's annual revenues [1]. Hence, avoiding fraud is an important challenges of modern enterprises [1]. In the recent past, self-learning classifiers have been increasingly used for detecting fraud in companies [2]. For testing and training such classifiers, a sufficient amount of authentic user data from a company's respective IT systems is necessary. These datasets should contain as many different fraud scenarios as possible in order to "literate" the classifier [3].

Due to security reasons, however, only a few companies are willing to hand over real-life datasets to researchers, especially, when the dataset is suspected to contain actual traces of fraudulent behavior [4].

As a possible remedy, we aim to develop a data generation tool, which simulates normal and fraudulent behavior. Due to the higher likelihood of being a target for fraudulent behavior, we chose the purchase-to-pay process of an SAP ERP system as simulation environment [5]. Compared to other data generation tools [6], we simulate user interactions within an SAP ERP system using respective automation technologies,

© Springer International Publishing Switzerland 2016
W. Abramowicz et al. (Eds.): BIS 2016, LNBIP 255, pp. 67–78, 2016.
DOI: 10.1007/978-3-319-39426-8_6

like Business Application Programming Interfaces (BAPIs) or Batch Input Maps. We have built our prototype directly in the ABAP programming language ABAP and also implemented a graphical user interface. The generated data can be used to train self-learning classifiers with known fraud cases as well as for testing existing fraud detection methods.

2   Method

We developed the fraud data generation tool based on the design science guidelines of Hevner et al. [7]. The prototype simulates fraudulent and non-fraudulent user interaction within the purchase-to-pay business process of an SAP ERP system.

To identify known fraudulent scenarios, we conducted a literature review following the guidelines of Webster and Watson [8]. In order to cover material from a wide number of scientific journals and conference proceedings, we utilized the databases EBSCOhost, ScienceDirect, IEEE Explore and Google Scholar.

Our search string was "fraud" or "white collar crime" or "misappropriation" or "corruption" or "conflict of interests" or "bribery" or "kickback" or "shell company" with the combination of "purchase-to-pay" or "procure-to-pay" or "accounts payable" or "procurement". While excluding all papers not dealing with fraud scenarios within the purchase-to-pay process, our search revealed 89 papers. From these papers, we extracted typical fraud scenarios, which we then implemented in our prototype.

3   Related Literature

First, we want to discuss related literature in the context of synthetic data generation. Synthetic user data can be defined as data, which is generated by a simulated user performing simulated actions in an artificial system [3]. Two approaches for data generation are available: the analytical and the constructive approach.

The *analytical approach* is based on the extrapolation of historic user data. It consists of five phases [3]: Data collection, analysis of historical data to identify parameters, building user profiles and modelling user and system behavior. The analytical approach's advantage is the dataset's authenticity. The main disadvantage is that the quality of the dataset is dependent on the sample dataset's quality. Noteworthy examples of this approach in the context of fraud detection can be found in Lundin et al. [9] or Barse et al. [3]. In the related field of intrusion detection we found the works of Chinchani et al. [10], Greenberg [11] or Schonlau [12] interesting.

The *constructive approach* models user behavior based on expert knowledge. First off, user interactions are defined in different abstraction levels within a transition system [10]. Afterwards, a three layer stochastic user simulation modeled as a Markov process chain is generated [4]. The advantage of this approach is that user behavior can be designed based on specific requirements. The main disadvantage is the mathematical complexity of the approach. The constructive approach can for example be found in Yannikos et al. [4].

In this paper, we aim to adopt the constructive approach, as our dataset should contain as many fraud scenarios as possible. Furthermore, we need a dataset with a disproportional high amount of fraud cases in order to prevent a classifier algorithm from marking all frauds as "rule compliant" (i.e. false negatives) [4].

An approach similar to ours can be found in Islam et al. [6], who simulate user behavior in an ERP system's purchase-to-pay business process. They fill respective tables with random data by means of a random generator [6]. The user-simulation does not conform to any modelling approach. Our approach is to identify typical fraud scenarios from literature and to implement them in the proposed simulator.

4   Requirements for the Fraud Data Generator

A typical example of automatic fraud detection algorithms are data mining algorithms like Decision Trees, Naïve Bayes, Neuronal Networks or Support Vector Machines. All four share the need for a sufficiently large training dataset. A large dataset is also necessary for evaluating existing fraud detection algorithms [3]. We thus have derived requirements for the proposed fraud data generator from related data generation and fraud-related papers. The requirements differ for training and test data.

General Requirements

- The generation of sufficiently large data amounts should be possible in order to test the scalability of the detection algorithms [13].
- Fraudulent and non-fraudulent (compliant) data has to be present in the generated dataset [3]. Some algorithms learn better, when the dataset is skewed towards malicious behavior patterns.
- A graphical user interface is necessary. The user of the prototype should be able to configure the ratio of each to-be-generated fraud scenario in relation to the "normal" data.
- The user conducting a process step in a simulation run should be identifiable [14].
- Normal behavior should be modelled close to reality, as this has influences on the hit rate of the detection algorithm [15].
- Random generation of parameters necessary throughout the process' steps like the amount of ordered goods or time stamps are required.

Training Data Requirements

- The audit record for a simulation run should contain a clear indication if a particular transaction was fraudulent or not [9].
- As many realistic fraud scenarios as possible should be included [9].

Test Data Requirements

- The amount and distribution of fraud activities should be as close to reality as possible [9].

To be able to model normal and fraudulent behavior, we first extracted the "standard" purchase-to-pay business process and known fraud scenarios from literature. This preliminary research was conducted to ensure the generated data's authenticity.

The purchase-to-pay business process is part of all businesses and represents the prime value chain [16–18]. The standard process' steps are shown in Table 1 and are based on [18–20].

Table 1. Standard purchase-to-pay business process (Source: own design)

1. Purchase requisition	A need for a good or service is identified within a department and a purchase requisition is created. Optionally, an authorized user should release the purchase requisition (depending on the customizing settings)
2. Vendor selection (e.g. framework contract)	The purchasing department selects a supplier. Framework contracts with a supplier may exist. A framework contract includes general agreements, which govern terms and conditions for making specific purchases (call-offs)
3. Purchase order	After the vendor selection, a purchase order is send to the supplier. Again, an authorized user releases the purchase order.
4. Goods receipt	The supplier delivers the goods. Usually, they are recorded in the system with reference to the purchase order
5. Invoice receipt	An invoice is send to the purchasing company. This invoice is recorded in the system with reference to the purchase order and goods receipt. Usually, the quantity and price of the delivered goods are checked automatically for discrepancies in the so-called *three-way match*
6. Invoice payment	The last step includes the invoice payment either through a payment run or through manual payment

For our data generation tool's development, we assume a single-level approval procedure for the purchase requisition and purchase order, as different attempts to bypass the approval process exist. We conducted a literature review to detect fraud scenarios within the purchase-to-pay business process. In the following, we shortly describe the most often mentioned fraud cases in literature, that we implemented and refer to the most relevant contributions [4, 6, 18, 21] (Table 2).

5   Technical Implementation of the Data Generator

In the following we describe how the data generation tool has been implemented in SAP ERP 6.06 IDES system. IDES is an SAP-own demo company with customizing settings and exemplary data for teaching and showcasing purposes. To gain an understanding of the implemented generation tool, we first describe typical SAP artifacts that we have used, followed by the tool's devised architecture.

Table 2. Implemented fraud detection scenarios (Source: own design)

	Description
Framework contract	To simplify business processes in a company, the approval process is not necessary when using a framework contract. This may be abused by a fraudster, for example, by using an invalid or expired framework contract. SAP ERP maintains framework contracts as amount or prize contracts [23]. Our data generation tool simulates the misuse of a quantity expired contract
Invoice without reference	Typically, the purchase-to-pay business process starts with a purchase requisition recognizing a demand. However, the purchase requisition and purchase order may be conducted in a separate system. Invoices are then posted in the ERP system as "Invoice without Reference". Such transactions are risky, as no check for correctness (quantity or price) is possible. Especially services have a high risk of fraud due to their immaterial character. Our fraud simulation tool is able to simulate missing purchase requisitions, purchase orders or goods receipts
CpD transaction	Conto pro Diverse (CpD) represents an anonymous collective account, used for one-time suppliers without master data. The CpD-account can be used for numerous creditors. For each payment, information about the supplier (address, account number etc.) has to be filled in manually. The CpD transaction's anonymity can be exploited, e.g. by processing purchases from a blocked supplier
Manual payment	Normally each payment is generated and triggered through the automatic payment run in an ERP system. An automatic payment run can only be started, if an invoice has been posted correctly, the master data of a creditor contains banking accounts and no payment block is set. Manual payments are possible without the mentioned restrictions and are more prone to misuse. Within the ERP system, different possibilities of manual payments are available. We conduct the manual payment through a posting in the general ledger. This results in an open payment without any liability in the accounts payable of the subsidiary ledger. The reference to the invoice receipt has to be created manually through the clearing of the invoice
Non-reference fraud	The scenario "Payment without reference" is similar to Manual Payment, as in both scenarios the payment is conducted manually. The difference lies in the order of the pertaining process steps. In this scenario, the payment is conducted before the invoice receipt or even before goods receipt
Double payment	The *double payment fraud* involves a good or service being paid twice. There may be different reasons for a double payment: – Multiple invoice receipts – one with reference to the goods receipt and one without – Double record of an invoice is posted due to a reminder – Double record of an invoice due to duplicate master data of a creditor – Double record of an invoice within a different accounting area Most ERP systems have implemented controls to detect double payments in general. However, paying in different currencies for example is not handled by the checks. Our tool is able to simulate double invoice payment (first automatic payment run, then manual payment)

(Continued)

Table 2. (*Continued*)

		Description
Redirect payment		In this scheme, the fraudster's aim is to redirect the payment to his own or an accomplice's bank account. Therefore, the bank account of the creditor master data is changed before the payment and changed back right after the payment to wipe traces
False invoice fraud		An overpriced invoice for goods receipt is posted in the system. With respective customization settings an automated payment block is set [24]. The block is released manually by the fraudster and the payment is triggered through the automatic payment run
Misappropriation fraud		The aim of this fraud is to steal company property, for example by purchasing personal goods on the company's account. This scenario is possible through erroneous or missing segregation of duties. In our example, one person has the rights to post the purchase requisition as well as the purchase order and to release both
Non-purchase payment		This scenario encompasses faking the reception of goods in order to generate a payment flow to a cooperating creditor. There are two variants of this fraud scenario: (1) The supplier sends an invoice without sending goods. This is usually blocked by the system. However, the fraudster unblocks it and the invoice is paid. (2) A goods receipt is posted by the fraudster although no physical goods have been received

5.1   Used SAP Artefacts

The aim of our fraud data generator was to simulate the already laid out fraud scenarios, as well as normal behavior. In our model, the scenarios are composed of discrete atomic steps, which either create or modify documents in the SAP ERP system. There are two basic ways to implement process steps programmatically: the use of Business Application Programming Interfaces (BAPI)-calls [24, 25] or Batch Input Maps (BDC) [26].

BAPIs are remote enabled RFC function modules [25]. They are part of the service-oriented architecture of SAP and can be seen as an API for the modification of business objects [25]. Unfortunately, the runtime environment does not provide BAPIs for all process steps, e.g. for triggering the automatic payment run or clearing activities. For simulating process steps not available as BAPIs, we therefore used Batch Input Maps (BDCs). A Batch Input Map can be seen as the recording of an SAP Dynpro execution, encapsulated in a function module, which can be filled with user-specified input parameters [27]. In our work, the use of BAPIs turned out to be more comfortable and robust. On the other hand, BDCs offer a more realistic simulation of the user behavior as they simulate data input via standard SAP Dynpros. Consequently, a BDC execution achieves the same consistency and undergoes the same authority checks as any regular execution of an SAP transaction. Furthermore, identical documents are created and posted.

5.2   Software Architecture of the Generation Tool

The software prototype of our fraud data generator was implemented using a 4-layer-architecture:

The first and lowest layer, i.e. the closest layer to the database, models the single user actions of the SAP purchase-to-pay process. Here, the creation of a purchase requisition, a purchase order or the execution of a payment run were implemented. The layer consists of a set of 14 ABAP Objects classes encapsulating the previously mentioned BAPI or BDC calls in a *perform_step()* method. Aside from the BAPI or BDC function call, the *perform_step()* methods contain logic for input parameter conversion and error handling. The execution of each action is triggered by a method call and results in the creation or modification of one or more SAP documents. The IDs of the created or modified documents are given back as output parameters. Table 3 shows a list of the used BAPIs and BDCs.

Table 3. BAPIs and BDCs used for process step simulation (Source: own design)

Process step	BAPI
Create purchase requisition	BAPI_REQUISITION_CREATE
Release purchase requisition	BAPI_REQUISITION_RELEASE_GEN
Create purchase order	BAPI_PO_CREATE1
Release purchase order	BAPI_PO_RELEASE
Create goods receipt document	BAPI_GOODSMVT_CREATE
Create incoming invoice document	BAPI_INCOMINGINVOICE_CREATE
Create incoming invoice document (without reference to purchase order)	BAPI_ACC_DOCUMENT_POST
Release blocked invoice	BAPI_INCOMINGINVOICE_RELEASE
Change creditor's bank details	BDC for transaction XK02
Automatic payment run	BDC for transaction F110
Manual payment run	BDC for transaction FB01
Clearing	BDC for transaction F-44

The second layer models normal and fraudulent user behavior by concatenating the single user actions (process steps) from the underlying layer. We have implemented an ABAP Objects class for each fraudulent scenario as well as normal process behavior, which can be called by the *perform_scenario()* method. Between the steps of a scenario, input and output data are transferred in order to assure that each process step works on the output document of the previously executed step. The input data either are taken from the output parameters of a previous step or are randomly generated (in layer three). As the self-learning classifiers need a label that indicates if a process instance is fraudulent or not, the second layer of our tool logs information like the process steps, a label (fraudulent or not), input and output parameters of the process' steps and (if applicable) error messages from BAPI or BDC executions.

The third layer of our architecture executes an arbitrary number of different fraudulent or non-fraudulent scenarios by repeatedly instantiating scenario classes of layer two and calling their *perform_scenario()* methods. We call a full set of *n* scenario executions a 'generation run'. Aside from the generation runs, the third layer uses a randomizer for selecting scenario types and random generation of parameter sets for the scenario execution. Before each scenario execution, the randomizer first determines the type of the scenario, as this information is needed for the generation of a suitable input parameter set. The scenario type selection is based on a discrete probability distribution, which can be selected in the GUI. In a second randomization step, the scenario parameters are sampled. Each of the scenarios requires (depending on its process steps) a special set of numeric and non-numeric parameters, e.g. material identifiers, quantities of ordered material, price variations in invoices, date and time of action execution and the IDs of the vendors or other organizational entities involved in the scenario. The sampling of consistent and unambiguous parameter sets was one of the most complex tasks of our work (e.g. depending on the configuration of the SAP system, not every material can be ordered for every organizational unit, by every user or respectively at every point in time). We chose an SAP system preloaded with an IDES dataset as environment for our generation runs as we required realistic and highly integrated master data [28].

The forth layer of our prototype represents the user interface, which we implemented as an SAP Dynpro transaction. Via the front-end mask, users can start new fraud data generation runs, configure the number of the generation rounds or select a discrete probability distribution defining the frequency of the different fraud scenarios. Figure 1 gives an impression of the GUI's configuration editor.

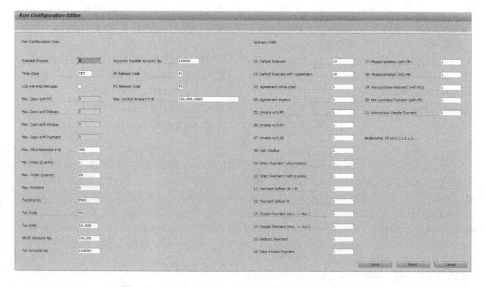

Fig. 1. Configuration editor (Source: own design)

Additionally, the user is able to stop runs currently executing and extract information about terminated runs (e.g. logging data). A run execution is implemented as an asynchronously executing SAP job, which decouples the data generation from front-end processing. The prototype adheres to the object-oriented programming style and can easily be extended by adding new process steps or composing new fraud scenarios. Besides the central classes for the process step and scenario implementation, we used many utility classes provided by the SAP system for database access, logging, error handling and data transfer between the process steps.

6 Validation

Following the guidelines of Hevner et al. [7], we have performed several tests to ensure the fraud data generation tool's functionality. The aim of the tests was to verify the fulfilment of the previously identified requirements. Within our first data generation run, we selected 100 process instances to be generated with all fraud scenarios having the same probability of occurrence. In the second run, we changed the probability of the different scenarios and decreased the number of runs to 50 executions.

After a performed run, the prototype displays a fraud distribution chart, visualizing how often each fraud scenario was actually executed and what parameters (e.g. Purchase Order Number or Invoice Number from the SAP ERP system) have been applied and received by the system. A screenshot displaying an output chart can be seen in Fig. 2.

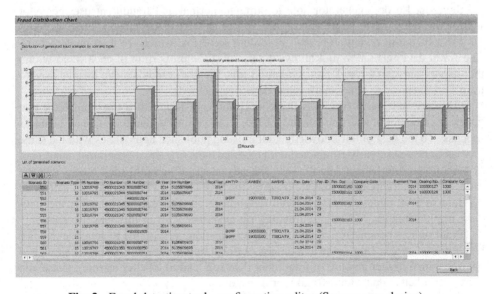

Fig. 2. Fraud detection tool - configuration editor (Source: own design)

We have determined the performance and error log of the execution of both tests, summarized in Table 4.

The error rate is quite small, although we had issues when feeding the random generator initial values. Some typical errors were, that a randomly chosen material and

Table 4. Fraud detection tool – performance KPIs (Source: own design)

Parameter	Test 1	Test 2
Number of rounds	100	50
Total duration	895 s	466 s
Average duration of a scenario	8,95 s	9,32 s
Number of errors (abs.)	1	2
Number of errors (rel.)	1 %	4 %

corresponding company code did not match, as not all materials are available in each company code. However, such errors, often referred to as noise, also happen in real-world companies (e.g. unexperienced users). Each fraud scenario was created at least once.

7   Discussion and Conclusion

The development of effective algorithms for automatic fraud detection is a challenging task as large amounts of data has to be analyzed, sometimes even on-the-fly. Self-learning classifiers present a reasonable approach to overcome this issues [29]. However, such algorithms need sufficiently large training and test datasets [9]. Authentic real-world datasets are usually not handed over to researchers [4]. Therefore, we developed a data generator able to simulate normal and fraudulent user behavior within the purchase-to-pay business process of an SAP ERP system. First, we conducted a literature review to reveal fraudulent scenarios, which we then implemented. The fraud data generation tool allows determining the proportion of each fraud within the dataset and start a fraud data generation run. The generated data can then be used to train classifiers or to benchmark other detection methods.

Our contribution to theory is the conception and design of a user simulator that generates fraudulent and non-fraudulent behavior within the purchase-to-pay business process. To the best of our knowledge, all existing published approaches with the same goal as ours are currently conducted with external tools (e.g. Java-programs) and do not systematically implement fraud cases. The advantage of an embedded tool is that the same software components that are used by the SAP GUI can be used to simulate data. As data in SAP tables are often stored redundantly in different tables, using the standard implementation of SAP assures that no inconsistencies in the database exist.

Practitioners benefit from the tool's implementation, as it is able to generate data for benchmarks of existing fraud classifiers or for training datasets. Using a simulated dataset with a higher amount of fraudulent data will lead to a better learning phase and therefore a better detection of real-world fraud.

For future research, we suggest to extend the fraud data generation tool with further fraud scenarios and to further relevant standard business processes. An interesting candidate is the order-to-cash process which describes the selling of goods [19]. Furthermore, the data produced by the data generator should be used to train classifiers, which then should be tested within a real-world example. In addition, existing classifiers should be benchmarked using the generated dataset. As we have labeled each process as fraudulent or not, it is possible to measure error rates.

References

1. ACFE: Report to the Nations on Occupational Fraud and Abuse (Association of Certified Fraud Examiners). Report, Austin, USA (2014)
2. Phua, C., Lee, V., Smith, K., Gayer, R.: A comprehensive survey of data mining-based fraud detection research. In: Intelligent Computation Technology and Automation (ICICTA), pp. 1–14. IEEE Press, Changsha, China (2010)
3. Barse, E.L., Kvarnström, H., Jonsson E.: Synthesizing test data for fraud detection systems. In: 19th Annual Computer Security Applications Conference (ACSAC), pp. 384–394. IEEE Press, Las Vegas, Nevada (2003)
4. Yannikos, Y., Franke, F., Winter, C., Schneider, M.: 3LSPG: forensic tool evaluation by three layer stochastic process-based generation of data. In: Sako, H., Franke, K.Y., Saitoh, S. (eds.) IWCF 2010. LNCS, vol. 6540, pp. 200–211. Springer, Heidelberg (2011)
5. Luell, J.: Employee fraud detection under real world conditions. In: Faculty of Economics, Doctoral dissertation, University of Zurich: Zurich, (2010)
6. Islam, A.K., Corney, M., Mohay, G., Clark, A., Bracher, S., Raub, T., Flegel, U.: Fraud detection in ERP systems using scenario matching. In: Rannenberg, K., Varadharajan, V., Weber, C. (eds.) SEC 2010. IFIP AICT, vol. 330, pp. 112–123. Springer, Heidelberg (2010)
7. Hevner, A.R., March, S., Park, J., Ram, S.: Design science in information systems research. MIS Q. **28**(1), 75–105 (2004)
8. Webster, J., Watson, R.: Analysing the past to prepare for the future: writing a literature review. MIS Q. **26**(2), xiii–xxiii (2002)
9. Lundin, E., Kvarnström, H., Jonsson, E.: A synthetic fraud data generation methodology. In: Deng, R.H., Qing, S., Bao, F., Zhou, J. (eds.) ICICS 2002. LNCS, vol. 2513, pp. 265–277. Springer, Heidelberg (2002)
10. Chinchani, R., Muthukrishnan, A., Chandrasekaran, M., Upadhyaya, S.: RACOON: rapidly generating user command data for anomaly detection from customizable template. In: 20th Annual Computer Security Applications Conference, pp. 189–202. Tucson, Arizona (2004)
11. Greenberg, S.: Using Unix: Collected Traces of 168 Users. Department of Computer Science, University of Calgary, Calgary (1988)
12. Masquerading User Data. http://www.schonlau.net/intrusion.html
13. Griffin, R.: Using big data to combat enterprise fraud. Financ. Exec. Int. **28**(10), 44–47 (2012)
14. Mercuri, R.T.: On auditing audit trails. Commun. ACM **46**(1), 17–20 (2003)
15. Maxion, R.A., Tan, K.M.C.: Benchmarking anomaly-based detection systems. In: International Conference on Dependable Systems and Networks (DSN), New York, pp 623–630 (2000)
16. Hall, J.A.: Accounting Information Systems. Cengage Learning, Mason (2011)
17. Porter, M.E.: Competitive Advantage: Creating and Sustaining Superior Performance. Free Press, New York (1998)
18. Bönner, A., Riedl, M., Wenig, S.: Digitale SAP-Massendatenanalyse. Erich Schmidt Verlag, Berlin (2011)
19. SAP TERP10: SAP ERP - Integration von Geschäftsprozessen. SAP AG, o.O. (2012)
20. SAP Freigabeverfahren. http://help.sap.com/erp2005_ehp_05/helpdata/de/75/ee14a355c811d189900000e8322d00/content.htm
21. Vendor Account Clearing in SAP. http://stabnet.blogspot.de/2012/05/vendor-account-clearing-in-sap-t-code-f.html
22. SAP Kundenkontrakt. http://help.sap.com/saphelp_erp60_sp/helpdata/de/dd/55fd53545a11d1a7020000e829fd11/content.htm

23. SAP Enhancements. http://help.sap.com/saphelp_nw70/helpdata/EN/bf/ec079f5db911d295 ae0000e82de14a/frameset.htm
24. SAP Jobeinplanung. http://help.sap.com/saphelp_erp60_sp/helpdata/de/ef/2c51389711087 2e10000009b38f889/content.htm
25. Wegelin, M., Englbrecht, M.: SAP-Schnittstellenprogrammierung. Galileo Press, Bonn (2009)
26. SAP Hintergrundverarbeitung. https://help.sap.com/saphelp_nw70/helpdata/de/ed/a9bb3f8e 236d3fe10000000a114084/content.htm
27. Batch Input – BDC. http://wiki.scn.sap.com/wiki/display/ABAP/Batch+Input+-+BDC
28. IDES - das SAP Modellunternehmen. https://help.sap.com/saphelp_46c/helpdata/de/af/fc4 f35dfe82578e10000009b38f839/frameset.htm
29. Jonsson, E., Lundin, E., Kvarnström H.: Combining fraud and intrusion detection - meeting new requirements. In: 5th Nordic Workshop on Secure IT-Systems (NORDSEC), p.o.S. Reykjavik, Iceland (2000)

Quantitative Analysis of Art Market Using Ontologies, Named Entity Recognition and Machine Learning: A Case Study

Dominik Filipiak[1]([⊠]), Henning Agt-Rickauer[2], Christian Hentschel[2], Agata Filipowska[1], and Harald Sack[2]

[1] Department of Information Systems, Poznań University of Economics,
Al. Niepodległości 10, 61-875 Poznań, Poland
{dominik.filipiak,agata.filipowska}@kie.ue.poznan.pl
[2] Hasso-Plattner-Institut, Prof.-Dr.-Helmert-Straße 2-3, 14482 Potsdam, Germany
{henning.agt-rickauer,christian.hentschel,harald.sack}@hpi.de
http://kie.ue.poznan.pl
http://hpi.de

Abstract. In the paper we investigate new approaches to quantitative art market research, such as statistical analysis and building of market indices. An ontology has been designed to describe art market data in a unified way. To ensure the quality of information in the knowledge base of the ontology, data enrichment techniques such as named entity recognition (NER) or data linking are also involved. By using techniques from computer vision and machine learning, we predict a style of a painting. This paper comes with a case study example being a detailed validation of our approach.

Keywords: Art market · Semantic web · Linked data · Machine learning · Information retrieval · Alternative investment · Digital humanities

1 Introduction

Due to the constantly growing interest in the alternative investment area, the art market has become a subject of numerous studies. By publishing sales data, many services and auction houses provide a basis for further research in terms of the latest data analysis trends. A closer look at available data shows missing information or inconsistency in many cases, though. An intense effort (see the next section) has been observed among scientists carrying out research on auction markets, especially in the field of constructing indexes. To the best of our knowledge, the problem of data quality has not been raised often in that field. To tackle this issue, we propose mixing standard econometric analysis with the usage of the latest solutions known from the computer science domain.

To address the issue of data quality (influencing the indexes being built) regarding paintings sold in auction houses, a framework for quantitative art market research using ontologies, named entity recognition (NER) and machine

© Springer International Publishing Switzerland 2016
W. Abramowicz et al. (Eds.): BIS 2016, LNBIP 255, pp. 79–90, 2016.
DOI: 10.1007/978-3-319-39426-8_7

learning has been introduced in this paper. This can be reached by combining various data sources and by employing recent developments in data science, such as semantic annotation of text and automated visual analysis. Visualising various trends or indicating economic incentives influencing the market is a possible outcome of this framework.

The remainder of this paper is organised as follows. The next section is a review of literature relevant to our framework. After that, we shed a light on the approach by explaining methods behind it. Since this paper introduces a new art market ontology, the following section is fully devoted to it. The next section contains a case study, being a detailed example of the usage of our method on a single Monet's painting. A short summary with future work closes this paper.

2   Related Work

Art Market Research. Quantitative analysis of art market data has been a subject matter for many studies, such as portfolio diversification [1] and measuring the volatility of the market [2]. The vast majority of conducted research relies on building art market indices, which are to show price movements of a standard artwork in a given period (typically on a year basis). The most popular types of indices for art market research are built on top of a hedonic regression (HR) and repeat-sales regression (RSR) [3]. No matter which type has been chosen, the problem of availability of data always arises. Models based on RSR rely on prices of artworks sold at least twice [4]. The famous Mei Moses Fine Art Index[1], an example of the employment of RSR, considers only lots sold in the two biggest auction houses - Christie's and Sotheby's. Due to its nature, repeat-sales regression can operate on significantly smaller datasets. Art is considered to be a long-term investment, so it may be a challenging task to collect a decent dataset of lots sold at least twice - in some auction houses it is even strictly forbidden. Models based on HR take into account all sold lots and rely on the relation between (typically) a hammer price of a given lot and all its features. An example of this model is the Two-Step Hedonic Regression [5]. HR-based calculations are prone to feature selection bias, however. To overcome the limitation of taking into account only sold lots, a probit estimation may be employed [6].

Knowledges Bases for Art Market Data. Our work proposes the use of semantic web technologies and linked open data principals [7] to store information about artworks sold in auctions. While the use of ontologies in the auction domain is rather small, different works exist in the art domain. The openART ontology [8] was developed to describe a research dataset about London's artworld. Its focus is on events related to artworks in that time, but some parts deal with sales data. Europeana is a research initiative to provide access to millions of cultural heritage objects (e.g., books, films, paintings). The Europeana Data Model (EDM) is a semantic web-based framework to describe cultural heritage objects. The DBpedia ontology [9] also covers parts of artwork descriptions

---

[1] http://www.artasanasset.com.

and can be used in art research [10]. Following the linked data principals we link these ontologies where necessary and reuse existing properties. In order to populate our auction knowledge base we apply state-of-the-art information extraction methods [11], in particular named-entity recognition and disambiguation [12–14].

Computer Vision Tools for Art Market Data. Typically, computer vision tools focus on the classification of real-world objects and scenes (e.g. "sunset", "faces" and "car"). Classification of paintings into different styles and art epochs has gained relatively little attention in the past. Several authors have evaluated the aesthetic quality of photographs. Datta et al. [15], designed algorithms to extract visual characteristics from images that can be used to represent concepts such as colourfulness, saturation, rule-of-thirds, and depth-of-field, and evaluated aesthetic rating predictions on photographs. The same methods were applied to a small set of Impressionist paintings [16]. Murray et al. [17] introduced the Aesthetic Visual Analysis (AVA) dataset, annotated with ratings by users of a photographic skill competition website. Marchesotti and Peronnin [18] proposed a method to predict aesthetic ratings by using data gathered from user comments published on the website. The attributes they found to be informative (e.g., "lovely photo," "nice detail") are not specific to image style though. Several authors have developed approaches to automatically classify classic painting styles, including [19,20]. These works consider only a handful of styles (less than ten apiece), with styles that are visually very distinct, e.g., Pollock vs. Dalí. These datasets comprise less than 60 images per style. Mensink [21] provides a larger dataset of artworks, but does not consider style classification. Finally, the authors in [22] publish a dataset of 85,000 paintings annotated with 25 genre labels ranging from Renaissance to modern art (e.g. "Baroque", "Roccoco" and "Cubism"). The authors present an approach for automatic classification of paintings into art epochs using latest computer vision methods and report per-class accuracies ranging from 72 % ("Symbolism", "Expressionism", "Art Nouveau") to 94 % ("Ukiyo-e", "Minimalism", "Color Field Painting").

3   Our Approach

Since numerous auction house websites are publishing data about sold lots in a well-structured manner, this information may be gathered using web crawlers. The quality and availability of information about a given artwork is the subject matter of this paper. There is a so-called *garbage in, garbage out* rule in data analysis and the art market is not different. Therefore, the process of collecting the data should be planned carefully. This should involve data refinement, due to the possible errors in collected data. A typo in the author's name may result in assigning artworks to two different creators. These problems can be mitigated by data cleansing techniques. On the other hand, some lots have missing values (for example, the style of an artwork is often not provided). We explain how to deal with this problem in further sections. Having gathered the data, standard linear regression (estimated by Ordinal Least Squares) may be used to indicate the relation between painting's price and its qualities:

$$\ln P_{it} = \alpha + \sum_{j=1}^{z} \beta_j X_{ij} + \sum_{t=0}^{\tau} \gamma_t D_{it} + \varepsilon_{it} \tag{1}$$

where $\ln P_{it}$ is the natural logarithm of a price of a given lot $i \in \{1, 2, ..., N\}$ at time $t \in \{1, 2, ..., \tau\}$; α (intercept), β and γ are regression coefficients for estimated characteristics included in the model. X_{ij} represents hedonic variables (numeric and dummies, explained in the next paragraph) included in the model, whereas D_{it} stands for time dummy variables. The last parameter, ε_{it} represents the error term.

Suppose there is a non-numeric parameter, like the presence of the signature. This information can't be included in a model in that form. To overcome this problem and to not lose the information, so-called dummy variables are introduced. They are characterized by the amount of possible levels l (in this case $l = 2$). A well-defined ontology (see Sect. 4) can help to summarize the possible levels. For the each level, an explanatory variable takes "1" if the condition is true (there is a signature) or "0" otherwise (lack thereof). Coefficients of dummy variables equal the average difference in impact on the model between these cases. Another example is the year of sale (D_{it}). For each sold lot, this variable is equal to "1" only if a given painting i was sold in a period t (otherwise it is equal to "0"). Supposing only paintings sold in 2014, 2015 and 2016 are considered, those years become dummy variables for representing time in the equation. Supposing a given observation is sold in 2016, only the last one takes the value of "1", the rest is equal to "0".

Although it will be hard to have more predictors than observations (so called $p > n$ setup) in our case, the correct selection of used explanatory variables is crucial and the overall result depends on it. Measures like statistical significance can be helpful here. It has to be remembered that one must not include l levels in the model, because this leads to multicollinearity. A simple way to avoid this situation is to use $l - 1$ levels.

The most important dummy variable is the one representing the year of sale. By calculating the coefficients, one can build an index to *measure* the art market and compare lots sold in a given year to other forms of assets (such as stocks). At the same time, this value is *separated* from the painting's qualities, such as its author or price. This allows to measure the impact of the given year on the hammer price without examining sold lots and regardless of their features (which typically involves experts with domain knowledge). A standard way to calculate the hedonic price index for a period t is:

$$Index_t = e^{\gamma_t} \tag{2}$$

where γs are taken from the Eq. (1). A more sophisticated way to calculate the index, like the Two-Step Hedonic Index considers following calculations:

$$Index_{t+1} = \frac{\prod_{i=1}^{n}(P_{i,t+1})^{1/n} / \prod_{i=1}^{m}(P_{i,t})^{1/m}}{\exp\left[\sum_{j=1}^{z}\beta_j\left(\sum_{i=1}^{n}\frac{X_{ij,t+1}}{n} - \sum_{i=1}^{m}\frac{X_{ij,t}}{m}\right)\right]} \tag{3}$$

No matter which way of building a hedonic index is used, the quality of information (i.e. observations in our case) always seems to play an important role. A higher number of explanatory variables allows examining the importance of each one more accurately. This also leads to a higher coefficient of determination (R^2) – it is not a good indicator of the quality of the model, though[2].

To improve the generalization of the model, the number of explanatory variables should be reduced. This can be achieved by e.g. univariate feature selection, recursive feature elimination or hill climbing solutions. Performing those methods makes sense only if a decent number of variables is available. As it was stated previously, some observations suffer from a lack of data. Therefore, we decided to extend the amount of available information.

The goal of data collection is the processing of different kinds of data sources and the extraction of data values and descriptions regarding lots sold in auctions. The extraction is performed in two steps. First, we crawl auction house websites and download all data on sold lots. Values that can be directly extracted using standard techniques (e.g., regular expressions) are stored as RDF data using our art auction ontology. Secondly, during data enrichment, extracted information are processed with named entity recognition and natural language processing in order to link them to external data sources, such as DBpedia or Europeana. This enables the acquisition of additional variables for art market analysis.

Finally, information automatically extracted from the visual domain may provide further information on the art object at hand. By using techniques from computer vision and machine learning we predict the style of a painting in order to further enrich extracted metadata. Our approach uses features extracted from a Deep Convolutional Neural Network (CNN) trained on the well-know ImageNet [23] dataset. The model architecture is based on the CNN architecture winning the 2012 ImageNet challenge [24] with some minor modifications (for the sake of increased training performance, the size of the full connect layer is reduced from 4,096 to 2,048 neurons). We follow the approach presented in [25] by taking the pen-ultimate (fully connected) layer of the trained CNN as a feature extraction layer yielding a 2,048 dimensional feature vector per image which is $L2$-normalized. Model training is conducted in a One-vs.-Rest-approach using a Support Vector classifier with a linear kernel. We optimize the cost parameter in a three-fold stratified cross-validation.

Groundtruth data is taken from the "WikiArt.org – Encyclopedia of fine arts" project[3] which contains a collection of paintings from different art movements, ranging from Renaissance to modern art. All paintings are manually labeled

---

[2] Due to the problem of overfitting.

[3] WikiArt.org – Encyclopedia of fine arts, http://www.wikiart.org.

according to the respective art movement. We used the dataset to generate groundtruth data by randomly selecting 1,000 images for training and 50 images for testing purposes. Since the number of images in the WikiArt.org paintings collection varies a lot (e.g., for some art movements less than 100 images are available while others provide more than 10,000 images), only those classes were selected that are supported by at least 1,050 images (we obtained a total of 22 classes).

4   Ontology

We designed an art auction ontology that is able to capture information gathered from different sources, such as auction house websites, linked data sets, and other art market databases. Currently, the ontology concentrates on the description of paintings that appeared in auctions, but the model is extensible for different kinds of artworks. We use Web Ontology Language (OWL) and Resource Description Framework Schema (RDFS) to specify the schema.

Figure 1 shows part of the ontology schema that is responsible for the sales data. The central entity is the *Lot*. In a lot a particular item is offered for sale, in our case *Artwork* items. The *Sale* event is organized by an *AuctionHouse* at a certain location. We link to the vCard ontology and reuse it for address description. The most important information for price indices is the *date* of the auction and final selling price. A lot is associated with different types of prices: *upperEstimate* and *lowerEstimate* are an appraisal of what price the lot will fetch. The *initialPrice* is the suggested opening bid determined by the auctioneer. *hammerPrice* is the final price at which a lot was sold (without any fees and taxes), and *premiumPrice* includes fees and taxes, such as buyer's premium and sales tax. *PriceSpecification* is done using the schema.org vocabulary (price value and currency). Additionally, the lot class is linked to the openART ontology [8] that provides additional properties for the artwork domain.

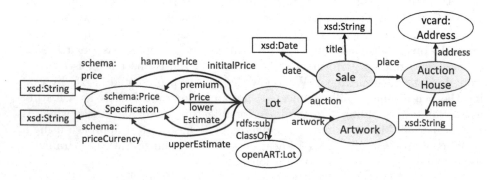

Fig. 1. Schema of the art auction ontology – sales data (excerpt).

The lot has a reference to the artwork as shown in Fig. 2. This part of the ontology is responsible for modeling the auction items and their creators in detail.

The *Artwork* class is the general description for different types of artworks, and has a creation date (in case it is known). The artwork class is linked to the DBpedia ontology [9] enabling reuse of its properties (*dbo* prefix). As mentioned before, our work currently operates on paintings data, thus one specialization *Painting* exists. Other kinds of specific artworks can be added later using the subclass relation. A painting is described by several datatype properties, such as name and a natural language description. The ontology is capable of describing the painting's size and information on the author's signature (present or not, and the location of the signature). An item can be categorized using the *artMovement* and *paintingTechnique* properties. We also provide properties to store art movement information that was classified automatically by image analysis (*artMovementClassified*, and *artMovementClassifiedConfidenceValue*, not shown in the Figure). A painting can be associated with provenance information describing previous owners, such as organizations and persons. Therefore, we link to the DBpedia *Agent* class. Similarly, exhibitions are described in which the painting appeared (not shown in the Figure).

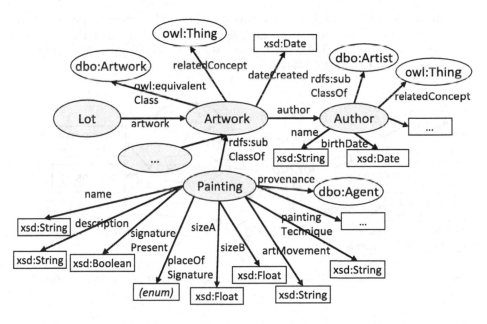

Fig. 2. Schema of the art auction ontology – artwork data (excerpt).

An artwork item is usually associated with an *Author*. We inherit from the DBpedia Artist class that offers a wide range of person properties, such as birth date, death date and nationality, as well as artwork-specific properties (e.g., *influencedBy*, *works*). In case the author is not known the ontology offers an origin property to be able to describe nationality or other indicators of the painting's source. Both the author and the artwork can be linked to other entities using the *relatedConcept* object property in order to store additional information, such as important cities, other artists or related artworks.

5   Case Study

This section applied our method on a real world example. We have chosen Monet's *Le Parlement Soleil Couchant*[4], sold at Christie's New York for $40,485,000 (hammer price plus buyer's premium) in 2015.

5.1   Data Collection

In our example, we illustrate the extraction of data values from Christie's website. We use techniques from information extraction [26] in order to transform data provided in an unstructured manner into a structured machine-readable format. In particular boilerplate removal (deletion of unnecessary HTML content), filtering, tokenization and regular expressions are applied to build a wrapper for the auction data. Figure 3 depicts an excerpt of the RDF visual representation that could be extracted from the example website. It shows that the painting "Le Parlement, soleil couchant" by Claude Monet was sold on 11 May 2015 for a premium price of USD 40,485,000.

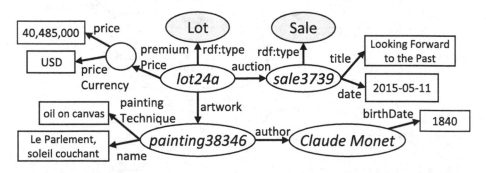

Fig. 3. Auction data in RDF format extracted from the auction website (excerpt).

5.2   Data Enrichment

We pursue two strategies to obtain additional data on artworks and artists. First, we employ named entity recognition [13] on text values and natural language descriptions. The identified entities are then disambiguated [14] and linked to other semantic data sets. Secondly, we use image classification to retrieve the style (art movement) of the painting which is often not provided.

---

[4] http://christies.com/lotfinder/paintings/claude-monet-le-parlement-soleil-couchant
-5895978-details.aspx.

Named Entity Recognition. The task of named entity recognition (NER) concerns the identification of words or phrases in natural language texts that refer to names. These names are associated with categories (e.g., person, location). NER on the already obtained name of the author and painting is relatively easy and used to verify the data collection task. We also apply NER on the long description of the lot (lot notes). From that we learn that London is the most important city related to the lot (mentioned 27 times), as well as the River Thames (natural feature, 10 times). Two organizations have been found: St. Thomas's Hospital and Saint James's Hospital (mentioned 16 times, also indirectly just with "hospital".) Paul Durand-Ruel could be identified as the most important person related to Monet.

Data Linking. Whereas NER can only identify position and category of a named entity, named entity linking performs disambiguation of the name and connects it to a specific instance in an existing knowledge base. Currently, we link to the DBpedia knowledge base [9]. In our case study, we are able to link the artist value (Claude Monet) to the DBpedia resource Claude_Monet[5]. The link allows the retrieval of additional structured data values (e.g., exact birth and death date, nationality, or author description). Furthermore, we are able to query additional information from the knowledge base using pre-defined properties from the DBpedia ontology and SPARQL language. For example, using the *dbo:movement* property we retrieve "Impressionism". In case this information or the artist is not available in DBpedia, we apply image analysis (see next Section). The other identified named entities (e.g., London, Thames) are linked to the lot using the *relatedConcept* property of our art auction ontology.

Art Movement Classification. Figure 4 shows the scores of our art movement classifiers when applied on Monet's *Le Parlement Soleil Couchant*. For reasons of brevity we visualize only positive classification scores. As can be seen, the SVM classifier for the art movement "Impressionism" produces the highest classification score.

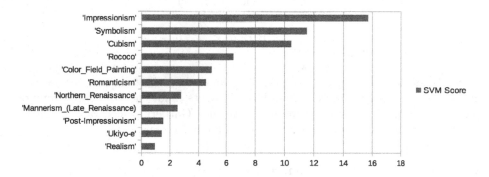

Fig. 4. Classifier scores for positively classified epochs.

---

[5] http://dbpedia.org/resource/Claude_Monet.

5.3   Data Analysis

Having enriched and refined a dataset, it is possible to perform the analysis en route to build the indices. The natural logarithm of the realized price (40,485,000) becomes the dependent variable. The currency is omitted, provided that all of the observations will be presented in USD. Table 1 shows the excerpt of possible explanatory variables for Eq. (1). *Signature* is an example of a dichotomous variable. Nominal variables, such as *Author* can't be employed in the equation in the original form. Therefore, *Claude_Monet* becomes a dummy variable (equal to one) among expected other ones, for example *Pablo_Picasso* which takes zero as the value. Ordinal variables, such as *Year_of_sale* may be transferred to the set of dummies representing each year. Concepts found by NER in text descriptions are also included as dummies (like *London*). These ones related directly to the painter (such as Paul Durand-Ruel) can't be included, due to the presence of the *Author* variable group, which may lead to multicollinearity. Because of the possibly large number of variables, a selection process is a must.

Table 1. Explanatory variables (excerpt) for *Le parlement, soleil couchant*

Group	Variable	Value
Author	Claude_Monet	1
	Pablo_Picasso	0
	Claude_Monet	0
Dichotomous	Signature	1
Continuous	Size_A	81.3
	Size_B	93
Year	XIX	1
Technique	Oil_on_canvas	1
	Tempera	0
Concept	London	1
	River_Thames	1
	Berlin	0

Such an approach applied for various artworks and various artists would enable us to construct indices describing markets, and lead to comparisons of markets in case of choosing between investments. Gathering such values enables also for other analyses which are subject to further research.

6   Conclusions and Future Work

This paper presented a method for quantitative art market research using ontologies, named entity recognition and machine learning. By connecting concepts to a given artwork or using linking to another data source we presented the value of data enrichment in this particular case. To suit our needs for data storing and provide a possibility to create a standard for presenting art market data, the art market ontology was developed. As for overcoming problems with missing style information, an artwork classifier was prepared, trained and employed in our research. A one-example use case illustrates the usage of the proposed solution. The next steps consider gathering data as a base for further research. Data refinement seems to be a challenge after that. It has to be decided which variables are playing a significant role regarding the description of a considered lot. This can be achieved by statistical analysis. Finally, we expect to build precise art market indices on a large-scale data sample.

Future work may consider extending the ontology to the other artworks (e.g. sculptures). This might even enable a possibility to carry out research on all of lots sold in auction houses – not only paintings. Adding other data enrichment sources is also a feature worth considering. Using an ontology enables the transfer of cultural knowledge or artworks itself more efficiently between different countries, so the further steps might consider publishing data and making a translational effort. An extensive study of the behavior or correlation of different indices on an enriched dataset is the next possibility.

References

1. Jureviciene, D., Savicenko, J.: Art investments for portfolio diversification. Intellect. Econ. **6**(2), 41–56 (2012)
2. Bocart, F.Y.R.P., Hafner, C.M.: Volatility of price indices for heterogeneous goods with applications to the fine art market. J. Appl. Econometrics **30**(2), 291–312 (2015)
3. Etro, F., Stepanova, E.: The market for paintings in paris between rococo and romanticism. Kyklos **68**(1), 28–50 (2015)
4. Ginsburgh, V., Mei, J., Moses, M.: The computation of prices indices. In: Handbook of the Economics of Art and Culture, vol. 1, pp. 947–979. Elsevier (2006)
5. Kräussl, R., van Elsland, N.: Constructing the true art market index: a novel 2-step hedonic approach and its application to the german art market (2008)
6. Collins, A., Scorcu, A., Zanola, R.: Reconsidering hedonic art price indexes. Econ. Lett. **104**(2), 57–60 (2009)
7. Heath, T., Bizer, C.: Linked data: evolving the web into a global data space, 1st edn. Morgan & Claypool, San Rafael (2011)
8. Allinson, J.: Openart: open metadata for art research at the tate. Bull. Am. Soc. Inf. Sci. Technol. **38**(3), 43–48 (2012)
9. Lehmann, J., Isele, R., Jakob, M., Jentzsch, A., Kontokostas, D., Mendes, P.N., Hellmann, S., Morsey, M., van Kleef, P., Auer, S., et al.: Dbpedia-a large-scale, multilingual knowledge base extracted from wikipedia. Semant. Web **6**(2), 167–195 (2015)
10. Filipiak, D., Filipowska, A.: DBpedia in the art market. In: Abramowicz, W., et al. (eds.) BIS 2015 Workshops. LNBIP, vol. 228, pp. 321–331. Springer, Heidelberg (2015). doi:10.1007/978-3-319-26762-3_28
11. Etzioni, O., Fader, A., Christensen, J., Soderland, S., Mausam, M.: Open information extraction: the second generation. In: Proceedings of the Twenty-Second International Joint Conference on Artificial Intelligence - Volume, IJCAI 2011, vol. 1, pp. 3–10. AAAI Press (2011)

12. Filipiak, D., Węcel, K., Filipowska, A.: Semantic annotation to support description of the art market. In: Joint Proceedings of the Posters and Demos Track of 11th International Conference on Semantic Systems - SEMANTiCS2015 and 1st Workshop on Data Science: Methods, Technology and Applications (DSci15) Colocated with the 11th International Conference on Sema, Vienna. CEUR Workshop Proceedings, pp. 51–54 (2015)
13. Usbeck, R., Röder, M., Ngomo, A.N., Baron, C., Both, A., Brümmer, M., Ceccarelli, D., Cornolti, M., Cherix, D., Eickmann, B., Ferragina, P., Lemke, C., Moro, A., Navigli, R., Piccinno, F., Rizzo, G., Sack, H., Speck, R., Troncy, R., Waitelonis, J., Wesemann, L.: GERBIL: general entity annotator benchmarking framework. In: Proceedings of the 24th International Conference on World Wide Web, WWW 2015, Florence, Italy, 18–22 May 2015, pp. 1133–1143 (2015)
14. Daiber, J., Jakob, M., Hokamp, C., Mendes, P.N.: Improving efficiency and accuracy in multilingual entity extraction. In: Proceedings of the 9th International Conference on Semantic Systems (I-Semantics) (2013)
15. Datta, R., Joshi, D., Li, J., Wang, J.Z.: Studying aesthetics in photographic images using a computational approach. In: Leonardis, A., Bischof, H., Pinz, A. (eds.) ECCV 2006. LNCS, vol. 3953, pp. 288–301. Springer, Heidelberg (2006)
16. Li, C., Chen, T.: Aesthetic visual quality assessment of paintings. IEEE J. Sel. Top. Sign. Proces. **3**(2), 236–252 (2009)
17. Murray, N., Marchesotti, L., Perronnin, F.: Ava: a large-scale database for aesthetic visual analysis. In: 2012 IEEE Conference on Computer Vision and Pattern Recognition (CVPR), pp. 2408–2415. IEEE (2012)
18. Marchesotti, L., Perronnin, F.: Learning beautiful (and ugly) attributes. In: Proceedings of the British Machine Vision Conference. BMVA Press (2013)
19. Keren, D.: Painter identification using local features and naive bayes. In: Proceedings of the 16th International Conference on Pattern Recognition, vol. 2, pp. 474–477. IEEE (2002)
20. Shamir, L., Macura, T., Orlov, N., Eckley, D.M., Goldberg, I.G.: Impressionism, expressionism, surrealism: automated recognition of painters and schools of art. ACM Trans. Appl. Percept. (TAP) **7**(2), 8 (2010)
21. Mensink, T., van Gemert, J.: The rijksmuseum challenge: museum-centered visual recognition. In: Proceedings of International Conference on Multimedia Retrieval, p. 451. ACM (2014)
22. Karayev, S., Trentacoste, M., Han, H., Agarwala, A., Darrell, T., Hertzmann, A., Winnemoeller, H.: Recognizing image style. In: Proceedings of the British Machine Vision Conference. BMVA Press (2014)
23. Russakovsky, O., Deng, J., Su, H., Krause, J., Satheesh, S., Ma, S., Huang, Z., Karpathy, A., Khosla, A., Bernstein, M., Berg, A.C., Fei-Fei, L.: ImageNet large scale visual recognition challenge. Int. J. Comput. Vis. **115**(3), 211–252 (2015)
24. Krizhevsky, A., Sutskever, I., Hinton, G.E.: Imagenet classification with deep convolutional neural networks. In: Advances in Neural Information Processing Systems 25, pp. 1097–1105. Curran Associates, Inc. (2012)
25. Chatfield, K., Simonyan, K., Vedaldi, A., Zisserman, A.: Return of the devil in the details: Delving deep into convolutional nets. CoRR abs/1405.3531 (2014)
26. Cimiano, P., Mädche, A., Staab, S., Völker, J.: Ontology learning. In: Staab, S., Studer, R. (eds.) Handbook on Ontologies. International Handbooks on Information Systems, pp. 245–267. Springer, Heidelberg (2009)

Search Engine Visibility Indices Versus Visitor Traffic on Websites

Ralf-Christian Härting[1(✉)], Maik Mohl[2], Philipp Steinhauser[1],
and Michael Möhring[1]

[1] Aalen University of Applied Science,
Competence Center for Information Systems in SME, Aalen, Germany
{ralf.haerting,michael.moehring}@hs-aalen.de,
Philipp.Steinhauser@kmu-aalen.de
[2] Metamove GmbH, Munich, Germany
maik.mohl@metamove.de

Abstract. "Search engine (optimization) visibility indices" or also called "SEO visibility indices" are a widespread and important key performance indicator in the SEO-Communities. SEO visibility indices show the overall visibility of a website regarding the search engine result page (SERP). Although search engine visibility indices are widespread as an important KPI, they are highly controversial regarding the aspect of a correlation between real website visitor traffic and search engine visibility indices. Furthermore, only a few online-publications examine this controversial aspect. Therefore, we designed a study, analyzing the correlation between organic visitor traffic and search engine visibility indices, the correlation amongst the indices themselves and the impact of Google Updates on the indices. The study is based on 32 websites of German enterprises from various business branches. Key findings imply that there is no high correlation between organic visitor traffic and search engine visibility indices, but a high correlation between the indices themselves. Furthermore, there is no identifiable pattern relating to the expected effect that Google Updates influence the search engine visibility indices.

Keywords: Search engine · Visibility · Indices · Traffic · Websites

1 Introduction

Based on emerging markets and a number of new technologies like mobile computing, cloud-computing and big data the importance of internet services is growing rapidly [2, 9, 26]. According to ITU [13], the volume of worldwide internet users surpassed 2.9 billion in 2014 – a growth by a factor of 8 over 14 years and is expected to continue its rise in the future. In the context of these developments, a widespread use of digital marketing activities appeared.

On B2C markets, Social Media Marketing and especially Search Engine Marketing became important [6, 37]. Especially online information search is a very important activity for customers in e-commerce [6, 10]. Google is the most used search engine with a global market share of 88.1 % (Jan 2015) [32] and registers 5.74 billion searches

© Springer International Publishing Switzerland 2016
W. Abramowicz et al. (Eds.): BIS 2016, LNBIP 255, pp. 91–101, 2016.
DOI: 10.1007/978-3-319-39426-8_8

averagely per day (2014) [33]. For enterprises it is very important to be ranked within the first few search results of Google. According to Mediative's recent eye tracking study 83 % of all people participating looked at the top organic listing of the SERP (search engine results page). 76 % of page clicks went on the top four organic listings of the SERP [24]. Enterprises being ranked within the first few ranks of the SERP have a higher chance of gaining web traffic as a result.

Current research shows the importance of Search Engine Optimization e.g. according to ranking quality, visitor satisfaction etc. [1]. In almost all enterprises on emerging markets, increasing budgets for digital marketing have to be stated [24]. In some industries the budgets have even exceeded the spending for traditional marketing.

Digital marketing does not only represent the planning and coordination of electronic supported marketing campaigns, it also means the monitoring of success. We can distinguish two approaches to control the success of digital marketing [16, 29]. The first approach is represented by off-the-page tools which check single digital marketing activities and especially the visibility rank of hosted websites. The second approach is represented by on-the-page Webanalytic-Tools (e.g. Google Analytics (GA)) which check the utilization figures and client data [25].

Regarding the first approach there are various companies offering a large range of different tools. A very important segment are SEO-Tools (e.g. [30, 31, 35]) which are calculating search engine visibility indices as KPIs (Key Performance Indicators). For our research we chose SISTRIX, XOVI and SEOlytics, one of the most important Tool-Providers of the German SEO-Community.

By enlarging the visibility, an increase of the web traffic is usually expected [7]. With focus on e-business a higher number of website visitors are expected. Therefore, it can be assumed that there is a high correlation between organic visitor traffic and search engine visibility indices. Although experts note that there could be a growing divergence between both KPIs based on the implementation of several Google algorithm updates by Google itself [34]. Google uses as many as 200 factors for its algorithm and there are more than 500 changes per year [38]. Important updates are the so called Penguin and Panda updates. Penguin 2.0 was launched in May 2013 (2.1 in Jan 2014) and relevant update Panda 4.0 in May 2014. Both updates were responsible for major changes in the ranking structure of SERPs.

Despite the divergence, the visibility indices are highly valued in the industry and the SEO community. Therefore, we designed a study to examine the research questions of the divergence between web traffic and the visibility index by comparing their correlations. The study is based on data gathered from 48 websites. Having eliminated incomplete and inconsistent data sets, 32 websites were used for data analysis. Each website data set contains real website visitor traffic (extracted from Google Analytics) and three search engine visibility indices (extracted from SISTRIX, XOVI and SEOlytics) over an extended period of 126 weeks, beginning in the second week of August 2012 and ending in the last week of December 2014.

Our paper proceeds as follows: First we discussed the basic properties of the foreign search engine visibility index. In Sect. 2 we introduce the function and the method of calculation of the search engine visibility index. In the Sect. 3 the research design and the process of data collection is presented. The results of our research are given in Sect. 4. To top off our paper, we present further thoughts on the topic in Sect. 5.

2 Search Engine Visibility Index

The search engine visibility is highly valued and important [11] and therefore tools for search engine visibility indices are widespread. The KPI is calculated by several tools offered by SEO-Tool providers and shows the visibility of a domain for the search engine result pages of Google. Within our examination we covered the visibility indices of three SEO-Tool providers, as already mentioned in chapter one (please refer to the introduction).

The visibility index of a website is generally created by a keyword pool in which each keyword is being ranked and weighed within the Google search results [19]. SISTRIX GmbH for example calculates the index through a keyword pool of one million keywords and keyword combinations weekly. 10 % of these keywords are formed by current and important occasions whereas 90 % always stay the same. Every week the top 100 positions in Google are being registered and analyzed for the specific keyword pool. The results are weighed regarding the aspects of position and anticipated search volume for each keyword [15]. Providers although vary their approaches and keep the exact algorithm for their calculation of the KPI a secret.

A huge impact factor on the development of search engine visibility indices are changes in rankings of keywords which can also be caused by updates of the Google algorithm. Each Google update targets different aspects of a website and each website is constructed differently. As a result, websites show different developments regarding their website visibility [21]. Regarding the SEO-Tool suppliers' statement on correlation developments between visibility indices and organic traffic, a drop of the visibility index should result in a drop in organic traffic. The following Figs. 1 and 2 show quite a different development for a selected enterprise website.

Fig. 1. Organic traffic by Google analytics

The three visibility indices show a significant drop after the Google Panda 2.0 update in May 2013 whereas the organic traffic develops differently. Current research focuses on aspects like the importance of Search Engine Optimization [1] and e.g. the constructions of visibility indices (e.g. [7, 27, 28]) as well as specific visibility factors and influences [18] of on- and off-the-page SEO-activities. There is no extensive research in the field of "established visibility indices in the real world versus visitor traffic on websites" based on a literature review [5] in scientific databases such as SpringerLink, IEEExplore, EbscoHost, Sciencedirect.

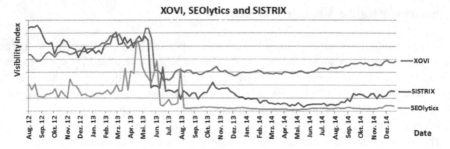

Fig. 2. Visibility indices by XOVI, SEOlytics and SISTRIX

3 Research Design

In order to investigate the assumed correlation between organic visitor traffic and search engine visibility indices in a deeper way, we designed an empirical observation study based on the following three hypotheses. Several sources of the German-speaking SEO-community indicate a high correlation between the organic visitor traffic and the search engine visibility index of several SEO-Tools [12]. To investigate this statement, we created our hypothesis 1:

H1: Search engine visibility indices are highly correlating with the organic (visitor) traffic on websites.

The next hypothesis discovers the relation of the search engine visibility indices among each other. According to [23] case studies, the visibility indices of different providers of SEO-Tools have similar performances, even if the exact characteristics of the indices are not published by the providers. Therefore, we designed the following hypothesis 2:

H2: Search engine visibility indices are correlating amongst each other.

A further field of our research is the impact of Google Updates on search engine visibility. Based on hypothesis 1 and in reference to [12] we can also indicate an influence on the correlation between organic visitor traffic and search engine visibility indices. Therefore we created hypothesis 3:

H3: Google updates influence the correlation between the search engine visibility indices and the organic traffic.

4 Research Methods and Data Collection

To investigate our empirical research model, we implemented an empirical study with real data based on extracted and collected SEO data of different commercial websites in Germany according to general empirical research guidelines [36].

We asked different leading internet marketing enterprises and their customers in 2015 to get real data for our research. In contrast to e.g. survey researches or laboratory settings, the analysis of real data can generate real insights in correlations as well as real world implications. We collected various search engine and SEO data (e.g. Google Analytics organic traffic, SISTRIX, XOVI, SEOlytics visibility indices) from 48 different enterprises to ensure a good quality of research. All enterprises are small or medium sized. Their business is mainly focused on BtoC-markets. E-commerce is very important.

After data cleaning (e.g. missing data), we got a final sample of 32. All websites are developed in German language. The timeframe of the collected data comprised the second week of August 2012 until the last week of December 2014 (126 weeks).

Table 1. Industry sector in percentage terms

Industry sector	Percentage
Online-retailer	46.875 %
Service industry	21.750 %
Recommendations sites	9.375 %
Offline-retailer	6.250 %
Real estate	6.250 %
Medicine	6.250 %
Education	3.125 %

Regarding the whole time frame and all 32 websites, the overall average was 100.55 visitors per website weekly. The investigated websites represent a wide area of different sectors (e.g. online retailer, service enterprises) according to Table 1.

For analyzing our hypothesis, we used correlation analysis [14] to investigate the linear relationship between the different factors. The correlation analysis is often used in research [37]. According to [4, 14, 17], the correlation coefficient r can be interpreted as follows (e.g. in behavioral sciences): Table 2

Table 2. Correlation coefficient r according to Cohen [4]

Correlation coefficient	Interpretation
$r = 0$	No correlation
$r > = |0.1|$	Weak correlation
$r > = |0.3|$	Moderate correlation
$r > = |0.5|$	Strong correlation

To ensure a high quality of our research, we tested the significance of the results of the correlation analysis according to general statistical guidelines [22]. All analysis were based on IBM SPSS 22 and Microsoft Excel [8, 20].

5   Results

According to our research model (hypotheses) we tested our collected empirical data via the research approach we stated in the last section. The first hypothesis explores the correlation between search engine visibility indices and the organic website traffic measured via Google Analytics (GA) [3, 15]. Based on the correlation analysis, we got the following results:

Table 3. Correlation matrix between GA and visibility indices

Enterprise	Correlation coefficient R / P-value		
	GA - XOVI	GA - SEOlytics	GA - SISTRIX
1	**0.70004967**	**0.6378532**	**0.69764889**
	p = 0	p = 0	p = 0
2	-0.4092155	0.3335812	0.15931457
	p = 0	p = 0	p = 0.075
3	**0.58921932**	0.47852415	**0.63475365**
	p = 0	p = 0	p = 0
4	0.13882503	0.1177219	0.1475417
	p = 0.121	p = 0.189	p = 0.099
5	0.12055815	0.00010472	0.06712095
	p = 0.179	p = 0.999	p = 0.455
6	0.42899444	-0.3959563	-0.35883255
	p = 0	p = 0	p = 0
7	0.09264655	0.33058752	0.10906778
	p = 0.302	p = 0	p = 0.224
8	-0.200939	0.1701752	0.04470587
	p = 0.024	p = 0.057	p = 0.619
9	**0.90105743**	0.14250452	**0.73440312**
	p = 0	p = 0.111	p = 0
10	0.46615647	0.20220485	**0.5954684**
	p = 0	p = 0.023	p = 0
11	−0.11268737	0.19409895	0.17169686
	p = 0.209	p = 0.029	p = 0.055
12	−0.5810689	**0.62416979**	**0.61793799**
	p = 0	p = 0	p = 0
13	**0.86117132**	**0.79395815**	**0.84526154**
	p = 0	p = 0	p = 0
14	**0.85507914**	**0.86238761**	**0.90542583**
	p = 0	p = 0	p = 0
15	**0.53465695**	0.13133665	**0.51189498**
	p = 0.143	p = 0,143	p = 0

(*Continued*)

Table 3. (*Continued*)

Enterprise	Correlation coefficient R / P-value		
	GA - XOVI	GA - SEOlytics	GA - SISTRIX
16	−0.29170908	−0.02862762	0.15882151
	p = 0.001	p = 0.75	p = 0.076
17	−0.3240669	−0.25480318	−0.38461682
	p = 0	p = 0.004	p = 0
18	**0.72030351**	**0.5494804**	**0.76038277**
	p = 0	p – 0	p = 0
19	0.28363493	0.30987148	0.37363679
	p = 0.001	p = 0	p = 0
20	0.29076247	0.30231945	**0.51135295**
	p = 0.001	p = 0,001	p = 0
21	0.43304034	−0.52622806	−0.66358449
	p = 0	p = 0	p = 0
22	0.46860094	0.1225066	0.09210708
	p = 0	p = 0.172	p = 0.305
23	−0.18371463	−0.12090527	0.20658401
	p = 0.039	p = 0.177	p = 0,02
24	**0.88745965**	0.47034633	**0.82426966**
	p – 0	p = 0	p = 0
25	**0.5618975**	**0.85226357**	**0.83268613**
	p = 0	p = 0	p = 0
26	0.41310372	0.36040141	0.48386392
	p = 0	p = 0	p = 0
27	−0.02235333	0.16923673	−0.07972043
	p = 0.375	p = 0.058	p = 0,375
28	0.48248523	0.24522369	0.40909071
	p = 0.006	p = 0,006	p = 0
29	**0.76633391**	0.4857807	**0.80676352**
	p = 0	p = 0	p = 0
30	−0.43770324	−0.50382589	−0.25085629
	p = 0	p = 0	p = 0.005
31	**0.79020062**	**0.69313593**	**0.76376549**
	p = 0	p = 0	p = 0
32	**0.89987365**	**0.67190309**	**0.87920129**
	p = 0	p = 0	p = 0

Regarding the correlation between GA and XOVI, there are 12 out of 32 (37.5 %) websites with a significant correlation of r > 0.5 (see bold results in Table 3). Only 8 of 32 (25 %) websites show a significant correlation of r > 0.5 for GA and SEOlytics. Examining the correlation between GA and SISTRIX 15 of 32 (46.875 %) websites show a significant correlation of r > 0.5. Based on these results we cannot confirm hypothesis 1 (*search engine visibility indices are highly correlating with the organic (visitor) traffic on websites*).

Hypothesis 2 explores the correlations among the three different visibility indices. The search engine visibility indices of XOVI and SEOlytics only show a moderate as well as a significant correlation (r = 0.48884375). XOVI and SISTRIX show a high significant average correlation of r = 0.60246875 regarding their search engine visibility indices. SEOlytics and SISTRIX also show a high average correlation of r = 0.574125. The exact index-characteristics are not published by the tool-providers. But the results suggest that the calculation of the KPI is based on similar criteria and keyword pools. Therefore, we can confirm hypothesis 2 (*search engine visibility indices are correlating amongst each other*).

Finally, hypothesis 3 explores the influence of the different Google Updates on the visibility indices vs. organic traffic via a correlation analysis (according to Fig. 3 with significant values p < 0.05):

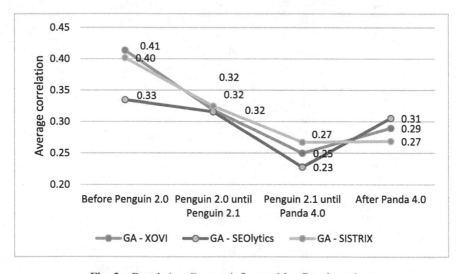

Fig. 3. Correlation diagram influenced by Google updates

Based on these results, we can confirm hypothesis 3 (*Google updates influence the correlation between the search engine visibility indices and the organic traffic*). On the one hand, each website is reacting individually to Google updates. The correlation between organic visitor traffic and search engine visibility indices can either increase, decrease or stay the same, depending on the website itself. On the other hand, there is a general downward trend for the average correlation of all three indices with the organic traffic until the introduction of the update Panda 4.0.

6 Conclusion

Our paper discovered the important practical and research topics influencing the search engine visibility indices and the real organic traffic measured with Google Analytics. After theory and hypotheses development, we designed a quantitative study of different German enterprises and analyzed these data with correlation analysis. Based on the results, we found a gap between the indices and the organic traffic. The indices mostly correlate with each other and there are different degrees of correlation between each of the three indices and the organic traffic.

Practical users can apply our results to choose an adequate index. They get a deeper understanding of the indices and are able to develop more accurate interpretations. Research can benefit from new knowledge about the indices and search engine behavior.

Our data set consists only of German enterprises in different industry sectors. Future research should enlarge the sample and integrate more countries.

Furthermore, there are additional explanations for a negative or weak correlation between the examined KPIs. One influencing aspect we investigated are Google updates. Google updates influence the search engine visibility indices and the correlation with organic visitor traffic. Correlations can increase, decrease or stay the same for each website itself. But our research showed a pattern for the average correlation of all data.

Another aspect for a weak correlation can be all traffic that is not being sufficiently considered by the SEO-Tool providers and their calculation of the search engine visibility indices. Examples are traffic from social media and bookmarks or type-in traffic. A further feature is the limited keyword-pool being used to calculate the indices. Keywords of niche industry branches are often not represented in the keyword pools. The same situation applies for strongly regional oriented websites.

Final aspects may be that SEO-Tool providers do not update their data records and keep the exact algorithm for the calculation of the search engine visibility indices secret. These aspects are up for discussion and would profit from further research.

References

1. Berman, R., Katona, Z.: The role of search engine optimization in search marketing. Mark. Sci. **32**(4), 644–651 (2011)
2. Bughin, J., Chui, M., Manyika, J.: Clouds, big data, and smart assets: Ten tech-enabled business trends to watch. McKinsey Q. **56**(1), 75–86 (2010)
3. Clifton, B.: Advanced web metrics with Google Analytics, 3rd edn. Wiley Inc., Indianapolis (2012)
4. Cohen, J.: Set correlation and contingency tables. Appl. Psychol. Meas. **12**(4), 425–434 (1988)
5. Cooper, H.M.: Synthesizing research: a guide for literature reviews, vol. 2, 3rd edn. SAGE Publications Inc., New York City (1998)
6. Dou, W., Lim, K.H., Su, C., Zhou, N., Cui, N.: Brand positioning strategy using search engine marketing. MIS Q. **34**(2), 261–279 (2010)

7. Drèze, X., Zufryden, F.: Measurement of online visibility and its impact on Internet traffic. J. Interact. Mark. **18**(1), 20–37 (2004)
8. Field, A.: Discovering statistics using IMB SPSS statistics, 4th edn. SAGE Publications Ltd., London (2013)
9. Gantz, J., Reinsel, D.: The digital universe in 2020: Big data, bigger digital shadows, and biggest growth in the far east (2012). http://www.emc.com/collateral/analyst-reports/idc-the-digital-universe-in-2020.pdf
10. Grefen, D., Straub, D.W.: The relative importance of perceived ease of use in IS adoption: a study of e-commerce adoption. J. Assoc. Inf. Syst. **1**(1), 8 (2000)
11. Goodman, A.: Winning Results with Google AdWords, 2nd edn. McGraw-Hill, New York City (2009)
12. Härting, R., Steinhauser, P.: Insights of the German SEO-Community regarding the relevance of Search Engine Visibility Indices (2015). http://www.kmu-aalen.de/kmu-aalen/?page_id=169
13. International Telecommunication Union: ICT Facts and Figures 2014, (2014). http://www.itu.int/en/ITU-D/Statistics/Documents/facts/ICTFactsFigures2014-e.pdf
14. Kendall, M.G.: Rank Correlation Methods. Charles Griffin & Company Limited, London (1948)
15. Kronenberg, H.: Wie wird der Sichtbarkeitsindex berechnet? (2013). http://www.sistrix.de/frag-sistrix/was-ist-der-sistrix-sichtbarkeitsindex/
16. Kumar, L., Kumar, N.: SEO technique for a website and its effectiveness in context of Google search engine. Int. J. Comput. Sci. Eng. (IJCSE) **2**, 113–118 (2014)
17. Weinberg, L., Knapp, S., Abramowitz, S.: Exploring Relationships Between two Variables. Data Analysis for the Behavioral Sciences Using SPSS. Cambridge University Press, Cambridge (2002)
18. Lim, Y.S., Han, W.P.: How do congressional members appear on the web? Tracking the web visibility of South Korean politicians. Govern. Inf. Q. **28**(4), 514–521 (2011)
19. Maynes, R., Everdell, I.: The Evolution of Google Search Results Pages and Their Effects On User Behaviour (2014). http://www.mediative.com/whitepaper-the-evolution-of-googles-search-results-pages-effects-on-user-behaviour/
20. McCullough, B.D., Wilson, B.: On the accuracy of statistical procedures in Microsoft Excel 2003. Comput. Stat. Data Anal. **49**(4), 1244–1252 (2005)
21. McGee, M.: Winners & Losers As Panda 2.0 Goes Global? eHow, Bing's Ciao.co.uk & More (2011). http://searchengineland.com/winners-losers-panda-goes-global-ehow-bings-ciao-more-72895
22. Miles, J., Shevlin, M.: Applying regression and correlation: a guide for students and researchers. SAGE Publications Ltd., London (2001)
23. Missfeldt, M.: Case Study: Sichtbarkeitsindex und Traffic im Vergleich (2013). http://www.tagseoblog.de/case-study-sichtbarkeitsindex-und-traffic-im-vergleich
24. Rivera J., Meulen, R.V.D.: Gartner Survey Reveals Digital Marketing Budgets Will Increase by 8 Percent in 2015, (2014). http://www.gartner.com/newsroom/id/2895817
25. Roebuck, K.: Web Analytics: High-impact Strategies -What you need to know: Definitions, Adoptions, Impact, Benefits, Maturity, Vendors. Emereo Publishing, Brisbane (2012)
26. Schmidt, R., Möhring, M., Härting, R.-C., Reichstein, C., Neumaier, P., Jozinović, P.: Industry 4.0 - potentials for creating smart products: empirical research results. In: Abramowicz, W. (ed.) BIS 2015. LNBIP, vol. 208, pp. 16–27. Springer, Heidelberg (2015)
27. Schmidt-Mänz, N., Gaul, W.: Measurement of Online Visibility. In: Ahr, D., Fahrion, R., Oswald, M., Reinelt., G. (ed.) Operations Research Proceedings 2003, pp. 205–212. Springer, Berlin Heidelberg (2004)

28. Schmidt-Mänz, N., Gaul, W.: Web mining and online visibility. In: Weihs, C., Gaul, W. (eds.) Classification - the Ubiquitous Challenge. Studies in Classification, Data Analysis, and Knowledge Organization, pp. 418–425. Springer, Heidelberg (2005)
29. Schroeder, B.: Publicizing your program: website evaluation. Des. Mark. Stratge. AACE J. **15**(4), 437–471 (2007)
30. SEOlytics GmbH: SEO Software for Professionals (2015). http://www.seolytics.com/
31. SISTRIX GmbH: The secret of successful Websites (2015). http://www.sistrix.com
32. Statista Inc.: Worldwide market share of leading search engines from January 2010 to January 2015 (2015). http://www.statista.com/statistics/216573/worldwide-market-share-of-search-engines/
33. Statisticbrain LLC: Google Annual Search Statistics (2015). http://www.statisticbrain.com/google-searches/
34. Tober, M.: Update: Google "Bad" SEO Update - Now named Penguin Update (2012). http://blog.searchmetrics.com/us/2012/04/25/google-bad-seo-update-a-first-earthquake-on-the-short-head/
35. XOVI GmbH: SEO and Social Monitoring made in Germany (2015). http://www.xovi.com/
36. Zikmund, W., Babin, B., Carr, J., Griffin, M.: Business Research Methods, 9th edn. South-Western College Pub, Nashville (2012)
37. Gruen, T.W., Summers, J.O., Acito, F.: Relationship marketing activities, commitment, and membership behaviors in professional associations. J. Mark. **64**(3), 34–49 (2000)
38. Killoran, J.B.: How to use search engine optimization techniques to increase website visibility. IEEE Trans. Prof. Commun. **56**(1), 50–66 (2013)

A Dynamic Hybrid RBF/Elman Neural Networks for Credit Scoring Using Big Data

Yacine Djemaiel$^{(\boxtimes)}$, Nadia Labidi, and Noureddine Boudriga

Communication Networks and Security Research Lab, Sup'Com,
University of Carthage, Tunis, Tunisia
ydjemaiel@gmail.com, nadia_abidi2003@yahoo.fr, noure.boudriga2@gmail.com

Abstract. The evaluation of credit applications is among processes that should be conducted in an efficient manner in order to prevent incorrect decisions that may lead to a loss even for the bank or for the credit applicant. Several approaches have been proposed in this context in order to ensure the enhancement of the credit evaluation process by using various artificial intelligence approaches. Even if the proposed schemes have shown their efficiency, the provided decision regarding a credit is not correct in most cases due to the lack of information for a provided criteria, incorrect defined weights for credit criteria, and a missing information regarding a credit applicant. In this paper, we propose a hybrid neural network that ensures the enhancement of the decision for credit applicants data based on a credit scoring by considering the big data related to the context associated to credit criterion which is collected through a period of time. The proposed model ensures the evaluation of credit by using a set of collectors that are deployed through interconnected networks. The efficiency of the proposed model is illustrated through a conducted simulation based on a set of credit applicant's data.

Keywords: Credit · Neural networks · Big data · Credit scoring · Collectors · Context · Decision

1 Introduction

Nowadays, credit scoring is considered among the most important tasks in financial institutions. In fact, the number of credit applicants is increasing through time which requires that financial institutions manage and predict financial risks to reduce losses that they can face in borrowing money to applicants.

Credit scoring consists of the assessment of risk associated with lending to an organization or a consumer (an individual) which is represented in most cases as a measure of creditworthiness of the applicant. It measures the default of credit based on credit applicant data such as the benefit value, the sector growth value, the guarantee value and the sales value. Moreover, the credit scoring helps financial institutions to make the good decision, reject or grant credit to applicants. As a consequence, financial institutions can reduce the percentage of loan not paid through the decrease of the number of non-performing loans. In fact, the credit

© Springer International Publishing Switzerland 2016
W. Abramowicz et al. (Eds.): BIS 2016, LNBIP 255, pp. 102–113, 2016.
DOI: 10.1007/978-3-319-39426-8_9

evaluation includes data reliable to human, technical, environmental and political. In the last years, several research works have studied the application of artificial intelligence models for credit risk management. The major problem of any financial institution or lender is to differentiate between "good" and "bad" applicants. Such differentiation is possible by using a credit-scoring model. In this paper, we propose a hybrid neural network for credit scoring that enhances the decision made by using big data that is defined as a huge volumes of data observed by several deployed collectors through interconnected networks. In [4], the development of Credit scoring models supports decision making mainly in the case of retail credit business. In this context, we focus our research on the evaluation of credit applicant data related to corporations where a set of collectors are selected according to their activities.

The paper contributions are four folds: (1) the definition of the context in the case of a hybrid neural network for credit evaluation that enhances the decision made during the credit evaluation process by using a hybrid model built upon combining Elman neural network and RBF (Radial basis function) neural network. The RBF neural network enables the classification in real time of credit applicant's data. The Elman neural networks ensure the enhancement of the decision made by the RBF neural network by considering the required data during a laps of time.

In order to ensure the evaluation of the credit in an efficient manner, the context is defined for each credit criterion which includes the set of needed attributes that enhances the classification of applications. (2) The enhancement of the decision made by the use of a hybrid neural network model that enables real time classification and the processing of the history through a period of time. (3) The identification and the use of observers that enable the collection of big data in relation with the credit applicant data. (4) The analysis of collected big data in order to extract knowledge needed for the defined contexts.

The remaining of the paper is organized as follows: Sect. 2 gives a survey of the proposed techniques for credit scoring and for decision making using neural networks. The next section discusses the need for big data to enhance the credit scoring by introducing the context for neural networks and the use of collectors to extract required knowledge from big data. Section 4 presents the proposed hybrid neural network that enables the enhancement of the credit scoring by considering the context based on available big data. Section 5 details the conducted simulation in order to evaluate the proposed hybrid neural network and to illustrate the importance of the context based on big data to enhance credit scoring. Section 6 concludes the paper and discusses some future prospects for the provided model.

2   Related Work

Several research works have used neural network to enhance the decision based on credit scoring. In this context, numerous studies have proven that Neural Network perform remarkably the best.

Artificial neural network (ANN) is used widely in intrusion detection systems but cannot restore the memory of past events and it takes a long time to train a Multilayer Perception neural network because of its non-linear mapping of global approximation. In [5], ANNs are able to classify normal traffic correctly and detect known and unknown attacks without using a large data. A detection rate of 88% on 50 attacks (known and unknown attacks) for four hidden layers is obtained while a false positive rate of 0% is achieved. Hence, all normal sessions were classified correctly as normal traffic. Their results are promising compared to other researches but they suggest that the limit in this case is the need to update data over the time to consider known and unknown attacks. Recurrent neural network (RNN) remedy the shortcoming of ANN which is the ignoration of time. In fact, this neural network represents a dynamic model that accepts for each timestamp an input vector, updates its hidden state via nonlinear activation functions, and uses it to make a prediction of its output. It's a rich model because their hidden states can store information as high dimensional distributed representations and their nonlinear dynamics can implement rich and powerful computations, allowing it to perform modeling and prediction tasks for sequences with highly complex structure [4]. In [3], the pH process is modeled based on RNN. The pH is one of the highly nonlinear dynamic system, hence it is difficult to identify and control it. The data set is then divided into 70% for training and 30% for testing data set. Given training data of input-output pairs where output is "pH of the solution" and input is "Base flow rate". The model structure selection consists of two sub problems: choosing a regressor structure and choosing network architecture. Neural networks utilized in modeling and identifying the pH process, have overcame successfully the nonlinear characteristic of the pH process. The recurrent neural networks models enhance the control performance of pH process, due to the important capacity in learning nonlinear characteristic. These results can be applied to model and identify not only the pH process but also other nonlinear and time-varied parametric industrial systems without considering changing in external environments.

Moreover, credit scoring model based on back propagation neural networks has been proposed in [1]. This model has enhanced the decision made regarding credit applicants' data which may be more enhanced if related big data is analyzed and considered as input to this network. In this way, combining two neural networks may be also a way to enhance the decision. At present, hybrid models that synthesizing advantages of methods is the focus of researchers, because they can lead to better results. This combination covers the weaknesses of the others and enhances the generated results. A hybrid RBF/Elman neural network model for both anomaly detection and misuse detection has been proposed in [6]. The hybrid neural network can detect temporally dispersed and collaborative attacks effectively because of its memory of past events. The RBF network is employed as a real-time pattern classification and the Elman network is employed to restore the memory of past events. When a classification result of an input is analyzed by RBF, the result will be feed forward and can be restored by Elman network connected to the output units of RBF network. The results demonstrate that the hybrid neural network, RBF/Elman neural network model, can detect intrusions with higher detection rate and lower false positive rate than current neural network. The use of the context in the case of this work has enabled better detection since the monitoring of the activity is considered for a period of time using the Elman network. Introducing context is helpful in this case but it is not sufficient since even for RBF, the classification may be also enhanced when introducing the context.

3   The Need of Big Data for Enhancing the Credit Accreditation Decision

Most credit scoring schemes may provide a false decision based on the provided set of values for credit criterion that does not reflect really the credit applicant capabilities to refund the requested credit. In this context, additional information related to the credit applicant should be collected based on the huge volumes of data in relation with the credit applicant profile. Facing this need, we introduce in this Section the context that should be defined for credit applicants' data in order to collect the required knowledge for the needed criterion using a set of collectors that are then introduced by detailing their specification. The collection of big data through time and techniques needed to process such data is then presented.

3.1   Context Specification

The evaluation of the credit is based on a set of parameters that are defined as credit applicant data. For each parameter, the evaluation is based on the provided values, it is difficult to evaluate credit applications in an efficient manner since this value may be incorrect or invalid for a laps of time. The context is defined based on a set of parameters that are related to the used credit criterion. The parameters are determined according to two methods. The first method is based on the elaboration of questionnaires that ensure the collection of information about personnel and professionals regarding the set of attributes that may be added to the context. The second method is a big data analytics that enable the extraction of knowledge from huge volumes of data in order to define the needed context parameters. This operation is performed for each credit criterion by considering the credit applicant's data and the credit type. A set of attributes are added to the context that are needed to enhance the evaluation of a criterion such as the benefit determined by information included in the balance sheet. These attributes may include: information about corporation's activities, the potential number of customers, etc. For each context parameter, a set of observers are defined in order to collect information. All attributes are considered in order to check if the provided value is correct or that should be updated considering the related big data collected from deployed observers. The time period used to collect big data needed for the defined context parameters is considered as an additional attribute for such context that may be varied depending to the degree of importance of the criteria for the evaluated credit type.

3.2   Big Data Collectors Specification

A collector or observer is a software module that is attached to running services through interconnected networks and that enables the collection of the needed data in relation with attributes associated to a credit criterion. The collection is performed based on the available big data generated by running services and stored in an attached storage devices. The collection of big data is ensured by querying data sources based on the defined values for the attributes in relation with the credit applicant data. For example, the collectors ensure the querying of the data sources (e.g. a database) that hold the users records that include information on their identity, their responsibilities and the generated amount of benefit even if the credit applicant recognizes more depreciation expense because of the ownership of assets more than is required by the business. In this case, the querying considers as input the credit applicant data that includes: benefit, guarantee, sector growth rate, etc. These criteria are used by the collector during the querying in order to enable the selection in first step of the organizations where the credit applicant may be solicited then to check the returned records (for structured big data) in order to collect useful information for the defined attributes (e.g. Assets value). Based on these querying criteria, a collector identifies the data sources that are available in the zone where it is deployed and it ensures the interrogation of remote collectors in a distribute manner in order to update the attribute value belong to an active context. For a specific criterion, many collectors may be able to provide needed data related to credit applicant's data with a varied degree of accuracy that may affect the final decision made by the neural network considering the context. In this way, a weight related to this criterion should be considered in the context and that should be considered when evaluating data collected from such observer.

3.3   Requirements for Decision Based on Collected Big Data Through Time

The decision regarding a credit may be affected by the available values for the credit criteria that are valid for an instant t but it may be not correct after that instant. The provided credit applicant's data are evaluated not immediately for the most cases but by considering a delay that leads to the use of incorrect value that may engender an incorrect decision since these values are considered during the evaluation process. Moreover, considering the additional context attributes at instant t may enhance the decision since the criteria attached to this context is updated but it is insufficient. This value may be enhanced if it is observed through a period of time that is defined in accordance with the credit applicant data. For example, the guarantee is evaluated for an amount at the time of filing the credit application but this value is updated at the moment of decision making, it can increase or decrease according to the market value. In this case, the attributes' values are updated for these periods of time which increase the probability of convergence for the real value that may contribute to the correct decision regarding the credit applicant's data.

3.4   Data Mining Techniques for Collected Big Data

These techniques are needed to extract knowledge from the collected big data using observers. These techniques are based on the use of keywords that characterize the credit applicant data that may be used to extract knowledge from the collected and stored big data. For example, the type of activity of the credit applicant and its specialty may be used as keywords to extract knowledge from databases, from stored files and data sheets in relation with the applicant. In addition, some additional techniques may be used to ensure the merge and fusion of data from multiple data sources that enables the processing of more rich records to generate more interesting analytics for an applicant. A querying of different databases based on the applicant data may enable the collection of different records from different databases that may be merged in order to collect more attributes attached to the applicant. In this case, the first record from the first fetched database enables the identification of the name of the corporation, the address, the activity sector however the second database provides additional fields that are attached to the applicant such as: the sector growth that may help to identify the additional potential data sources that may be used to collect data needed for the identified contexts.

4   A Model for Credit Evaluation Using RBF/Elman Neural Networks

In this section, a hybrid RBF/Elman neural network model for credit scoring using big data through defined context is introduced.

Model Principles. The processing of the credit applicant data should be performed based on the collected information for the different criterion related to the credit applicant's data. The processing of such information should be performed according to two steps. The first step aims to check the validity of the provided information at the instant of the demand processing. The provided decision should be then refined by considering the history for a defined period for each investigated criterion. This verification step is based on the collected big data related to the processed criteria. The collection of needed big data is performed using a set of observers that ensures the collection and the analysis in order to validate the provided credit applicant data. This principle is implemented by the use of a hybrid model that enables the real time processing of the provided data in order to ensure the classification and then to refine the decision based on credit applicant history related to the checked criterion.

4.1   Model Hypothesis

The proposed model ensures the evaluation of credit applicant's data if a set of the following assumptions are considered: (1) Sufficient credit criterion are considered as input for the credit evaluation process; (2) the processing of the provided input should be performed first by the RBF neural network then refined by the Elman neural network. The first step ensures the evaluation of the credit criteria using an RBF neural network. The second step processes the RBF output by an Elman neural network; (3) the evaluation for the both steps should consider the attached contexts in addition to the defined criterion, and (4) The number and location of collectors should be dynamically defined according to the complexity of the credit applicants' data.

4.2   Big Data-Based RBF/Elman Neural Network Model

The evaluation of the credits is performed based on a hybrid model that is composed of an RBF neural network which is a kind of feed forward neural network, ensuring the real time classification of credit applicant's data using collected big data that are valid for the current processing time of the credit applicant's data. The provided decision is then refined using an Elman neural network that keeps memory of the information in relation of the checked credit criteria that are collected and extracted from huge volumes of data.

An input vector is defined for the hybrid model that is expressed as follows: $V_a = \{v_{a_1}, v_{a_2}, ..., v_{a_k}\}$. This vector represents the possible values associated to each credit criterion where k represents the number of criteria used to evaluate a credit for an applicant a. Each credit criterion v_i, is represented through a set of additional variables where the possible values for each variable are defined as follows: $v_i = \{v_{i1}, v_{i2}, ..., v_{in}\}$ where n represents the number of additional variables that are used to evaluate a criteria. The first decision provided by the RBF neural network is modeled as: $R_i = \{r_{i1}, r_{i2}, ..., r_{ij}, ..., r_{iN}\}$ where N represents the number of output units for the RBF neural network. The weights that connect the outputs of the RBF network to the input of the Elman network is: $W^R = (w_1^R, w_2^R, ..., w_i^R, ..., w_N^R)$. The recurrent weights of the Elman neural network are: $W^E = (w_1^E, w_2^E, ..., w_i^E, ..., w_N^E)$. The values in context units of RBF are modeled as follows: $C^R = [c_1^R, c_2^R, ..., c_i^R, ..., c_N^R]^T$. Each element of the provided context vector represents the collected big data in relation with the defined criteria as input to the RBF neural network defined in the vector V. For the Elman neural network, the context is defined as follows: $C^E = [c_1^E, c_2^E, ..., c_i^E, ..., c_N^E]^T$. The weights connect the outputs of the RBF to the inputs of Elman is $W = (w_1, w_2, ..., w_N)^T$. The recurrent weights of the Elman are expressed as follows: $W' = (w_1', w_2', ..., w_N')^T$. The output of the RBF network is expressed using the following formula: $r_{ik} = \sum_{[l=1]}^{n} v_{ik} \cdot w_{ik} + c_i^R \cdot w_i^R$. The output of the Elman network is modeled as follows: $O_i = [o_{i1}, o_{i2}, ..., o_{ij}, ... o_{iN}]^T$. Each element of the vector O_i is deduced using the following expression: $o_{ij} = r_{ij} * w_i + c_i^E * w_i^E$. The hidden nodes are used to activate the context nodes when it is required to enhance the provided decision by the RBF neural network. The expression of Elman non-linear state space is as follows:

$$\begin{cases} x(t) = f(W^A x_c(t) + W^B u(t-1)) \\ x_c(t) = x(t-1) \\ y(t) = g(W^C x(t)) \end{cases}$$

where $x(t)$ is the output of hidden layer in general, $y(t)$ is the output of the output layer, $u(t-1)$ is the input of Elman network. W^A is the weight of the connection between context units and hidden layer, W^B is the weight of connection between input layer and hidden layer, W^C is the weight of connection between hidden layer and output layer, $f()$ and $g()$ are respectively the activation functions for hidden and the output layer. The Gaussian activation function used as a basis function has the following expression: $\phi(X_k, t_i) = G(\|X_k - t_i\|) = exp(-\frac{1}{2\sigma_i^2}\|X_k - t_i\|) = exp(-\frac{1}{2\sigma_i^2}(x_{km} - t_{im})^2)$ where $t_i = [t_{i1}, t_{i2}, ..., t_{iM}]$ is the central of the Gaussian activation function, and σ_i is the variance of gaussian activation function. Three parameters should be learned for the RBF network that are: the center of radial basis function, the variance of radial basis of function and the weight. Each context node receives input from a single hidden node and sends its output to each node in the layer of its corresponding hidden node. The activation function is defined as a sigmoid function that is given by [2]: $f(x) = \frac{1}{1+e^{-ax+b}}$ and $g(x) = kx$

The proposed hybrid neural network is illustrated by Fig. 1.

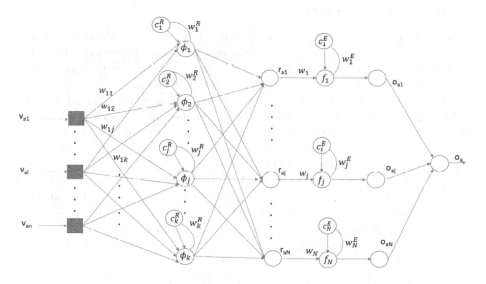

Fig. 1. Proposed hybrid neural network for credit scoring

The context elements are dynamically identified based on the values of the different weights. This process is activated by considering a set of thresholds that are checked to either activate or deactivate the collection of big data for a defined context element. The threshold vector is defined respectively for RBF and Elman neural networks, as follows: $T^E = \{t_1^E, t_2^E ......, t_i^E ......, t_n^E\}$ and $T^R = \{t_1^R, t_2^R ......, t_i^R ......, t_n^R\}$. The obtained score is computed from the neural network outputs using the following expression: $\frac{\sum_{i}^{N} o_i}{N}$[1]. The sensitivity of the proposed model can be modified when the recurrent connection weight and the threshold of the output nodes are varied. Moreover, the hybrid RBF/Elman neural network keeps a memory of the collected data in relation with the defined parameters for a credit applicant. When the RBF network identifies a new value for the monitored credit parameters, the Elman network will accumulate a large value in its context unit. Based on this processing, the proposed hybrid neural network emphasizes the modification made on the given parameters. The weights associated to both contexts for RBF and Elman neural networks are updated based on the results obtained from the collectors and associated to the possible effects detailed in Sect. 5.1.

5   Simulation

In order to illustrate the efficiency of the proposed model, a simulation is conducted for the proposed model using the matlab neural networks toolbox.

5.1   Scenario Description

Using the matlab neural network toolbox, we have built the hybrid neural network, illustrated by Fig. 2, according to the proposed model detailed in Sect. 4.2. Four classes of observers are defined to collect the needed big data associated to the provided values for the context parameters that are used as a way to evaluate credit applicant's data. The configuration is made for the neural network considering 4 neurons for each layer, except the last layer which is composed of single neuron in order to minimize the mean square error. 70 % of data is used for training the proposed neural network and 30 % are used for the testing data set. The Return On equity (ROE), the Sector growth rate and the guarantee value are the three input variables ($V = \{v_1, v_2, v_3\}$) to the studied neural network. The aforementioned variables are defined according to the following expressions: $ROE_t = \frac{Netincome_t}{ShareholderEquity_t}$, Sector growth rate$_t = \frac{Prod_t - Prod_{t-1}}{Prod_{t-1}}$, where $Prod_t$ represents the Gross Product of Sector for the year t and Guarantee in percentage of credit value$_t = \frac{GuaranteeValue_t}{CreditValue}$. Three transfer functions are used respectively for the four layers: a Gaussian, a linear, a sigmoid and a linear functions. Moreover, 30 credit applicants' data are considered as input for the proposed neural network. Weights are initialized randomly for the RBF,

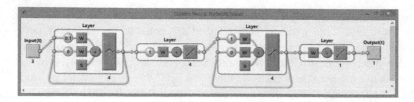

Fig. 2. The hybrid Elman/RBF neural network using matlab neural networks toolbox

Elman and defined contexts for layer 1 and layer 3 respectively for RBF and Elman neural networks. Three effects of context on the credit applicant's data are defined as follows: (1) A positive effect is introduced when an update is available through the analysis of the collected big data related to the set of input variables, (2) A negative effect is applied when an erroneous value is provided by the credit applicant for the input variables, and (3) Neutral effect is applied when no change to be made to the provided input values regarding the credit applicant data. According to each effect, context weights values are then updated. The context in this case aims to enhance the decision, by checking the set of deployed collectors through networks in order to analyze related big data.

A subjective evaluation of credit applicant's data is also performed based on a set of rules that are defined for the possible values for each considered variable in order to perform a comparison between the obtained scores by the traditional technique (subjective score) and the scores provided by the proposed hybrid neural network. As an example for the defined rules for the subjective technique, we can mention the following rule: $score_{ROE} = 1\,if\,ROE \in [0, 10\,\%]$. For this rule, the score associated to the ROE provided as an input for a credit applicant is equal to 1 when provided value as input is between 0 and 10 %.

5.2   Results Interpretation

After training the proposed neural network using a set credit applicant's data samples and by defining the appropriate targets associated to these data, the simulation is then conducted for 30 cases that represent the credit applicant's data as input. First, the simulation is performed by disabling the context while considering the hybrid RBF/Elman neural network. The obtained scores related to this step in addition to the subjective scores are illustrated by Fig. 3. As shown by this figure, two respective score values are defined respectively for a rejected and an accepted credit applicant's data that are: 0 and 1. By considering the proposed model, the generated scores are fluctuating between these two aforementioned values where a threshold, which is equal in this case to 0.5, is considered to decide whether the credit applicant data is accepted or rejected. For example, credit applicant #5 in Fig. 3 for the second curve, has a score that is lower than this threshold then it is considered as rejected. For the applicant #25, the score is greater than this threshold, as a consequence the credit applicant data is accepted. This result is coherent with the decision provided by the subjective approach. The provided scores using the neural network help to decide better in this case to distinguish the weak accept and reject among the obtained decisions. In addition, there is some few cases where the proposed technique detects a false made decision such as for the applicant #20.

The next step is to enable the context for the score evaluation by using the results collected by the set of deployed four collectors to explore big data related to the three variables considered as credit applicant's data. In order to illustrate the effect of the context on the generated decision by the proposed hybrid neural network, the same input data is processed by the simulated network. The set of defined collectors have deduced a negative effect for the applicant #15 data

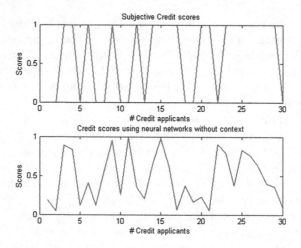

Fig. 3. Illustration of obtained scores using the hybrid neural network without context and subjective evaluation

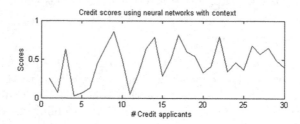

Fig. 4. Illustration of the proposed hybrid neural network with context enabled

that is illustrated through Fig. 4 by a new score that is less than 0.5. For the applicant #28, the context has enabled the update of the provided values for the defined three variables which has induced a positive effect, illustrated by Fig. 4.

6   Conclusion

In this paper, we have introduced a novel hybrid neural network that enables the evaluation of the credit applicants' data that may enhance the decision based on the defined context using big data. The proposed hybrid model has enabled the classification of credit applicant data based on the use of RBF and the decision is enhanced based on Elman neural network by considering the collected big data through a period of time. The set of defined weights for each layer are updated in a dynamic manner considering the related big data identified by the set of deployed collectors through interconnected networks. The proposed model has shown its efficiency through the conducted simulation using the neural network toolbox to enhance the credit scoring compared to conducted subjective evaluation. Some enhancements still possible for the proposed neural network by integrating weights associated to the collectors' efficiency that will be considered for credit scoring processing in addition to the identification of the most appropriate period of time in order to converge to the best credit scoring. The selection of required collectors should be enhanced by considering the dependency between the available data sources and the level of trust associated to the deployed collectors. The application of the proposed model for additional financial problems will be also investigated.

References

1. Al Douri, B., Beurouti, M.: Credit scoring model based on back propagation neural network using various activation and error function. Int. J. Comput. Sci. Netw. Secur. **14**(3), 16–24 (2014)
2. Edelman, D.B., Crook, J.N., Thomas, L.C.: Recent developments in consumer credit risk assessment. Eur. J. Oper. Res. **183**(3), 1447–1465 (2007)
3. Lee, T.C., Chen, I.: A two stage hybrid credit scoring using artificial neural networks and multivariate adaptive regression splines. Expert Syst. Appl. **28**, 743–752 (2005)
4. Martens, J., Sutskever, I.: Learning recurrent neural network with hassian-free optimization. In: The 28th International Confrence on Learning Machine, Bellevue, WA, USA (2011)
5. Pradhan, S.K., Pradhan, M., Sahu, S.K.: Anomaly detection using artificial neural network. Int. J. Eng. Sci. Emerg. Technol. **2**(1), 29–36 (2012)
6. Wang, Z., Tong, X., Yu, H.: A research using hybrid RBF/Elman neural networks for intrusion detection system secure model. Comput. Phys. **180**, 1795–1801 (2009)

Smart Infrastructures

Situation Awareness for Push-Based Recommendations in Mobile Devices

Ramón Hermoso[1]([✉]), Jürgen Dunkel[2], and Jan Krause[2]

[1] University of Zaragoza, Zaragoza, Spain
rhermoso@unizar.es
[2] Hannover University of Applied Sciences and Arts, Hannover, Germany
juergen.dunkel@hs-hannover.de, jan.krause@stud.hs-hannover.de

Abstract. The paper presents an innovative architecture for push-based Context-aware Recommendation Systems (CARS) that integrates different description and reasoning approaches. Complex Event Processing (CEP) is applied on live data to provide situation awareness. Ontologies and semantic rules are used to define domain expertise that allow individualized and domain-specific recommendations. A case study of a museum serves as a proof of concept of the approach.

Keywords: Context-aware recommender systems · Mobile recommender systems · CEP · Environment

1 Introduction

During the last years, with the rise of information technologies, many different sources of data may be used in order to support decision making processes. Besides, this enormous quantity of information might make the user feel overwhelmed and incapable of making appropriate decisions. For that reason, different techniques have been used in order to help users refine information in different ways to focus only on relevant data to make more accurate decisions. Some examples of these techniques are Information Fusion [17], Big Data analysis [6] or Complex Event Processing (CEP) [14]. In this paper, we focus on recommender systems as specific instances of decision-making processes. In particular, we are interested in the so-called Context-Aware Recommender Systems (CARS) [2] that use context (e.g. location of the users, weather, time, mood, etc.) to assess recommendations on items or activities the user is potentially interested in. We claim that significant progress in the field of CARS has not yet been fully accomplished, as issues related to the high dynamicity in some scenarios, which may involve a continuous change in the locations of the users, and changes in other contextual elements, are still hard challenges [8]. Besides, users typically interact with mobile devices using wireless communications and the concept of environment should be carefully defined and exploited to articulate an appropriate recommendation process. The final goal is to enable context-aware and

© Springer International Publishing Switzerland 2016
W. Abramowicz et al. (Eds.): BIS 2016, LNBIP 255, pp. 117–129, 2016.
DOI: 10.1007/978-3-319-39426-8_10

adaptive information systems that pro-actively recommend interesting items or activities to mobile users.

The massive proliferation of mobile devices yields an enormous amount of events that can be exploited for improving recommendation processes. In this paper, we propose an innovative push-based CARS architecture that integrates CEP and semantic rules for providing the following benefits: (i) *situation and context awareness* by taking the current situations and contexts of all users into account (e.g. locations, movements, surrounding objects, time); (ii) *flexibility by declarative rules* (in particular CEP and semantic rules) that define all triggering and reasoning in the recommendation processes, respectively; and (iii) *real-time processing* by using CEP that can handle high-frequent data streams from mobile phones and sensors in real-time.

In summary, our approach provides a flexible CARS architecture that can give individualized recommendations for users, based on the context assessed along with the users preferences, in real-time.

The paper is structured as follows: in Sect. 2, we present the reference model we adhere to build push-based context-aware recommender systems. Then Sect. 3 shows our architectural approach. We explain it in more detail by using a case study of a museum scenario in Sect. 4. In Sect. 5, we consider related work. Finally, Sect. 6 summarizes our work and sketches some future avenues.

2   Reference Model

We present in this section an architectural model for articulating push-based recommendation processes based on contextual information. We adhere to the model presented in [11].

2.1   Contextual-Based Model

In this section we introduce the elements forming part of the model namely *contexts, environments, agents, events* and *activities*.

Context. A *context* must be considered as a purpose unit for the recommendation process. A context delimits the scope or purpose of a recommendation. For example, in the context of an *art museum*, which artwork to go next is a valid recommendation outcome, while suggesting a route to get to work results meaningless. Formally we define a context as follows:

Definition 1. *A context c is a tuple $\langle \mathcal{E}, \delta \rangle$, in which $\mathcal{E} = \{e_1, ..., e_n\}$ is the set of environments in which the user is active and δ is the purpose of the recommendation process.*

Environment. This entity allows encapsulating a recommendation process and its associated contextual information, as well as the communication among different entities. Formally:

Definition 2. *An environment $e_c = \langle U, \Theta, Act \rangle$, belongs to context c, and is a common area, physical or virtual, in which users U (the set of users currently active in the environment) coexist to perform a set of activities Act under certain environmental constraints $\Theta = \{\theta_1, \theta_2, ..., \theta_n\}$.*

Constraints in Θ are physical or virtual boundaries used to unambiguously delimit the environment (e.g. physical location, time delimiting the duration of the environment, the max number of agents accepted, etc.). Moreover, note that contexts with the same purpose but different associated environments might result in different recommendation outcomes.

Agents. They represent actors in the recommendation process. Agents are divided into two subgroups namely: *users* and *Environmental Managers (EMs)*. The former are either the receivers of the recommendation or representatives of a third party while the latter are special agents associated to specific environments, being in charge of controlling the membership of users and communication issues in that environment.

A user u_i belongs to an environment e_c iff he/she fulfils all the constraints Θ of that environment. Formally, let $e_c = \langle U, \Theta, Act \rangle$ be an environment, then $belong(u_i, e_c) \leftrightarrow \forall_{\theta_j \in \Theta} \, fulfills(u_i, \theta_j)$. Consequently, the user u_i leaves an environment he/she belonged to iff any of the constraints in Θ is no longer fulfilled by him/her.

Events. We divide the set of possible events into two non-overlapping sets: physical and communicative. The former are perceived by physical sensors and represent uncontrollable phenomena that can occur due to the inherent nature of the environment, such as weather conditions, time or location. Contrary, communicative events are released by agents in order to inform users in the environments about different issues. For example, a museum might broadcast a guided tour for a short period of time to those users already in the environment.

Activities. Activities are denoted by the set $Act = \{act_1, ..., act_n\}$ and represent the actions the user can carry out in an environment. For example, a museum visitor can be recommended to go to join a guided tour, to go to a certain painting, depending on the circumstances.

2.2 Management of Environments

In order to obtain recommendations in a certain environment a user needs to be part of that environment. This membership management is handled by the EM of each environment. EMs poll the users with periodic messages indicating which constraints they must fulfil to be active in the environment. Once a user receives a poll message, his/her device checks whether the constraints are (or are not) satisfied; if they are it replies to the EM with an ACK message, and so becoming a member of the environment.

3 Situation-Aware Recommendation Process

In this section, we present our architectural approach. In particular, we discuss how User Agents and Environmental Mangers exploit their specific knowledge for deriving individualized recommendations for certain users. An overview of our architecture is given by Fig. 1.

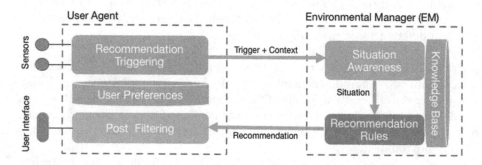

Fig. 1. Architectural approach of a recommendation system

User Agents (UAs) implement the recommendation process on user side by providing the recommendation system with the current situations of the present users. Furthermore, UAs foster privacy, because they transfer only information that users explicitly permit forwarding to the recommendation system.

To understand the users' personal situation, each UA is monitoring continuously its sensor data, e.g. the current GPS coordinates (in outdoor locations) or the strength of detected beacons signals (if the user is in indoor facilities). Monitoring this data, the UAs have to deal with two different use cases:

1. *Recommendation triggering:* If the user is in an appropriate situation to receive a new recommendation, its UA contacts the EM and requests a recommendation. We can distinguish two types of such situations:
   (a) The user is entering a new environment: according to the monitored sensor data, the UA detects that the user fulfills all environmental constrains $\Theta = \{\theta_1, \theta_2, ..., \theta_n\}$.
   (b) The users' situation has significantly changed (e.g. her location) and furthermore she is in an appropriate situation for receiving new recommendations. For instance, the user is not occupied and it has passed some time since the previous recommendation has arrived.
   For triggering a new recommendation, the UA has to pass context information to the EM. This includes all the preferences the user wants to reveal and additional information about the actual situation.
2. *Post Filtering:* When a UA receives new recommendations from the EM, it processes a filtering step. According to the users' private preferences and her current situation the UA selects the most appropriate recommendations

and filters out what is not acceptable due to the users' personal constraints. Afterwards, the UA shows the selected recommendations on the agents' user interface. Note that this step cannot be done by the EMs, because they lack the users' private preferences.

Environmental Managers (EMs) implement the recommendation process on the environment side. When an EM receives a users' request for a new recommendation, it processes the following steps:

1. First, the EM evaluates the *current situation of that user* by exploiting its detailed knowledge about the environment.
2. Then, it infers the *global situation of the environment* by taking the situations of *all* present users, as well as the domain knowledge into account.
3. On base of the users' local and the environments' global situation, the EM applies domain-specific recommendation rules to deduce appropriate and *personalized recommendation* for the user.

The proposed architecture presented in Fig. 1 yields a situation-aware recommendation system due to the following aspects:

- First, each UA infers the actual situation of the user by exploiting its sensor data. With the knowledge about the current situation it decides, if the user wants a new recommendation. Then, the situation context is propagated to the EM for enabling personalized recommendations.
- Secondly, an EM makes situation-aware decisions, because it takes the situations of all known users into account. So the manager can find a compromise between the conflicting desires of users with regard to the requirements and constraints of the environment.

Situation-aware recommendation systems have to exploit two different types of knowledge: structural and situational knowledge.

- **Structural Knowledge** describes the expert knowledge about a certain domain by means of an ontology [13]. An ontology contains an TBox describing terminological or conceptional knowledge, and an ABox defining the given facts or the assertional knowledge. Structural knowledge can be characterized as stable knowledge that does not change over time (as the TBox) or at least in a very low frequency as the facts defined by the ABox.
- **Situational Knowledge** defines the current state in an environment by exploiting live data. Usually, live data is produced by a continuous stream of sensor data. Such sensors could be in-built sensors of the users' smartphones, or permanently installed in the environment (e.g. movement sensors, cameras or smoke detectors). Situational knowledge can be seen as dynamic knowledge with a high change frequency. Each data set corresponds with a particular event in the environment. The stream of events must be evaluated in realtime to achieve situation awareness.

4   System Design

4.1   Case Study: A Museum Scenario

A recommendation system for museum visitors serves as a case study to explain the proposed architecture in detail, with the following actors involved:

User Agents: The UAs are software agents running on the smartphones of all museum visitors. Each UA represents the intentions and desires of a particular visitor. For instance, it contains data about preferences of certain artists or art styles, but also specific restrictions and constraints, such as spoken languages or coming-up appointments outside the museum. Furthermore, User Agents exploit sensors such as GPS, beacons[1] and acceleration sensors to infer the current position and behavior of its user. Museum guides have their own specific UA running on their smartphones. This UAs know about the guides' expertise and their current situation, e.g. in which museum room they are.

Environmental Manager: The EM has a detailed *structural knowledge* about the museum, which is necessary to make appropriate recommendations: for instance, it knows about the floor plan, the locations of all artworks and the positions of all installed beacons. Furthermore, the EM has a general understanding of art. It knows about different art styles, which artists belong to certain styles, and how to relate visitors' preferences to certain artworks. Furthermore, the EM exploits *situational knowledge* provided by all UAs: i.e. it knows the current situations of all visitors and guides. Therefore, the EM can adjust its recommendations to reduce the walking distances of visitors and to prevent overcrowded museum rooms. For simplification reasons we assume a single EM.

Cooperation: The UAs and the EM cooperate due to the following process:

1. As soon as an UA detects via GPS that its user is entering the museum, it contacts the EM sending the location, as well as the users' personal preferences for triggering a recommendation.
2. The EM evaluates the current situation and context of the new visitor, the occupancy of rooms and the situations of appropriate museum guides to infer some individualized recommendations that it sends back to the UA.
3. When a UA receives recommendations from the EM, it processes a post-filtering step by taking private preferences of the user into account before presenting the results to its user.

All UAs are continuously sending their positions to the EM, in particular, when they are in the mood of receiving new recommendations.

---

[1] Beacons are sensors that send continuously an unique ID, which can be used by smartphones for indoor localization.

4.2   Design of the User Agent

The detailed architecture of the User Agent is based on various technologies as illustrated in Fig. 2.

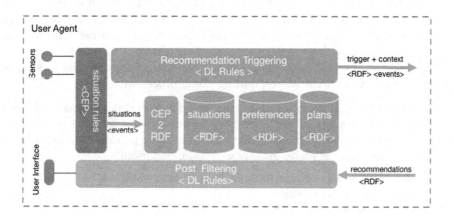

Fig. 2. Design of the User Agent

Knowledge Base: The knowledge of each UA can be described by Semantic Web Technologies [13]. The UA basically contains only facts (or ABox) that can be described by RDF [1] triples, but has no specific terminological knowledge. Using RDF provides numerous advantages: RDF is the most general format for exchanging data in open systems. Furthermore, it is easily integrated with ontology languages such as OWL [12] and rule systems like Jena Rules[2]. The following knowledge base is written in RDF and gives an idea of a possible knowledge base content.

```
:User1  :likesArtist      :Kandinsky;
        :likesStyle       :Expressionism;
        :speaks           :Spanish;
        :isLocatedIn      :EntranceHall;
        :hasAppointmentAt :"12:30".
```

There are different types of facts. We can distinguish the users' *preferences* (liking Kandinsky and expressionism) and *capabilities* (speaking Spanish), her future *plans* (appointment at 12:30), and the current *situation* (located in the entrance hall).

Situational Rules: CEP rules capture the dynamic aspects in the recommendation system and are applied to provide the current situation of the UAs' user. The following CEP rule correlates two different iBeacon events to infer that its user has changed her location.

---

[2] https://jena.apache.org.

```
CONDITION:  iBeaconEvent AS B1 → iBeaconEvent AS B2
            ∧ B1.id ≠ B2.id
   ACTION:  create ChangedLocationEvent(from:=B1, to:=B2)
```

Note that this rule creates a new complex event of type `ChangedLocationEvent` that must be integrated in the knowledge base. The integration is provided by the CEP2RDF component that maps this event to the `:User1 :isLocatedIn :B2` RDF fact. Because the role 'isLocatedIn' is functional, a former 'isLocatedIn' role assertion must be deleted from the knowledge base.

CEP rules can be expressed in an Event Processing Language (EPL) such as Esper, and executed on the corresponding event processing engine.

Recommendation Triggering: DL rules such as Jena Rules access the knowledge base to detect a situation that is appropriate for requesting a new recommendation from the EM. The following rule written in pseudo code describes a situation, when a user is ready for a new recommendation:

```
IF  currentTime>timeOfLastRecommendation+15min
    ∧ changeLocation(?user)
    ∧ notInMuseumShop(?user)
    ∧ noUpcomingAppointment(?user,currentTime+20min)
THEN triggerRecommendation(?user)
```

The rule formulates four conditions to trigger a recommendation: at least 15 min have been elapsed since the last recommendation, the user has changed her location, she is not in the museum shop and has no upcoming appointment within the next 20 min. Note that this rule combines different types of facts: the facts about the users' location are situational. They are provided by CEP rules and the corresponding RDF facts (produced by the CEP2RDF component). The appointment fact is from the part of the knowledge base dealing with plans and could be provided by the users calendar.

Post Filtering: Finally, there might be some post filtering rules of DL type, which deal with incoming recommendations. For instance, the user might not want to reveal to the EM, which artists she dislikes. Therefore, recommendations for affected artists must be filtered out by the UA.

4.3   Design of the Environment Manager

Figure 3 shows the essential components of the Environment Manager, and which technologies they use.

Situation Awareness: This component monitors the data streams arriving from the UAs by using CEP. Appropriate CEP rules derive the current situation of all museum visitors, as well as all the global situation in the museum. The following rule infers that a particular user stays in a certain room by using the `ChangedLocationEvent` event received from that user. This `UserPositionEvent` event can be mapped to corresponding RDF fact stored in a related triple store.

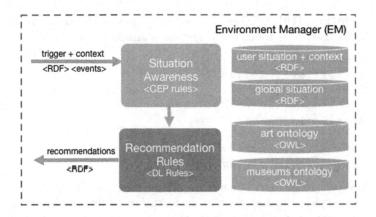

Fig. 3. Design of the Environment Manager

```
CONDITION: ChangedLocationEvent AS L
           ∧ L.to.location_type = 'room'
           ∧ L.user_type = 'user'
  ACTION: create UserPositionEvent(room_id:=L.to.id,
                                   user:=L.user)
```

The following rule captures all UserPositionEvents to transform them into a RoomOccupancyEvent: it aggregates all UserPositionEvents of the last 5 min in a certain room and counts them.

```
CONDITION: (UserPositionEvent AS P )[win:time:5min]
           ∧ group_by(P.room_id)
           ∧ count(P) AS C
  ACTION: create RoomOccupancyEvent(room_id:=P.room_id,
                                    nb_users:=C)
```

The RoomOccupancyEvent events reflect the occupancies of the museum rooms, and can also be stored as RDF facts in the EM knowledge base.

Knowledge Base: Besides the situational knowledge produced by CEP rules, the EM contains an OWL ontology describing terminological and assertional knowledge about the domain. The following part of an ontology defines some concepts (Painter, Expressionism,..) and facts about artists and paintings.

```
:Painter        rdfs:subClassOf :Artist.
:Expressionism  rdfs:subClassOf :ModernArt.

:Kandinsky  :a          :Painter;
            :style       :Expressionism;
            :hasPainted  :The_Rider.

:Macke  :a          :Painter;
        :style       :Expressionism;
        :hasPainted  :Promenade.
:The_Rider  :hangingIn  :Room_13.
:Promenade  :hangingIn  :Room_15.
```

Recommendation Rules: These rules infer a personalized recommendation for a particular user. The following rule gives a simplified example.

```
IF  likesArtist(?user, ?artist1)
  ∧ style(?artist1, ?style)
  ∧ style(?artist2, ?style)
  ∧ hasPainted(?artist2, ?painting)
  ∧ hangingIn(?painting, ?room1)
  ∧ located(?user, ?room1)
  ∧ nearBy(?room1, ?room2)
THEN recommend(?user, ?painting, ?room1)
```

If a user likes ?artist1, who has a certain style and there is another ?artist2 with the same style, who has a painting in the museum, then it is concluded to send the user to that painting. Note that this rule correlates facts from the art ontology about artists, their styles and paintings with facts of the museum ontology about the location of relevant paintings.

5   Related Work

Complex-Event Processing has become a major paradigm to identify situations in environments in which multiple events may take place in a short period of time [14]. This is particularly relevant in context-aware systems, since identification of complex events must bring about appropriate decision-making. There exist approaches combining CEP and context-awareness in different application domains, spanning from medical scenarios [4,18] to dynamic business process adaptation [10]. As a special case of decision-making support, recommender systems have also benefited from the rise of CEP. Authors in [5] present StreamRec, a recommender system designed as a CEP application taking into account different types of events. Although StreamRec can work as both pull and push-based modes, its main difference with our approach relies on the use of contextual information. In particular, StreamRec does not rely on domain knowledge specified by semantic web technologies and DL rules.

In line with push-based approaches, there exist some advances in ontology definition for recommender systems. [7] presents an ontology to allow the representation of event-based rules that trigger recommending processes.

An interesting work supporting our idea of using event processing for push-based processes is described in [3]. In there, authors present a context-aware information push service for the tourism domain by using ECA rules. In contrast to our work, users must subscribe to certain types of messages in order to allow the push-based mechanism to work. We consider our approach is more flexible and transparent for the user, since no subscription is needed, as well as it has a stronger recommender system flavor, as it takes into account user preferences to assess recommendations. Furthermore unlike CEP rules, conventional ECA rules are not suited for processing sensor data streams in real-time.

Tourism domain is a recurrent case of study when attempting to test the validity of a mobile recommender system [9]. Authors in [15] present a mobile recommender system for visiting museums, in which user's preferences as well as museum's constraints are taken into account to provide individualized recommendations so improving the quality of the experience. Here, recommendations are mainly focused on minimizing walking time in the museum (by checking other visitors routes) as well as satisfying (if possible) the user personal interests. However, this work does not take into account much relevant contextual information that could potentially lead to a more accurate recommending process. Other approaches such as [16] propose a context-aware recommender system in which knowledge is represented by semantic web languages to recommend objects according to a user profile and the gathered context information. However, context in this work is merely the location of the user collected by the mobile built-in sensors. In our approach, context can contain arbitrary information about the user and her situation, e.g. time restrictions or preferences. Moreover, our approach uses CEP in order to process events what brings about more flexibility in scalable environments, in terms of events.

Another important issue is to state the position of our proposal in terms of the existing literature about CARS. Our work is based on a knowledge-based approach instead of a collaborative filtering or content-based flavours other authors have addressed. We claim both approaches are complementary and can be used together depending on the constraints of the domain, i.e. the privacy of the data the users have to share.

6   Conclusions

In this paper, we have presented an innovative architecture for Context-aware Recommendation Systems (CARS) that integrates different description and reasoning approaches for particular purposes: (i) Complex Event Processing is applied to achieve situation awareness, (ii) ontologies are used for describing structural domain knowledge and (iii) semantic rules for specifying individualized recommendations.

Our approach has various advantages: it provides situation and context awareness: both the current behavior of particular users and the state of the entire system can be inferred in real-time to entail more personalized (and probably more accurate) recommendations. Furthermore, the general usage of rules and semantic web technologies endows the system with enough flexibility to be adapted to different domains/scenarios. Moreover, the proposed architecture supports privacy: private information that the user does not want to disclose, is considered in the post-filtering phase on user side only.

To describe our approach in some detail, we have presented a museum scenario as a case study, in which we are currently testing our work.

There are different future lines of research. First, we want to integrate some mechanisms to learn automatically recommendation rules by correlating between user behavior. For instance in the museum scenario, visitors who linger a long time in front of a certain artwork, could be nearly always interested in a particular other piece of art. As future work we intend to cover some important open issues, such as adding the concept of uncertainty to the information inferred from the events, since they often offer inaccurate information. This is of particular interest, since we could take advantage of advances on sensor uncertainty treatment as well as on promising works on incorporating uncertainty into CEP.

Acknowledgments. This work has been supported by the projects DGA-FSE, EU-COST Action IC1302 and TIN2015-65515-C4-4-R.

References

1. RDF 1.1 primer: Technical report, World Wide Web Consortium (2014)
2. Adomavicius, G., Tuzhilin, A.: Context-aware recommender systems. In: Ricci, F., Rokach, L., Shapira, B., Kantor, P.B. (eds.) Recommender Systems Handbook, pp. 217–253. Springer, Heidelberg (2011)
3. Beer, T., Rasinger, J., Höpken, W., Fuchs, M., Werthner, H.: Exploiting E-C-A rules for defining and processing context-aware push messages. In: Paschke, A., Biletskiy, Y. (eds.) RuleML 2007. LNCS, vol. 4824, pp. 199–206. Springer, Heidelberg (2007)
4. Bruns, R., Dunkel, J., Billhardt, H., Lujak, M., Ossowski, S.: Using complex event processing to support data fusion for ambulance coordination. In: 17th International Conference on Information Fusion, FUSION 2014, Salamanca, Spain, 7–10 July 2014, pp. 1–7 (2014)
5. Chandramouli, B., Levandoski, J.J., Eldawy, A., Mokbel, M.F.: StreamRec: a real-time recommender system. In: Proceedings of the 2011 ACM SIGMOD International Conference on Management of Data, pp. 1243–1246. ACM (2011)
6. Chen, M., Mao, S., Zhang, Y., Leung, V.C.: Big data analysis. In: Big Data, pp. 51–58. Springer, New York (2014)
7. Debattista, J., Scerri, S., Rivera, I., Handschuh, S.: Ontology-based rules for recommender systems. In: SeRSy, pp. 49–60 (2012)
8. del Carmen Rodríguez-Hernández, M., Ilarri, S.: Pull-based recommendations in mobile environments. Comput. Stand. Interfaces **44**, 185–204 (2015). doi:10.1016/j.csi.2015.08.002
9. Gavalas, D., Konstantopoulos, C., Mastakas, K., Pantziou, G.: Mobile recommender systems in tourism. J. Netw. Comput. Appl. **39**, 319–333 (2014)
10. Hermosillo, G., Seinturier, L., Duchien, L.: Using complex event processing for dynamic business process adaptation. In: 2010 IEEE International Conference on Services Computing (SCC), pp. 466–473. IEEE (2010)
11. Hermoso, R., Ilarri, S., Trillo-Lado, R., del Carmen Rodríguez-Hernández, M.: Push-based recommendations in mobile computing using a multi-layer contextual approach. In: The 13th International Conference on Advances in Mobile Computing and Multimedia. ACM (2015)
12. Hitzler, P., Krötzsch, M., Parsia, B., Patel-Schneider, P.F., Rudolph, S.: OWL 2 Web Ontology Language: Primer. W3C Recommendation, 27. http://www.w3.org/TR/owl2-primer/
13. Hitzler, P., Krötzsch, M., Rudolph, S.: Foundations of Semantic Web Technologies. Chapman & Hall/CRC, London (2009)
14. Luckham, D.: The Power of Events, vol. 204. Addison-Wesley, Reading (2002)

15. Lykourentzou, I., Claude, X., Naudet, Y., Tobias, E., Antoniou, A., Lepouras, G., Vassilakis, C.: Improving museum visitors' quality of experience through intelligent recommendations: a visiting style-based approach. In: Intelligent Environments (Workshops), pp. 507–518 (2013)

16. Ruotsalo, T., Haav, K., Stoyanov, A., Roche, S., Fani, E., Deliai, R., Mäkelä, E., Kauppinen, T., Hyvönen, E.: Smartmuseum: a mobile recommender system for the web of data. Web Semant. Sci. Services Agents World Wide Web 20, 50–67 (2013)

17. Xiong, N., Svensson, P.: Multi-sensor management for information fusion: issues and approaches. Inf. Fusion 3(2), 163–186 (2002)

18. Yao, W., Chu, C.-H., Li, Z.: Leveraging complex event processing for smart hospitals using RFID. J. Netw. Comput. Appl. 34(3), 799–810 (2011)

A New Perspective Over the Risk Assessment in Credit Scoring Analysis Using the Adaptive Reference System

Gelu I. Vac[1(✉)] and Lucian V. Găban[2]

[1] Babeş-Bolyai University, 400084 Cluj-Napoca, Romania
geluvac@yahoo.com
[2] 1 Decembrie University, 510009 Alba Iulia, Romania
luciangaban@yahoo.com

Abstract. The main goal of this paper is the development of a platform which can insure the effectiveness and the simplification of the loan granting process performed by financial credit institutions and banks oriented to small and medium enterprises. The factors considered include employee's education, experience, philosophy, self-beliefs and self-understanding of the bank's target and values and his self-commitment to the bank's objectives. This paper proposes a platform which implements a statistical model, containing financial indicators. The model is flexible, being able to include, besides financial indicators, some emotional ones, considered as model corrections pertaining to the decision maker. The latter indicators are important in borderline decisions. Our platform has been validated on samples containing financial data for Romanian small and medium sized enterprises.

Keywords: Adaptive Reference System · Risk assessment · Risk Culture · Credit risk

1 Introduction

The credit risk taken by these institutions constitutes a major problem, largely debated by the specialized international communities. In accordance with the rules of lending, as specified in Basel II (2006) [7] and Basel III (2010) [8], the objective of the banking institutions is to identify factors having a significant impact on the probability of granting a loan, and to construct credit risk models. Basel II encourages banking institutions to implement their own internal models for measuring financial risks. The credit risk remains one of the major threats which financial institutions face, so therefore it is essential to model it. Consumer finance has become one of the most important areas of financial credit institutions and banks because of the amount of credits being lent and the impact of such credits on global economy. Lately, an increasingly large number of scientists have obtained remarkable results in estimating the credit risk [32–34]. A significant number of international journal articles have addressed different credit

© Springer International Publishing Switzerland 2016
W. Abramowicz et al. (Eds.): BIS 2016, LNBIP 255, pp. 130–143, 2016.
DOI: 10.1007/978-3-319-39426-8_11

scoring techniques. The problems resulting from the non-performing loans can cause financial distress in banks, which in turn creates economic and social turmoil. As we are facing a constant increase of credit solicitors, the analysis of the credit requests becomes a major problem for the creditor.

Our work is based on two models: the Adaptive Reference System (ARS) [35], a general framework dedicated to evaluate and improve the behavior and proficiency of an decision maker, and a logistic regression model for the credit risk assessment.

The main goal of this paper is the development of a platform which can insure the effectiveness and the simplification of the loan granting process, available for both the financial institutions and the ones who want to be given a loan. As we live in a high speed era, we are trying to propose through ARS a publicly exposed platform where potential credit consumers could create profiles and upload their financial data which could save time for the financial institutions while pre-filtering their loan granting customers. While doing so, all data being centralized, we can automatically evaluate the proficiency of each financial institution, of each of their agencies and provide both comparison means and also suggest improvements for the lower performers.

The structure of the paper is as follows. After this introductory section, the next one gives a literature review, including a brief description of ARS. The rest of the paper contains our original work, which finally leads to an application of the general methodology specific to ARS in a particular field, namely credit assessment. This way, Sect. 3 presents the statistical methodology involved (i.e. the logistic regression model), while Sect. 4 describes the physical platform implementing ARS methodology in the above-mentioned field. Final section contains conclusions and future work.

2 Related Work

The focus of this paper is a split between the evaluation of the risk assessment, the proficiency of a decision maker and the implementation of a tool (software platform) that handles them. So, the areas for the background documentation are consequently.

The main goal of the financial institutions is to ensure the flow of resources from sectors with an excess of funds to those with a deficiency in funds. Our paper fits into the trend of finding advantageous solutions for the credit risk management. The growth in consumer lending has known during the past decades could not have been possible without a formal and automated approach to assessing the risk that a loan to a consumer will not be repaid. The papers [22,29] are relevant in this respect.

There are many papers and reviews dedicated to the evaluation of agencies and their proficiencies, like Fitch, Moody's and Standard&Poor's [26]. Out of the results we have read, when it comes to credit assessment, their conclusions remain in the area of financial indicators.

Also, there are many studies concerning the relevance and efficiency of the decision maker and there are quite a few methodologies to follow:

- One study we found relevant is about the usage of expected utility theory versus the use of prospect theory by which it has been demonstrated that probabilities should be replaced by decision weights [20].
- Another researched approach is the model of herd behavior, by which the decision maker looks at the decisions made by previous decision makers [6].

As the banking system is consistently constrained by the confidentiality of data, we couldn't find any public tools nor platforms available for research. Neither could we find any description of algorithms for computing the credit assessment.

In terms of managing the risk, our aim is not the study of the risk management directly, but better yet offer valuable output about the consequences of the potential of conflicts [19,21] induced by ignoring (at the management level) the cultural differences between remote agencies of the same organization [27,30]. In this case, the conflicts take the shape of lower performance in credit assessment in comparison with the context specified by the financial institution.

3 The Problem Context

The decision to either grant a loan or not, is a multi-criteria decision [36], involving objective factors (client data, usually financial indicators), as well as subjective factors, specific to the decision maker.

3.1 Input Data

There is a wide range of mathematical models which combine financial indicators. The best performing equations used to compute the risk assessment are the non-linear ones which dis-associates the degree of dependency between the indicators.

Our study has demonstrated that the analysis of the financial indicators alone in credit scoring can easily reach a gray area of decision (i.e. borderline decisions) either to grant the loan or not. When they do so, the bank tends to ask for extra information about the client's financial status which could give the bank an additional perspective over that specific company.

We believe that there should be an additional indicator, computed by the theory of planned behavior and enclosed by the name of Risk Culture [3], computed out of several cognitive criteria which carefully selected and evangelized across the bank, could influence the bank's overall proficiency in its credit engagement. This indicator is a subjective one, belonging to the decision maker [10].

3.2 Adaptive Reference System

An automated data processing technique which gives an accurate description over the current status of future clients as well as an accurate perspective over the future behavior of current clients, requires a broader approach which is suitable for the Adaptive Reference System (ARS) [35], described shortly below.

The Adaptive Reference System (ARS) is used to evaluate organization's performance. ARS can adapt easily to any organizational environment change by tracking three very specific values: Standard Value ($ARS(Cr, SV)$), Expected Value ($ARS(Cr, ExV)$) and Actual Value ($ARS(Cr, AV)$). These values need to be acknowledged and fairly well-identified by the decision maker and/or well-computed inside each industry field in join with each evaluation criterion. Each such assembly of these values in the context of an evaluated criteria set grouped by a very specific Area of Interest (AoI) constitute a powerful *key performance indicator, KPI* we can use to track the organization's performance level by the chosen AoI which can be an actual department of the organization (Sales, Accountancy, etc.). Each AoI has its specific set of $KPIs$. The adaptability capacity of the ARS resides in its historical stored context of all $KPIs$ which can be tracked in time to evaluate the *decision performance* by individual context (natural state).

3.3   Financial Indicators Considered

AoI in our case is credit risk department of an financial institution. Having considered the important role of the financial institutions in the economic activity development, we consider as relevant the choice of such an institution for an partial implementation of ARS that can help a Relevant Decision Maker (RDM) [35] to better evaluate his alternatives. The problem of either acquirement/non-acquirement of loan can be viewed as a binary classification problem where applicants for credit are predicted to be either good or bad risk. The practical problems of consumer credit modeling relate to how to measure the performance of the future client, how to build a score for a new credit based on the existing data, how to improve the monitoring of the score and how and when to adjust and rebuild score or change the operating policy. It was always required a considerable emphasis in consumer credit modeling on ensuring that the data is timely, valid and free from error.

The goal of this paper is to improve the quality of loan granting decisions taken by an RDM in financial credit institutions and banks. Having in consideration the above statements, we can definitely say that the viability of a credit granting institution is depending on the scientific grounding of decisions in the Risk Department. The criteria underlying the estimation of credit risk, accredited in the scientific literature, are based on the following financial ratios described below. According to ARS, these variables are resembled as KPI [23, 25] and are grouped as follows:

1. Profitability KPIs
   - Return on assets (ROA), Return on equity (ROE), Net profit margin (NPM) and Earnings Before Interest, Taxes, Depreciation and Amortization ($EBITDA$)
2. Leverage KPIs
   - Leverage ratio (LEV) and Stability ratio (SR)
3. Liquidity and solvency KPIs

- Current liquidity ratio (CL), Flexibility ratio $(FLEX)$, Cash-flow (CWR) and General solvency $(SOLV)$
4. Activity KPIs
    - Assets turnover $(ASTU)$, Current assets turnover $(CASTU)$ and Current debts turnover $(CDTU)$

The ARS's KPIs measure the company's economic proficiency, by several market perspectives. The many observations, the more relevant the accuracy of interpretation. And since data computing automation is commoner and no longer a blocker in the credit scoring process, we should consider as many as we find to be statistically and economically relevant.

4   Statistical Methodology

This section presents some theoretical references considered necessary regarding statistical methodology implemented in ARS. A special place is granted to evolutionary computing techniques, necessary for an integrating approach regarding: classification, variable selection, and parameter optimization, specific to the credit risk assessment as introduced by Marques et al. (2013) [24]. The regression approach in credit risk assessment of an applicant or current borrower is logistic regression, now being the most common approach, presented by L.C. Thomas et al.(2005, 2010) [33,34]. This approach allows one to indicate which are the important questions for classification purposes, as described in L.C. Thomas (2000) [32]. Further we will describe shortly the defining elements of a binary outcome model. The dependent variable (denoted VD) is a binary variable as in: $VD = 1$ for companies that have accessed credits, representing the default event; $VD = 0$ for companies that have not accessed credits yet. The binary outcome models are used for identifying the probability of success (granting credit), which is modeled to depend on regressors. One such model explains an unobserved continuous random variable y^*, further on called latent variable. What needs to be observed is the binary variable y, which may take the value 1 or 0, as y^* exceeds a threshold (in our case 0.5). Let y^* be an unobserved variable, and regression model for y^* is the index function model (see Cameron and Trivedi (2005) [12]).

$$y^* = X'\beta + u \tag{1}$$

where the regressor vector X is a $K \times 1$ column vector, the parameter vector β is a $K \times 1$ column vector, and the error vector u is a $K \times 1$ column vector. Then $X'\beta = \beta_1 X_2 + \beta_2 X_2 + \ldots + \beta_k X_k$. The parameter vector β from model (1) cannot be estimated, because y^* is not observed. We have

$$y = \begin{cases} 1 & if \quad y^* \geq 0.5 \\ 0 & if \quad y^* < 0.5 \end{cases} \tag{2}$$

where 0.5 is the threshold. The threshold is being chosen randomly by the institution, according to its own credit granting claims. From (1) and (2) we have

$$Pr(y = 1/X) = F(X'\beta) \tag{3}$$

where $F(X')$ is the cumulative distribution function (c.d.f.) of $-u$. If u is logistically distributed is obtained the logit model. The probability of default (the probability of success), is related to the application characteristics (the ARS's Key Performance Indicators) $X_1, X_2, \ldots, X_k$ by (3). Let a vector of data for KPI, denoted as $X_i = (X_{1i}, \ldots, X_{ki})$, where $i = 1, \ldots, N$ from N observations, then from the model (3), the conditional probability is given by

$$p_i \equiv Pr(y = 1/X) = F(X_i\beta). \tag{4}$$

The conditional probability p_i is a positive sub-unitary value which measures the credit risk for entity i, who's performance gets synthesize by the KPI data vector. The maximum likelihood estimator (MLE) is the estimator for binary models. Given a sample (y_i, X_i), where $i = 1, \ldots, N$ of N independent observations, we obtain a ML estimation vector $\hat{\beta} = (\hat{\beta}_0, \hat{\beta}_1, \ldots, \hat{\beta}_k)$. The estimated coefficients $\hat{\beta}_i$ help us to weight the impact of each of the independent Key Performance Indicator i on the estimated probability of default (credit risk). The $\hat{\beta}_i$ estimates are forming the weights I_{Cr_i} needed to compute the Risk Culture Indicator of the Credit Department. The change on the probability $Pr(y = 1)$, respectively Risk Culture Indicator, uses the following equation:

$$(\partial Pr(y_i = 1/X_i))/(\partial x_{ij}) = F(X_i\beta)\beta_j, \tag{5}$$

and represents marginal effects of X_i. In conclusion the weight $\hat{\beta}_i$ measures the impact of criterion i over the credit risk (the Risk Culture Indicator) specific to the Risk Department. The relationship of calculation for the Risk Culture Indicator, adapted accordingly to the financial institution, becomes the following:

$$RC_j = \left[\sum_{i \in D} (I_{Cr_i} \times KPI_i)\right] \times I_D, \tag{6}$$

where we noted with RC_j the Risk Culture Indicator associated to the client j, and I_D represents the Department's weight inside the financial or banking institution. The range of values for: the weights I_{Cr_i}, KPI_i and RC_j can be inferred from the context of application [16]. The formula (6) is linear for the performance indicators KPI_i, but in some cases we find the interaction effects which leads to non-linearity. Next we refer to practice on how to estimate the weights I_{Cr_i}, on the database needed to compute these estimates and last but not least to compute predictions regarding the client's economical behavior. These elements are all required in order to compute the Culture Indicators (RC_j), of the Risk Department, necessary decision making for the future decisions of an RMD.

We find mandatory to submit the relevance of a database and its update. According to Berger et al. (2007) [9] the credit risk is an instrument of estimating the request of a loan customer based on his/her basic characteristics and past experiences with credits. It is obvious that risk estimation requires an historical database regarding credit behavior of both existing and future loan customers. The estimation model of the credit risk is based on quantitative information gathered in a database regarding the factors related to the business of the customer when applying for a credit. Once the Credit Risk indicators are identified, using a binary outcome model one can build a statistical model that estimates the weight I_{Cr_i} of each KPI_i. The KPI_i which have a significant impact (statistically significant) on the repayment behavior are weighted more heavily in score. At the first step, based on the customer's historical data, we compute the I_{Cr_i} estimates. Based on this information, one can analyze and develop an evaluation module that gives a score that describes the data accumulated on loan applicants, and a probability of granting a loan can be estimated. As a consequence, in the second phase, using the Eq. (6) updated by the values of each KPI_i, one can evaluate the credit risk RC_j for each loan applicant. The purpose of the credit risk model is to assign (or not) a loan to customers. Choosing the variables [15] used in building credit risk models depends on the quality of the data and the availability of those who supply them. There is no preset number of indicators KPI_i that should be used in building scoring models. The data used affect the quality of the model. Based on some periodical economic analysis, some companies will enter in the database and some others will be deleted [13]. This risk model allows risk making predictions using the right weights for the KPIs. This way ARS can signal periodically the potential risk of each client. By improving the database, the predictive classification power of a model should be more realistic, and the standard tests of statistical significance should be more compelling. The implemented model is based on sample data containing financial data for small and medium sized enterprises from Romania. The companies come from various fields of activity that have been analyzed by the bank in order to access a credit with a positive result. Taking into account the economic and financial data, we have implemented a model which estimates the likelihood of granting a loan. The analyzed financial statements taken into account by the bank refer to the activity of the company.

5   The ARS Implementation

One can build hard numbers and evaluate accounting data as linear or nonlinear as one could, but the behavior or misbehavior of a loan consumer is also driven by the empathy and social and moral ascending between bank employees and bank customers [1].

We believe that a smart system which could easily combine accounting and culture values should get much closer to behavior predictability. Also, we believe that the risk culture indicator should have its own coefficient to be multiplied by and participate together to compute a more accurate score for credit eligibility [5, 14, 36].

Apart from the actual computation of the credit eligibility score, we would also like to compute the tendency of the Credit Department to grant loans - and this should include the most of the context by which they become likely to grant the loan. Starting from this point of view, we would like to track the core values and beliefs of the bank employees and implement (teach) the most performing structure of such values into all credit granting responsible employees [17,18,31].

ARS is a concept using three actors in order to build several perspectives of the same indicators which's values are rendered inside its axis: the Standard Value (SV), the Expected Value (ExV) and the Actual Value (AV). In the case of Risk Assessment for Credit Scoring there are currently no Standard Values set (as per industry) [35].

The arguments in favor of using ARS are given by the possibility of also computing external economical factors inside the same loan granting model not just the creditor's expectations and the client's economical results.

The advantages of also considering the Standard Values in credit scoring are tremendous since depending on the geographical area where the client resides, there might be a question of long or medium term stability of a certain industry field. For example, considering Romania for applying this model it is well known that the Cluj county offers greater stability for *the IT industry field* while the Salaj county offers almost nil stability, while considering *the agriculture industry field*, the situation is complete opposite [23].

Even further, as at this moment each banking institution chooses its own way of computing the Credit Score, so they don't even share their client's historical behavior to each other until it is too late and the client is reported to the Banking Risk Committee (BRC), while one of the KPIs to follow could be the degree of difficulty of a client as a client per say, and therefore implies a big deal of effort to manage it. And we strongly advice such dissemination of indicators to complement the risk culture indicator of the Credit Department, see (Mikusova 2010) [25].

Nevertheless, for this paper we only consider using the creditor's expectations for the Expected Values (ExV) as follows:

- The creditor describes *the indicators* it wants to evaluate (the current thread proposes only three significant KPIs, namely EBITDA, CR and ROI explained earlier in this article);
- The creditor computes *the weight* I_{Cr} (the ARS's importance mark) of each accounting indicator (which it refers to as *the coefficient value*) as part of the creditor's Risk Culture;
- The creditor declares *the expected value* of the non-linear output (computed from the ARSs linear score using the above mentioned indicators and weights) which we call Eligibility Threshold (e.g. an Eligibility Threshold set to 0.5 means that any resulted value above 0.5 means that the company becomes eligible to receive financial aid from the creditor, while any resulted value lower than 0.5 means that the company is risk full to receive financial aid);

which are computed against any applicant company's real data extracted from the company's historical economical and financial exercise for the Actual Values (AV). Then provide means to overcome the variations between them in order to facilitate the applicant's eligibility to contract a credit by suggesting economical measures which could enforce its Credit Scoring. (e.g. either increase EBITDA, ROI or CR separately or all together to meet a better probability to become eligible for a credit).

5.1   The ARS Tool

Our web community portal www.adrefsys.org [2] has already focused on these aspects and handles both assistance cases:

1. better elaborate the values of coefficients based on historical loans (already granted and therefore able to evaluate their impact for the bank) and
2. evaluate current customers based on the freshly computed coefficient values.

For the sake of building a strong argumentation of ARS, we have built an open platform called www.adrefsys.org which provides the actual tools to measure real data in real time and build to test as many scenarios as any subject creditor should have.

In order to comply a Credit Assessment into ARS, we have built as strategy to start from defining a working **Scenario** in which we settle the context for what to expect. During this setup process we have foreseen to give the current scenario as much data as needed to uniquely identify it in order to address it later on (a suggestive name, the Company Name it addresses the exercise to, the period we intent to evaluate (e.g. 1st of January 2015 until 20th of December 2015) and the Eligibility Threshold as the minimum probability of becoming eligible for financial aid (e.g. any probability value greater than 0.5 means good chances of eligibility)).

Using this strategy we can later overlap various Scenarios in order to obtain the overall KPIs like *the credibility of the model*, or *the potential misbehavior* of the client. But of course, this decision belongs to the Credit Department, see (Austin 1996) [4].

While setting the period we actually validate further monthly inputs so it becomes valid as a scenario we gave the suggestive name to. For each scenario we will define the creditor's credit assessment culture, such as:

– the list of indicators he is willing to investigate,
– each indicator's coefficient value,
– the eligibility threshold.

Remark 1. If any of the indicators participates in the equation in any of its altered forms (e.g. squared EBITDA, natural logarithm of ROI, etc.) we suggest to create a separate indicator with an according name and symbol and fill-in its corresponding Coefficient Value.

A full **Indicator** definition also includes the specification of its *Name* (long description), *Symbol* (short description), *MinThreshold* and *MaxThreshold* which are values we use to validate future inputs against.

This far, we have managed to define the base context of a Credit Risk Assessment or in our case the ARS's Expected Values set. The second part is about applicant data which we will refer to as the set of Actual Values.

We will constitute the applicant's data into sessions, namely these can be either monthly, quarterly, semester or annually accounting extracts. Defining a **Session** implies providing its name (suggestive words), the calendar date of the accounting extract and an observations field where we can update information as we go through the evaluation process of the current scenario. [28]

For each Session, we will define a set of Actual Values, that would be one value for each indicator we have defined in the context of the current scenario.

When having all these data updated in ARS, we are ready to compute the values of the score (denoted by X) as the linear/nonlinear output. The score is computed using logistic regression. The Eq. (6), for ID = 1 becomes

$$X = \left[\sum_i (I_{Cr_i} \times KPI_i)\right] \tag{7}$$

The probability of score, $p(X)$, is computed using the following equation

$$p(X) = \frac{e^X}{(1 + e^X)} \tag{8}$$

For the case study, presented in the Sect. 5.2, only three KPIs, namely EBITDA, CR and ROI have a relevant impact over the score X. For each evaluation session we compute X and $p(X)$ and the resulting graphs contain the evolution of these two indicators over all evaluation sections. Finally these graphs give the decision makers an educated guess of the applicant's chances to qualify for the credit amount he is aiming.

Remark 2. the evaluation sessions do not necessarily need to be actual data from live accounting, but simply exercise data a financial consultant can play with in order to assess the applicant with the proper advice that will suite the applicant's goal of scoring well for the credit amount.

5.2   Case Study

In order to give a practical example for the above statements, we have chosen three Romanian companies which are willing to solicit loans from the bank which has built the database required to estimate the values for the coefficients I_{Cr_i} (according to the regressive model (7)). In the following we have built the charts for both score and probability (computed according to Eq. (8)) in order to show better observed behavior of probability versus credit score over that period, considering only the financial indicators. For each company we have elaborated predictions for the next 2 months (P2), 4 months (P4) and 6 months (P6) for the EBITDA, CR and ROI indicators. As you can easily see below, the scoring chart becomes more harmonized by computing its probability chart. A growth in scoring does imply a growth in eligibility, but not directly (not as linear). In the probability chart below, you can also see the Eligibility Threshold (set to 0.5) and suggested by the red support line

In Fig. 1a and b we show the evolution of the probability and score for the first company. It is a company which's evolution is in favor to grant the solicited loan. The Fig. 2a and b show us a company which doesn't evolve in the direction desired

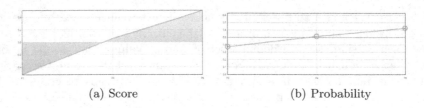

(a) Score   (b) Probability

Fig. 1. Study results for company A (Color figure online)

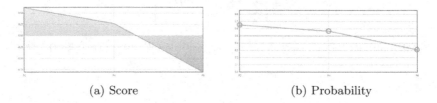

(a) Score   (b) Probability

Fig. 2. Study results for company B (Color figure online)

by the bank. EBITDA is the indicator which decreases triggers an alarming shrink-age of the probability to grant a loan. The graphs from the Figs. 3 and 4 reflect a company which cannot make any claims to loan a credit in the next six months. Although it shows a probability to be granted a loan of little above 0.5 (our thresh-old), the company doesn't have any eligible guaranties for the bank. Even if the company is not credible, for the 6 months period of time, still it requires a careful analysis for a longer period. Here is where we believe that our system will be able to make a difference. If for the companies A and B the situations were to obvious in favor or against granting a loan, for the company C where the financial indica-tors are not conclusive, the residual influence of the risk culture indicator could trigger the bank's exposure to a good deal or a high risk.

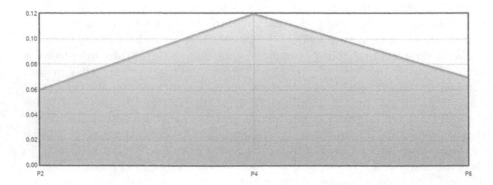

Fig. 3. Score for Company C

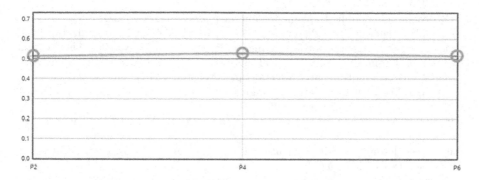

Fig. 4. Probability for Company C (Color figure online)

6 Conclusions

The economical purpose of any credit institution is to sell credits and the purpose of contracting a credit is to produce economical growth. By using the ARS methodology to evaluate a (Creditor, Applicant) context is to improve the odds of making business and accelerate economical growth for each.

The strength of ARS is that it follows the creditor's key indicators individually for smaller periods of time, which gives better transparency than computing the same key indicators by a cumulative manner.

Therefore, we have identified each of the requested indicators by their constituents as they can be identified in the legally accepted accounting system and compute them in a loop for as many times as the time series's items provided by the applicant for analysis.

Graphically, the Expected Value is going to be a straight line (e.g. the support line of 0.5 probability), expressed as a reference threshold for the applicant and aims to guide the applicant in which areas should he stimulate his business, so that he can become eligible for a credit of the desired amount.

Furthermore, in order to accomplish an agile implementation of the Risk Assessment in Credit analysis using ARS, we will try to envision a system using a company's monthly balance as a physical accountancy extract in a computer generated format file (e.g. CSV, TXT), where the monthly data extract will stand for the time series. And having the applicant piped live to the ARS portal, he could see in real time weather he complies for the targeted credit amount.

ARS is not only a formal system, but the grounds for a powerful tool which can intermediate communication between credit institutions and credit consumers and provide better means for each to cope for economical growth.

References

1. Adam, F., Humphreys, P.: Encyclopedia of Decision Making and Decision Support Technologies. University College Cork, Ireland. London School of Economics and Political Science, UK (2008)

2. ARS community public implementation. http://www.adrefsys.org
3. Ashby, S., Palermo, T., Power, M.: Risk culture in financial organizations: an interim report. Accounting & Finance with Plymouth University, The London School of Economics and Political Science. http://www.lse.ac.uk/accounting/CARR/pdf/risk-culture-interim-report.pdf. November 2012
4. Austin, R.D.: Measuring and Managing Performance in Organizations. Dorset House Publishing, New York (1996)
5. Bagnasco, A., Chirico, M., Parodi, G., Marina, S.A.: A model for an open and flexible e-training platform to encourage companies' learning culture and meet employees' learning needs. Educational Tech. Soc. **6**(1) (2003). DIBE, University of Genoa, ISSN 1436–4522
6. Banerjee, A.V.: A simple model of herd behavior. Q. J. Econ. **107**(3), 797–817 (1992). The MIT Press
7. Basel II, International Convergence of Capital Measurement and Capital Standards: A Revised Framework Comprehensive Version, Bank for International Settlements, Press & Communications, CH-4002 Basel, Switzerland, ISBN print: 92-9131-720-9, ISBN web: 92-9197-720-9 (2006)
8. Basel Committee on Banking Supervision. Basel III, A Global Regulatory Framework for More Resilient Banks and Banking Systems, Bank for International Settlements (2010)
9. Berger, A., Barrera, M., Parsons, L., Klein, J.: Credit Scoring for Microenterprise Lenders. The Aspen Institute/FIELD (2007). http://fieldus.org/projects/pdf/CreditScoring.pdf
10. Bradberry, T., Greaves, J.: Emotional Intelligence 2.0. TalentSmart, San Diego (2009). ISBN 978-0-9743206-2-5
11. Business Dictionary. http://www.businessdictionary.com
12. Cameron, A.C., Trivedi, P.K.: Microeconometrics: Methods and Applications. 1058 p. Cambridge University Press, Cambridge (2005)
13. Cameron, A.C., Trivedi, P.K.: Microeconometrics Using Stata, 692 p. A Stata Press Publication, StataCorp LP, College Station, Texas (2009)
14. Collins, J., Porras, J.I.: Built to Last: Successful Habits of Visionary Companies. HarperBusiness, New York (1994). ISBN 0887306713
15. Fernandes, J.E.: Corporate Credit Risk Modeling: Quantitative Rating System and Probability of Default Estimation, SSRN working paper (2005). www.ratingexpress.net/content
16. Gokhale, M.: Use of Analytical Hierarchy Process in University Strategy Planning. University of Missouri-Rolla, Master of Science in Engineering Management (2007)
17. Heathfield, S.M.: Culture: your environment for people at work. Am. Soc. Quality's J. Quality and Participation (2000). http://humanresources.about.com/od/organizationalculture/a/culture.htm
18. Hillston, J.: A Compositional Approach to Performance Modeling. Cambridge University Press, New York (1996). University of Edinburgh, ISBN: 0-521-57189-8, http://www.dcs.ed.ac.uk/pepa/book.pdf
19. Hofstede, G.: Culture's Consequences: Comparing Values, Behaviors, Institutions and Organizations Across Nations, 2nd edn. Sage Publications, Thousand Oaks (2001)
20. Kahneman, D., Tversky, A.: Prospect theory: an analysis of decision under risk. Econometrica **47**(2), 263–292 (1979)
21. Issa, K.T.: Princess sumaya university for technology. Int. J. Bus. Soc. Sci. **3**(6), 105–111 (2012). (Special Issue - March 2012)

22. Li, X.-L., Zhong, Y.: An overview of personal credit scoring: techniques and future work. Int. J. Intell. Sci. **2**, 181–189 (2012)
23. Macdonald, B., Rust, C., Thrift, C., Swanson, D.: Measuring the Performance and Impact of Community Indicators Systems: Insights on Frameworks and Examples of Key Performance Indicators. The International Institute for Sustainable Development (2012)
24. Marques, A.I., Garcia, V., Sanchez, J.S.: A literature review on the application of evolutionary computing to credit scoring. J. Oper. Res. Soc. **64**, 1384–1399 (2013)
25. Mikusova, M., Janeckova, V.: Developing and implementing successful key performance indicators. World Acad. Sci. Eng. Technol. 42, p. 983 (2010)
26. Special, N., Consultant Phoebus, J., Dhrymes, N.S., Consultant Tsvetan, N., Beloreshki.: CREDIT RATINGS FOR STRUCTURED PRODUCTS A Review of Analytical Methodologies, Credit Assessment Accuracy, and Issuer Selectivity among the Credit Rating Agencies by NERA Senior Vice President Andrew Carron (2003)
27. Pabian, A., Felicjan, B., Kuceba, R.: Role of cultural differences in international small business development, In: 28th Annual SEAANZ Conference Proceedings, Small Enterprise Association of Australia and New Zealand, Melbourne, pp. 1–3, July 2015
28. Patton, M.Q.: Qualitative Research Evaluation Methods. Sage Publishers, Thousand Oaks (1987)
29. Pavlidis, N.G., Tasoulis, D.K., Adams, N.M., Hand, D.J.: Adaptive consumer credit classification. Int. J. Intell. Sci. **2**, 181–189 (2012)
30. Ranf, D.E.: Cultural differences in project management. Annales Universitatis Apulensis Series Oeconomica **12**(2), 657 662 (2010)
31. Sahota, M.: An Agile Adoption and Transformation Survival Guide: Working With Organizational Culture. Published by InfoQ, Enterprise Software Development Series, ISBN 978-1-105-73572-1 (2012). http://pearllanguage.org/images/4/4f/Agile_Transition_Survival_Guide.pdf
32. Thomas, L.C.: A survey of credit and behavioral scoring: forecasting financial risk of lending to consumers. Int. J. Forecast. **16**(2000), 149–172 (2000)
33. Thomas, L.C., Oliver, R.W., Hand, D.J.: A survey of the issues in consumer credit modeling research. J. Oper. Res. Soc. **56**, 1006–1015 (2005)
34. Thomas, L.C.: Consumer finance: challenges for operational research. J. Oper. Res. Soc. **61**, 41–52 (2010). doi:10.1057/jors.2009.104
35. Vac, G.I.: Adaptive reference system: a tool for measuring managerial performance. Studia Univ. Babeş-Bolyai Informatica **LX**(1), 63–78 (2015)
36. Velasquez, M., Hester, P.T.: An analysis of multi-criteria decision making methods. Int. J. Oper. Res. **10**(2), 56–66 (2013)

Drivers and Inhibitors for the Adoption of Public Cloud Services in Germany

Patrick Lübbecke[1]([⊠]), Markus Siepermann[2], and Richard Lackes[2]

[1] Institute for Information Systems at the DFKI,
Saarland University, Saarbrücken, Germany
Patrick.Luebbecke@iwi.dfki.de
[2] Department of Business Information Management,
Technische Universität Dortmund, Dortmund, Germany
{Markus.Siepermann,Richard.Lackes}@tu-dortmund.de

Abstract. In this paper, we present an empirical study on the factors that influence companies in their decision regarding the adoption of Cloud Computing services. While this issue has been subject to a number of studies in the past, most of these approaches lack in the application of quantitative empirical methods or the appreciation of the inherent risk. With our study we focus on the factors that promote and inhibit the adoption of cloud services with a particular consideration of application risks. Our findings show that decision makers are significantly influenced by the risk of data loss in the first place but also by the risk that comes along with the service provider or technical issues that might occur during the use of the service. On the other side, the attractiveness could be identified as an important driver for the adoption of cloud services.

Keywords: Cloud computing · Adoption · Cloud readiness

1 Introduction

Although the concepts used with Cloud Computing (CC) were predominant for a long time, the emergence of Amazon Web Services in 2006 brought CC to the mind of mainstream corporate and private users (Regaldo 2011). After available services and business models evolved for nearly a decade, CC became an alternative to traditional approaches of service delivery such as Application Service Providing (ASP) or on-premise hosting. CC offers many advantages. For example, Cloud Service Providers (CSPs) offer flexible "pay-as-you-use"-service plans where users only get charged for the actual consumed units. Furthermore, no investments are necessary to ramp up on server infrastructure as the traffic of the cloud services increases (Armbrust et al. 2010). In addition, cloud service plans often allow short-term contracts from a couple of days to hours in extreme cases which again improve the flexibility for the user.

Despite mostly positive experiences of companies who actually use cloud services (Narasimhan and Nichols 2011), many companies are still reluctant towards the use of CC for certain reasons. In Germany, only 12 % of all companies use cloud services (Statistisches Bundesamt 2014). Fear from cyber-attacks, problems with data security and the uncertain legal situation are among the most stated reasons for companies not

© Springer International Publishing Switzerland 2016
W. Abramowicz et al. (Eds.): BIS 2016, LNBIP 255, pp. 144–157, 2016.
DOI: 10.1007/978-3-319-39426-8_12

to use public cloud services (Boillat and Legner 2014; Carroll et al. 2011). According to these results, we emphasize the role of risk when it comes to the adoption decision (research question RQ1) and consider the attractiveness of CC as the opposite (RQ2):

RQ1: What are primary risk factors that influence the adoption of cloud services and how strong is their impact on the decision?
RQ2: To what extent does the attractiveness of a cloud service oppose the negative effect of risk in an adoption decision?

Following these research questions, the goal of this paper is to identify the determinants for a company's decision to use a cloud service or not (cloud service adoption). The focus is set on public cloud services as private cloud services have a strong resemblance to the traditional on-premise or ASP service type. In the first step, we will identify possible decision factors using established theories in the field of acceptance research (e.g. transaction cost theory, theory of core competence). Based on these factors a research model will be developed. In addition, the specific nature of CC with its unique characteristics like dynamic resource allocation or location independent network access is also taken into account. In the second step, we empirically test our model and present the results of the survey.

2 Literature Review

In 2014, Schneider and Sunyaev (2014) comprehensively reviewed the CC literature published before April 2014 in conferences and journals of the top 50 AIS ranking. They identified 88 papers treating IT outsourcing and adoption of CC services. Only 13 of these papers used quantitative empirical methods to identify the drivers and barriers of CC adoption, providing profound in-depth analyses to gain an understanding on the relationship between certain decision factors and the adoption as well as on the relationship among the factors themselves. Based on these results, we searched the common scientific databases (ACM, Business Source Premier, Emerald, Google Scholar, IEEE Xplore, Sciverse, Springerlink) for the terms "cloud", "cloud computing", "cloud service" in combination with "adoption", "introduction", "adoption factors", "determinants" and found nine additional papers (see Table 1 for all papers).

Half of the papers (e.g. Benlian et al. 2009; Wu 2011) concentrate on the adoption of SaaS, neglecting IaaS and PaaS, while the other half (e.g. Alharbi 2012; Blaskovich and Mintchik 2011) investigates CC in general. The latter do not distinguish between SaaS, PaaS and IaaS but pose only questions generally concerning CC. Only one paper (Heinle and Strebel 2010) investigated the influence factors of IaaS but used expert interviews instead of a structured questionnaire. Some papers (e.g. Opitz et al. 2012) focus only on adoption drivers, neglecting the barriers. Others focus on characteristics of the service user (e.g. Low et al. 2011) instead of the characteristics of the cloud service and the causality which is the focus of this publication. The goal of this paper is to extract relevant adoption factors considering the three different cloud service types and focusing on the risk that comes with the application of cloud services. Therefore, explicit questions are used to examine how the different service models are assessed by the interviewees instead of looking at CC only in general.

Table 1. Results of the literature review

Author	Object	Interviewees	Basic theory	Findings
Alharbi (2012)	CC	IT professionals of Saudi Arabia	Technology Acceptance Model	Age, education influence attitude towards CC. Younger and higher skilled people have a more positive attitude
Asatiani et al. (2014)	SaaS	SMEs in Finland	Global Information-Intense Service Disaggregation; Transaction Cost Theory	Frequency, need for customer contact have a negative influence on adoption while information intensity, asset specificity and uncertainty have a positive one
Benlian (2009)	SaaS	SMEs in Europe	Transaction Cost Theory	Environmental uncertainty, application specify, usage frequency hinder adoption
Benlian and Hess (2011)	SaaS	German IT Executives	Theory of Reasoned Action; Perceived Risk Framework	Cost advantages are strongest driver while security risks are the strongest barriers
Benlian et al. (2009)	SaaS	German IT Executives	Transaction Cost Theory; Theory of Planned Behavior; Resource Based View	Low specificity applications are more often outsourced than high specificity applications. Size of an enterprise plays no role for the adoption decision
Blaskovich and Mintchik (2011)	CC	Accounting Executives	Resource Based View; Institutional Theory	Skills of CIO's negatively affect the adoption
Gupta et al. (2013)	CC	SMEs in APAC Region	–	Cost savings are not the most important factors. Ease of use and convenience are more important for SMEs
Heart (2010)	SaaS	Managers in Israel	–	Trust in the vendor community positively affects adoption
Heinle and Strebel (2010)	IaaS	German IT Executives	–	Provider characteristics (size, market share, etc.) had the strongest impact on the decision of moving to a cloud environment or not
Kung et al. (2013/ 2015)	SaaS	Manufacturing and Retail Firms	Institutional Theory; Diffusion of Innovation; Technology-Organization-Environment Framework	Interaction effects between mimetic pressure and perceived technology complexity affect the adoption
Lee et al. (2013)	SaaS	Korean IT Consultants	–	Reduced costs and distrust in security are the main driver and inhibitor
Lian et al. (2014)	CC	Taiwan Hospitals	Technology-Organization-Environment Framework; Human-Organization-Technology Fit	Technology is the most important dimension. Data security is the most important factor
Low et al. (2011)	CC	High Tech Industry	Technology-Organization-Environment Framework	Relative advantage has a negative influence. Top management support, firm size, competitive pressure, and trading partner power have a positive influence

(Continued)

Table 1. (*Continued*)

Author	Object	Interviewees	Basic theory	Findings
Narasimhan and Nichols (2011)	SaaS	North American IT Decision Makers	–	Adopters become convinced quickly and expand adoption quicker than non-adopters. Agility is a more influencing driver than costs are
Nkhoma and Dang (2013)	CC	Secondary Analysis of Technology Decision Makers	Technology-Organization-Environment Framework	An enterprises general adoption style influences the adoption of CC
Opitz et al. (2012)	CC	German CIOs of Enterprises Listed in Stock Indexes	Technology Acceptance Model	Social influence, subjective norm, and job relevance are important factors for adoption
Repschläeger et al. (2013)	CC	Technology and Web 2.0 Start-Ups	–	Identification of 5 customer segments and their preferences concerning cloud services
Safari et al. (2015)	SaaS	IT Professionals of IT Enterprises	Diffusion of Innovation; Technology-Organization-Environment Framework	Top influencing factors are relative advantage, competitive pressure, security and privacy, sharing and collaboration culture, social influence
Saya et al. (2010)	CC	IT Professionals	Real Option Theory	Growth and abandonment options have an effect on adoption, while deferral options do not.
Tehrani and Shirazi (2014)	CC	SMEs in North America	Diffusion of Innovation; Technology-Organization-Environment Framework	Level of knowledge about CC is the most influential factor for adoption decisions
Wu (2011)	SaaS	Taiwanese IT/MIS Enterprises	Technology Acceptance Model	Expert opinions, doing things faster with SaaS, security of data backups are the most influential factors for adoption
Wu et al. (2011)	SaaS	Taiwanese SMEs	Trust Theory, Perceived risks and benefits	"Easy and fast to deploy to end-users", "seems like the way of future" were the most influential factors for adoption. Data locality and security and authentication and authorization issues were the most influential concerns
Yigitbasioglu (2014)	CC	Australian IT Decision Makers	Transaction Cost Theory	Legislative uncertainty influences the perceived security risk that negatively affects the adoption. Vendor opportunism also hinders the adoption

3   Research Framework

The *readiness of enterprises to adopt* cloud services is determined by two opposing phenomena. On the one hand, there are factors that create *attractiveness* like cost savings (Benlian and Hess 2011; Boillat and Legner 2014), increased flexibility (Asatiani et al. 2014), or the concentration on core competencies (Gupta et al. 2013). But there are also factors that represent a barrier for the adoption like missing data security (Yigitbasioglu 2014), vendor lock-in (Armbrust et al. 2010), or loss of control (Alharbi 2012). In the case of cloud services, one can generally speak of *application risks* of various kinds. The characteristics and severity of these relationships are crucial for determining whether a company will adopt a cloud service or not. Furthermore, the factors and their severity need to be considered relatively to existing solutions used by the firm. Although a cloud service may be more attractive in direct comparison with an on-premise solution in terms of some criteria, this does not necessarily have to lead to the adoption of the cloud service. For instance, risks in other areas might be perceived higher and therefore prevent from the adoption of the service. On the other hand, an adoption can take place if the service has a high attractiveness and the risk is considered to be moderate.

Therefore, the two factors *application risk* and *attractiveness* are not considered as two extreme points of a one-dimensional variable that are mutually exclusive, but as two independent dimensions of an adoption decision. Thus, an allusion is made to the hygiene factors and motivators of Herzberg et al. (1967). In the context of job satisfaction, the authors argue that there are two variables that can produce job satisfaction. The so-called "hygiene factors" prevent from the state of dissatisfaction. If there are no hygiene factors available, a person will be unhappy with his or her situation. The second dimension is called "motivators" and creates the feeling of satisfaction. If a person lacks in motivators, he or she will not be "dissatisfied", but rather "not satisfied". In the context of CC, we assume specific risks to function as hygiene factors and the attractiveness of CC as a motivator.

3.1   Hygiene Factor: Risk

Firesmith (2004) identified nine essential safety requirements that have a significant impact on the quality of IT-based services naming the integrity as an essential criterion. He divided integrity into the aspects data integrity, hardware integrity, personnel integrity, as well as software integrity. Data integrity is the degree of protection of the data from deliberate destruction or alteration. The hardware integrity describes protection from deliberate destruction of the hardware while the personnel integrity describes the risk of the employees to be compromised. The protection of data and hardware components against intentional destruction or falsification, however, is only one aspect that determines the level of security of the data and physical server infrastructure. In addition to an intentional destruction of data or hardware by third parties, an unintended data loss or failure of the server hardware can occur as well.

The probability with which these negative events can occur determines the level of *functional and data integrity*. The direction of functional and data integrity is formulated as hypothesis H1 as follows:

H1: The higher the technical reliability in the form of functional and data integrity of a public cloud is perceived, the smaller the perceived application risk will be.

The employee integrity is considered in the model as a separate construct. To include organizational effects that go beyond the reliability of the employee, a construct called *provider integrity* is used. For example, there could be risk that providers follow their own interests (analysis or resale of data) (Benlian and Hess 2011; Nkhoma and Dang 2013). In addition, the provider may not properly delete the data after the termination of the contract (Yigitbasioglu 2014). It is assumed that provider integrity as well as functional and data integrity influence the application risk as follows:

H2: The higher the provider integrity is, the lower the application risk will be.

Both of the constructs provider integrity and function and data integrity relate to the use of cloud services for the execution of certain business processes. In order to detect the extent of the application-specific overall risk within the measurement model, the two constructs will be merged in a third construct with the name *application risk*. It is also assumed that the application risk has a direct influence on the adoption readiness (Asatiani et al. 2014):

H3: The lower the risk of the application of a public cloud services is perceived, the higher is the readiness of a user to adopt the service.

The construct functional and data integrity represents the technical risk of data loss or alteration of data. In addition, the legal question of who is responsible for the protection of the data is an important issue in CC (Lin and Squicciarini 2010). Often, the data transfer from the user to the service provider results in devolution of control over the data to the service provider (Alharbi 2012). If these are personal data, the data protection law of the respective country must be taken into account. The legal situation is further complicated by the fact that often cloud services are used across national borders. For instance, the service provider may be headquartered outside the European Union (EU) while the service user is located inside. Here, specific legislation applies for the outsourcing of personal data determined by the data protection laws of the member states of the EU. Therefore, a cross-border use of cloud services poses high demands on data protection (Lübbecke et al. 2013).

Despite the loss of control mentioned, the outsourcing company remains responsible for the proper handling of personal data from a legal perspective (Opitz et al. 2012). This rule is derived from the data protection laws of the countries in the EU which are now largely harmonized. This creates a risk that the service provider could act contrarily to the instructions of the outsourcing company. In addition, the location where the data is actually stored or processed is often unknown and is not necessarily the same country where the headquarters of the service provider are located. These risks are collectively subsumed under the term *data risk* and have an impact on the service users' readiness for the adoption:

H4: The lower the data risk of a cloud application is, the higher the readiness for the adoption will be.

3.2 Motivators: Attractiveness

A major advantage of cloud services is the ability to acquire only the amount of resources that is actually needed. This allows preventing from an underutilized server infrastructure at the in-house data center and therefore supports an economical provision of IT resources (Armbrust et al. 2010; Zhang et al. 2010). Companies can avoid large investment in hardware and software while being still able to use IT services as needed. The organizational hurdles such as the development of specialists for creating custom IT services can also be avoided. This ultimately leads to a shift from fixed to flexible costs. Subsuming these factors under the construct *economic efficiency*, we hypothesize:

H5: A higher economic efficiency of public cloud services compared to an in-house solution has a positive effect on the attractiveness of the services to the user.

Another characteristic of cloud services is the higher *flexibility* in comparison to traditional IT services. Computing capacity in the cloud can be added or reduced dynamically and on demand. This is not readily possible when running own server infrastructure (Armbrust et al. 2010). Ideally, such an adjustment can take place without any direct interaction with the provider, usually in the form of self-service portals on the provider's website. As cloud services are usually offered through standardized web technologies, they can be accessed from anywhere and with different terminal types (e.g. PC, smart phone, tablets). The location independency is of great importance in some scenarios such as for sales representatives that can access Customer Relationship Management applications on the road. However, the flexibility concerns not only the technical aspects but also organizational and administrative issues. Because of the short duration of contracts, one can realize advantages in terms of price/performance by changing providers (Durkee 2010) as long as the switching costs do not exceed this price advantage. Because of this wide range of advantages, the flexibility in the model is taken into account as follows:

H6: A higher flexibility of cloud services leads to a higher attractiveness.

Finally, it is assumed that adoption of cloud services will only take place if a sufficiently large degree of attractiveness is present.

H7: The attractiveness of cloud services has a positive impact on the adoption readiness.

The constructs and hypotheses result in the structural equation model of Fig. 1.

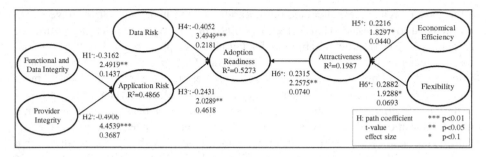

Fig. 1. Research model and results of PLS algorithm

4 Evaluation of the Structural Equation Model

4.1 Data Collection

The creation and evaluation of the structural equation model is based on data collected between late 2013 and early 2014. The data was collected without focusing on a specific industry. We consulted a corporate database to contact 200 companies without any restriction in terms of size or domain. From the companies that we contacted, we identified 183 that had at least a basic understanding of cloud computing. A questionnaire with 35 questions measured in a 5-point Likert-scale was sent to these 183 companies as paper and online version. The persons who responded to the poll mostly came from lower and middle management of the IT department. All responders were familiar with the core concept of CC and its service types (SaaS/IaaS/PaaS). A total of 67 companies participated in the survey. After eliminating incomplete records, 53 completed answer sheets remained. Although the number of only 53 samples is considered small, the PLS-approach is capable of dealing with such a small sample size (Chin 1998; Hair et al. 2013). Table 2 shows the participants of the survey with their respective size and intention towards the adoption of cloud services.

Table 2. Participants of the survey

Number of employees	Share	Adopters	In the future	Non-adopters
<50 (small)	35.19 %	26.32 %	26.32 %	47.37 %
50–250 (medium)	33.33 %	27.78 %	33.33 %	38.89 %
>250 (large)	31.48 %	23.53 %	41.18 %	35.29 %
No answer	6.89 %			
Total		25.93 %	33.33 %	40.74 %

4.2 Evaluation of the Measurement Model

The evaluation of the measurement model in structural equation modeling is performed in two steps for reflective indicator-construct relationships. In a first step, a verification of the indicators and their ability to represent the related constructs properly is performed.

In the case of reflective indicators, the reliability of the items and other convergence criteria need to be tested (Henseler et al. 2009). In this paper, the Average Variance Extracted (AVE), construct reliability (CR), Cronbach's alpha, Fornell-Larcker-criterion, cross-correlations between the items and the Stone-Geisser-criterion (Q^2) were tested. Indicators that did not meet the quality thresholds were removed from the measurement model in an iterative process. An item was considered as sufficient when the reliability of the item exceeded a critical value of 0.5 (Henseler et al. 2009). In some cases, however, a critical value of 0.4 is called in the literature if the remaining convergence criteria are met at the same time (Krafft et al. 2005). In addition to the reliability, the t-values for a significance level of 0.95 must be above 1.66. At a significance level of 0.90, a t-value of 1.29 would be acceptable.

To determine the item reliability, the bootstrap-method with 5000 samples and 53 cases was carried out with SmartPLS which is the proposed setting suggested by Hair et al. (2010). The elimination procedure led to the removal of 11 of the initial 31 reflective items because either the item reliability or the t-values did not exceed the threshold (Table 3). The values of the AVE, Fornell-Larcker criterion and construct reliability passed the necessary threshold of 0.5 (AVE) or 0.7 (CR) (Bagozzi and Yi 1988). The Fornell-Larcker criterion was met for all constructs, as the squared AVE of each construct was higher than the correlations with each of the remaining constructs (Fornell and Larcker 1981). The items did not have any cross-correlations. The Cronbach's Alpha of the construct flexibility missed the threshold of 0.7 slightly; the construct attractiveness missed the threshold significantly with a value of 0.3248.

Table 3. Results of the measurement model

Construct	Item	Loading/weight	Reliability	t-values
Provider integrity	Service provider follows its own agenda	0.9012	0.812	23.6285
	Dependency on service provider	0.8020	0.643	23.3471
Application risk	Data base and data storage (PaaS)	0.8020	0.643	15.8861
	Development environment (PaaS)	0.7114	0.506	6.8395
	Ready-to-use software (SaaS)	0.7941	0.631	12.5492
	Virtual server infrastructure (IaaS)	0.7193	0.517	8.4036
Flexibility	Easy adoption of resources	0.8417	0.708	5.6061
	Easy and instant alteration of resources	0.7696	0.592	4.5017
	Independence of location and devices	0.6984	0.488	4.7492

(*Continued*)

Table 3. (*Continued*)

Construct	Item	Loading/weight	Reliability	t-values
Attractiveness	Attractiveness of PaaS	0.8813	0.777	7.4703
	Overall opinion on public cloud	0.6344	0.402	3.1723
Adoption readiness	Data base and data storage (PaaS)	0.7900	0.624	10.2919
	Development environment (PaaS)	0.6750	0.456	4.3533
	Ready-to-use software (SaaS)	0.6431	0.414	5.1935
	Virtual server infrastructure (IaaS)	0.8349	0.697	17.0029
Functional and data integrity	Data errors during data transfer	0.9187	0.844	35.1452
	Data loss/mutilation	0.8303	0.689	16.8613
Economic efficiency	Cost reduction	0.8138	0.662	7.6603
	Financial benefits	0.8632	0.745	6.9636
	Administrative effort	0.8831	0.780	7.4336
Data Risk (formative)	Data exposal to unauthorized persons	−0.0611		0.3885
	Data safety	0.7008		3.0393
	Unknown place of data storage	−0.0793		0.3783
	Loss of control	0.5130		1.9979

Formative indicators do not only represent a result of the associated construct but are responsible for the expression of the construct that they are supposed to describe. For this reason, it is not recommended to remove items during the evaluation because this would lead to a significant change in the core meaning of the according construct (Jarvis et al. 2003). Henseler et al. (2009) suggest testing for external validity and significance of the item weights. The item weights provide information on the extent to which a construct is influenced by the associated items. In formative models, the focus is on the weight of the items rather than on the loadings of the indicators. In order to make a significant contribution to the characterization of the construct, the corresponding t-values should exceed 1.96 at a significance level of 0.95 and 1.66 at a significance level of 0.90. At least, two indicators exceeded the threshold of 1.96. The Fornell-Larcker criterion for the construct data risk was also met.

We tested the formative construct Data Risk for multi-collinearity with SPSS 24 through the Variance Inflation Factor (VIF). All VIF values turned out to be below 1.7 which indicates almost no collinearity among the items.

4.3   Evaluation of the Structural Model

The hypotheses are evaluated for the postulated effect relationships between the constructs. The path coefficients between the constructs serve as test criteria. Chin (1998) suggests a threshold of 0.2 for the path coefficients in order to attest a meaningful relationship. The values of t-statistics of the path coefficients must exceed the threshold of 2.57 at a confidence interval of 0.99. At a 0.95 confidence the coefficients must exceed the value of 1.98 and 1.66 at a maximum error probability of 10 percent. During the evaluation of this study, the hypotheses H2 and H4 could be accepted at an one percent probability of error. The hypotheses H1, H3, and H6 could be accepted at a five percent significance level and H5 and H6 at a 10 percent level.

For the coefficient of determination (R^2) Chin (1998) denotes values above 0.67 as "substantial", above 0.33 as "moderate", and above 0.19 as "weak". In our model, R^2 is at a moderate level for the constructs *adoption readiness* and *application risk* and at a weak level for *attractiveness*. That means, that 52.73 %, 48.66 % and 19.87 % of the variance can be explained. Effect sizes f^2 above 0.35 indicate a strong effect, between 0.35 and 0.15 a mean effect and sizes below 0.15 but above 0.02 indicate only a slight effect. In our case, H2 (.3687) and H3 (.4618) have a strong effect, H4 (.2181) a mean effect and the remaining hypotheses only a slight effect.

5   Conclusion

The presented structural equation model provides insights to answer the question, what factors influence the adoption decision of public cloud services. The results of the survey are satisfactory. All hypotheses could be confirmed. 52.7 % of the variance of the construct *adoption readiness* could be explained by the other constructs used in the model. The construct *attractiveness* missed the threshold for Cronbach's alpha significantly, *flexibility* missed it slightly and the variance explained for *attractiveness* is only 19.97 %. In conclusion, we can state that in our model the adoption readiness is mostly influenced by the risk side (RQ2). Answering RQ1, we found that data risk has the strongest influence on the adoption readiness followed by the application risk. The application risk is influenced stronger by the provider integrity than the functional and data integrity.

That means that technical issues play a much less important role to firms when deciding on the adoption of cloud services than concerns about data security and provider integrity do. As a consequence, cloud service provider should focus on a better data protection to strengthen the customers' trust in the provider. They should work as transparently as possible and provide insights into their processes so that customers can monitor the safety and security of their data. As Heart (2010) shows, this would positively affect the adoption process.

Limitations occur concerning the sample of the study. The sample size is quite small and comes from only one country. Therefore, it may not be possible to generalize the results. Future studies could focus on the impact of cultural aspects on the adoption of cloud services. For example, the adoption rate in Germany is quite low in comparison to Australia (Yigitbasioglu 2014) or other APAC-countries (Gupta et al. 2013).

Because of the small sample, we also did not separate the results based on the company size or based on the user's extent of knowledge with Cloud Computing. As Boillat and Legner (2014) emphasize, there is a difference between asking enterprises that have already adopted CC and those who intend to do or rather think about moving to the cloud. A distinction of these two groups could provide cloud service providers with useful insights on (a) how to address firms which are new to CC and (b) what are the most important characteristics of CC for real users. However, the number of responders of our survey who have already moved to the cloud is too small so that this analysis must be postponed to future studies. With a larger sample, the results could be interpreted in more detail according to the company size (small, medium, large) or the adoption state. Then, individual results could be generated for each cluster which could help service providers to tailor their service to a certain target group.

References

Alharbi, S.T.: Users' acceptance of cloud computing in Saudi Arabia: an extension of technology acceptance model. Int. J. Cloud Appl. Comput. (IJCAC) 2(2), 1–11 (2012)

Armbrust, M., Fox, A., Griffith, R., Josef, A.D., Katz, R.H., Konwinski, A., Lee, G., Patterson, D.A., Rabkin, A., Stoica, I., Zaharia, M.: A view of cloud computing. Commun. ACM **53**, 50–58 (2010)

Asatiani, A., Apte, U., Penttinen, E., Rönkkö, M., Saarinen, T.: Outsourcing of disaggregated services in cloud-based enterprise information systems. In: 47th Hawaii International Conference on System Sciences (HICSS). IEEE (2014)

Bagozzi, R.P., Yi, Y.: On the evaluation of structural equation models. J. Acad. Mark. Sci. **16**, 74–94 (1988)

Benlian, A., Hess, T.: Opportunities and risks of software-as-a-service: findings from a survey of IT executives. Decis. Support Syst. **52**(1), 232–246 (2011)

Benlian, A.: A transaction cost theoretical analysis of software-as-a-service (SAAS)-based sourcing in SMBs and enterprises. In: Proceedings of the 17th European Conference on Information Systems, Verona, Italy (2009)

Benlian, A., Hess, T., Buxmann, P.: Drivers of SaaS-adoption – an empirical study of different application types. Bus. Inf. Syst. Eng. **5**(1), 357–369 (2009)

Blaskovich, J., Mintchik, N.: Accounting executives and IT outsourcing recommendations: an experimental study of the effect of CIO skills and institutional isomorphism. J. Inf. Technol. **26**(2), 139–152 (2011)

Boillat, T., Legner, C.: Why do companies migrate towards cloud enterprise systems? A post-implementation perspective. In: 16th IEEE Conference on Business Informatics (CBI). IEEE (2014)

Carroll, M., Van der Merve, A., Kotze, P.: Secure cloud computing: benefits, risks and controls. In: Information Security South Africa (ISSA) (2011)

Chin, W.W.: The partial least squares approach to structural equation modeling. In: Marcoulides, G.A. (ed.) Modern Methods for Business Research. Lawrence Erlbaum Associates, New Jersey (1998)

Durkee, D.: Why cloud computing will never be free. ACM Mag. Queue - Emulators **8**, 1–10 (2010)

Firesmith, D.: Specifying reusable security requirements. J. Object Technol. **3**, 61–75 (2004)

Fornell, C., Larcker, D.F.: Evaluating structural equation models with unobservable variables and measurement error. J. Mark. Res. **18**, 39–50 (1981)

Gupta, P., Seetharaman, A., Raj, J.R.: The usage and adoption of cloud computing by small and medium businesses. Int. J. Inf. Manage. **33**(5), 861–874 (2013)

Hair, J.F., Black, W.C., Babin, B.J., Anderson, R.E.: Multivariate Data Analysis. Prentice-Hall, Englewood Cliffs (2010)

Hair, J.F., Hult, G.T.M., Ringle, C., Sarstedt, M.: A Primer on Partial Least Squares Structural Equation Modeling (PLS-SEM). Sage, New York (2013)

Heart, T.: Who is out there? Exploring the effects of trust and perceived risk on SaaS adoption intentions. ACM SIGMIS Database **41**(3), 49–68 (2010)

Heinle, C., Strebel, J.: IaaS adoption determinants in enterprises. In: Proceedings of the 7th International Conference on the Economics of Grids, Clouds, Systems and Services, Ischia, Italy (2010)

Henseler, J., Ringle, C., Sinkovics, R.: The use of partial least squares path modeling in international marketing. Adv. Int. Mark. **20**, 277–319 (2009)

Herzberg, F., Mausner, B., Snyderman, B.B.: Motivation to Work. Wiley, New York (1967)

Jarvis, C.B., Mackenzie, S.B., Podsakoff, P.M.: A critical review of construct indicators and measurement model misspecification in marketing and consumer research. J. Consum. Res. **30**, 199–218 (2003)

Krafft, M., Götz, O., Liehr-Gobbers, K.: Die validierung von strukturgleichungsmodellen mit hilfe des partial-least-squares (PLS)-ansatz. In: Bliemel, F., Eggert, A., Fassott, G., Henseler, J. (eds.) Handbuch PLS-Pfadmodellierung – Methode, Anwendung, Praxisbeispiele. Schäffer-Poeschel, Stuttgart (2005)

Kung, L., Cegielski, C.G., Kung, H.J.: Environmental pressure on software as a service adoption: an integrated perspective. In: Proceedings of the 19th Americas Conference on Information Systems (AMCIS), Chicago, USA (2013)

Kung, L., Cegielski, C.G., Kung, H.J.: An integrated environmental perspective on software as a service adoption in manufacturing and retail firms. J. Inf. Technol. 30(4), 352-363 (2015)

Lee, S.G., Chae, S.H., Cho, K.M.: Drivers and inhibitors of SaaS adoption in Korea. Int. J. Inf. Manage. **33**, 429–440 (2013)

Lian, J.W., Yen, D.C., Wang, Y.T.: An exploratory study to understand the critical factors affecting the decision to adopt cloud computing in Taiwan hospital. Int. J. Inf. Manage. **34**(1), 28–36 (2014)

Lin, D., Squicciarini, A.: Data protection models for service provisioning in the cloud. In: Proceedings of the 15th ACM Symposium on Access Control Models and Technologies (SACMAT 2010), Pittsburg, USA (2010)

Low, C., Chen, Y., Wu, M.: Understanding the determinants of cloud computing adoption. Ind. Manage. Data Syst. **111**(7), 1006–1023 (2011)

Lübbecke, P., Anton, T., Lackes, R.: Cross-border risk factors of cloud services: risk assessment of IS outsourcing to foreign cloud service providers. In: Proceedings of the 5th IEEE International Conference on Cloud Computing Technology and Science (CloudCom), Bristol, UK (2013)

Narasimhan, B., Nichols, R.: State of cloud applications and platforms: the cloud adopters' view. Computer **3**, 24–28 (2011)

Nkhoma, M.A., Dang, D.: Contributing factors of cloud computing adoption: a technology-organisation-environment framework approach. Int. J. Inf. Syst. Eng. (IJISE) **1**(1), 38–49 (2013)

Opitz, N., Langkau, T.F., Schmidt, N.H., Kolbe, L.M.: Technology acceptance of cloud computing: empirical evidence from German IT departments. In: 45th Hawaii International Conference on System Science. IEEE (2012)

Regaldo, A.: Who Coined 'Cloud Computing'? In: Pontin, J. (ed.) MIT Technology Review: Business in the Clouds (2011). http://www.technologyreview.com.br/printer_friendly_article. aspx?id=38987. Accessed October 2011

Repschläger, J., Erek, K., Zarnekow, R.: Cloud computing adoption: an empirical study of customer preferences among start-up companies. Electron. Markets 23(2), 115–148 (2013)

Safari, F., Safari, N., Hasanzadeh, A.: The adoption of software-as-a-service (SaaS): ranking the determinants. J. Enterp. Inf. Manage. 28(3), 400–422 (2015)

Saya, S., Pee, L.G., Kankanhalli, A.: The impact of institutional influences on perceived technological characteristics and real options in cloud computing adoption. In: International Conference on Information Systems (2010)

Schneider, S., Sunyaev, A.: Determinant factors of cloud-sourcing decisions: reflecting on the IT outsourcing literature in the era of cloud computing. J. Inf. Technol. 31, 1–31 (2014)

Statistisches Bundesamt: 12% der Unternehmen setzen auf Cloud Computing, Pressemitteilung Nr. 467/14 (2014)

Tehrani, S.R., Shirazi, F.: Factors influencing the adoption of cloud computing by small and medium size enterprises (SMEs). In: Yamamoto, S. (ed.) HCI 2014, Part II. LNCS, vol. 8522, pp. 631–642. Springer, Heidelberg (2014)

Wu, W.W.: Mining significant factors affecting the adoption of SaaS using the rough set approach. J. Syst. Softw. 84, 435–441 (2011)

Wu, W.W., Lan, L., Lee, Y.: Exploring decisive factors affecting an organization's SaaS adoption: a case study. Int. J. Inf. Manage. 31, 556–563 (2011)

Yigitbasioglu, O.: Modelling the intention to adopt cloud computing services: a transaction cost theory perspective. Australas. J. Inf. Syst. 18(3), 193–210 (2014)

Zhang, Q., Cheng, L., Boutaba, R.: Cloud computing: state-of-the-art and research challenges. J. Internet Serv. Appl. 1, 7–18 (2010)

Parallel Real Time Investigation of Communication Security Changes Based on Probabilistic Timed Automata

Henryk Piech and Grzegorz Grodzki[✉]

Czestochowa University of Technology, Dabrowskiego 73, 42201 Czestochowa, Poland
{henryk.piech,grzegorz.grodzki}@icis.pcz.pl

Abstract. The proposition is connected with the research of the security or threats referring to message decryption, user dishonesty, a non-fresh nonce, uncontrolled information jurisdiction, etc. (that means security properties - attributes), in network communication processes [3]. Encrypted messages are usually sent in the form of protocol operations. Protocols may be mutually interleaving, creating the so called runs, and their operations can appear as mutual parallel processes. The investigation regards both particular security attributes and their compositions referring to more general factors, such as: concrete users, protocols, public keys, secrets, messages, etc. Probabilistic timed automata (PTA) and Petri nets characterize the token set and the complex form of conditions which have to be fulfilled for the realization of transition [5]. The abovementioned situation forms a conception pertaining to the parallel strategy realized with the help of the Petri net that includes the set of security tokens (attributes) in each node.

Keywords: Tensor analysis · Protocol security · Auditing system · Probabilistic timed automata

1 Introduction

To clearly state problem we propose pay attention on a new specific approach based on searching communication threads not by identifying directly the dangerous situations but on utilizing long time prognosis and systematically building information for trends of integrated group of parameters. Therefore, we cannot compare the results of out strategy with the results for the existing short time prognosis. According to our system we supplement the current information dynamically, what is obviously realized on time. The parallel approach give us possibility not only to accelerate the investigation procedures but also to effectively realize them in reference to a given group of users. The set of security communication attributes is presented in [5]. Here we present the inferring mechanism. Selected elements are logically combined in the form of rules. For rules creation we have chosen the BAN and Hoare [3] logic. Rules have a traditional form *"if conditions then conclusions"*. Conditions are represented by protocol

© Springer International Publishing Switzerland 2016
W. Abramowicz et al. (Eds.): BIS 2016, LNBIP 255, pp. 158–168, 2016.
DOI: 10.1007/978-3-319-39426-8_13

actions, whereas conclusions by attributes. Attribute corrections (modifications) are realized according to rules. Our research based on tensor formalism gives us the pictures of complexity scale of monitoring algorithm and points on dependencies and among security parameters. Such projecting strategy facilitates the building of independent parallel processes and their separate execution. The distributed form of investigation, according to chosen security factors, suggests to use the parallel composition of the PTA node structure [2,5]. In general, the organization of the calculation process is presented in the section devoted to the thread creation and the dynamics of their new designation [12]. Our presentation starts from semantic formalisms (Sect. 2) which permit us to collect global information to support the creation of a parallel implemented structure (Sect. 3). Then we show the system shell scheme of a parallel configuration (Sect. 4). Finally we give selected intermediate and efficiency result.

2   Semantic Formalisms for the Presentation of the Communication Security State

2.1   Communication Protocol Run Characteristics in Tensor Grammar Description

Security state is the compendiousness of security attribute levels in aspect of cryptography authentication. The value of attribute security at_i is calculated in probability space. In this case, the probability space has dimension n, where the number of all attributes is regarded in the security analysis, and is defined in the following way: Measure space [1] P with the measure $\mu : M \rightarrow [0,1]^n$ which holds the following conditions:

1. $0 \leq \mu\left(At\right) \leq 1$ for a given attribute set At $\in$ M
2. $\mu\left(\cap_{i \in \{1,2,...,la\}} at_i\right) = \prod_{i=1}^{la} \mu\left(at_i\right),$

where: n - the number of all attributes, M - the set of all attributes, At - the selected subset of attributes, la - the number of selected attributes, $\mu(At)$ is a *probabilistic security space*. Attributes are included in the structure of security module components. Essentially, the following type of modules can be named as protocols, messages, secrets, intruders, nonces [3]. The detailed information about the freshness of keys and nonces, the degree of encryption, sources, the number concerning the usage of each key, the number of honest users and intruders is contained in security attributes [5,8]. The value of attributes is corrected by authentication rules and the time of their (attributes) activation. Let us start from attributes. Each module is described by a specific set of attributes. The simple sum of the m one-dimension probabilistic spaces (the measure of attributes has a probabilistic form) is treated as a tensor of valence 2 and dimensions $m \cdot n$ and consists of n - attributed components (state vectors), where m - the number of security modules, n - attribute number. This tensor Ta is presented in the following form [1]:

$$Ta^{(2)} = \begin{pmatrix} at_{1,1} & at_{1,2} & \cdots & at_{1,m} \\ at_{2,1} & at_{2,2} & \cdots & at_{2,m} \\ \vdots & \vdots & & \vdots \\ at_{n,1} & at_{n,2} & \cdots & at_{n,m} \end{pmatrix}, \tag{1}$$

where $at_{i,j} \in [0,1]$ - the i-th attribute in the j-th security module.

Attributes are modified by rule and time correction functions. The rule correction in the simplest form is realized by the multiplication of the attribute by the correction coefficient:

$$at_{i,j}(t) = c_{i,k} at_{i,j}(t-1),$$

where $k \in \{1, r\}$ - the code of the rule which implicates correction.

The rule regarding the correction tensor Tc, with dimensions $r \cdot n$, is presented in the following way:

$$Tc^{(2)} = \begin{pmatrix} c_{1,1} & c_{2,1} & \cdots & c_{n,1} \\ c_{1,2} & c_{2,2} & \cdots & c_{n,2} \\ \vdots & \vdots & & \vdots \\ c_{1,r} & c_{2,r} & \cdots & c_{n,r} \end{pmatrix}, \tag{2}$$

where $c_{i,k} \in [1, r]$ - the i-th attribute correction referring to the k-th rule. After the correction realized by the multiplication of the attribute $a_{i,j}$ by the coefficient $c_{i,k}$, the reduction tensor will be described as a tensor product $Tp_{r \cdot m \cdot n} = Tc_{r \cdot n} \otimes Ta_{n \cdot m}$. In the expression (3), the exemplary exposed tensor sector (frame - granule - component) refers to the set of attributes evocated by the 1-st rule in the 2-nd security communication module (e.g. 2-nd protocol).

$$Tp^{(3)} = \begin{pmatrix} c_{1,1}at_{1,1} & c_{1,1}at_{1,1} & \cdots & c_{1,1}at_{1,m} \\ c_{2,1}at_{2,1} & c_{2,1}at_{2,1} & \cdots & c_{2,1}at_{2,m} \\ \vdots & \vdots & & \vdots \\ c_{n,1}at_{n,1} & c_{n,1}at_{n,1} & \cdots & c_{n,1}at_{n,m} \\ c_{1,2}at_{1,1} & c_{1,2}at_{1,2} & \cdots & c_{1,2}at_{1,m} \\ c_{2,2}at_{2,1} & c_{2,2}at_{2,2} & \cdots & c_{2,2}at_{2,m} \\ \vdots & \vdots & & \vdots \\ c_{n,2}at_{n,1} & c_{n,2}at_{n,2} & \cdots & c_{n,2}at_{n,m} \\ \cdot & \cdot & & \cdot \\ \cdot & \cdot & & \cdot \\ \cdot & \cdot & & \cdot \\ c_{1,r}at_{1,1} & c_{1,r}at_{1,2} & \cdots & c_{1,r}at_{1,m} \\ c_{2,r}at_{2,1} & c_{2,r}at_{2,2} & \cdots & c_{2,r}at_{2,m} \\ \vdots & \vdots & & \vdots \\ c_{n,r}at_{n,1} & c_{n,r}at_{n,2} & \cdots & c_{n,r}at_{n,m} \end{pmatrix}, \tag{3}$$

The presented grammar refers to a general approach, i.e. when the same attribute can be corrected independently in different security modules. In this approach, for each module, the appointed attribute can be modified in a different and independent way. In the simpler variant, the single attribute is reduced only in one way for all modules. In this case the formal description will be significantly easier.

Attribute vectors (tensor with rank 1) would have the feature (4). In this approach, the attribute modification way, which is based on averaging the correction coefficients of different authentication rule proveniences, will be more convenient.

2.2 The Conception of Time Attribute Correction - Tensor Description

Generally, the influence of the correction strength tensor $T\alpha$ is described in the following way:

$$T\alpha^{(2)} = \begin{pmatrix} \alpha_{1,1} & \alpha_{2,1} & \cdots & \alpha_{n,1} \\ \alpha_{1,2} & \alpha_{2,2} & \cdots & \alpha_{n,2} \\ \vdots & \vdots & & \vdots \\ \alpha_{1,m} & \alpha_{2,m} & \cdots & \alpha_{n,m} \end{pmatrix}.$$

We obtain the tensor of attribute time reduction Tp'' as a result of the tensor product $Ta \otimes Ta$, with dimensions $m \cdot m \cdot n$. After narrowing tensor operations according to non-adequate attributes and time correction coefficients, a simplified version of this tensor Tp''', with dimensions $m \cdot n$, has the form:

$$Tp'''^{(2)} = \begin{pmatrix} a_{1,1}ct_{1,1} & a_{1,2}ct_{2,1} & \cdots & a_{1,m}ct_{m,1} \\ a_{2,1}ct_{1,2} & a_{2,2}ct_{2,2} & \cdots & a_{2,m}ct_{m,2} \\ \vdots & \vdots & & \vdots \\ a_{n,1}ct_{1,n} & a_{n,2}ct_{2,n} & \cdots & a_{n,m}ct_{m,n} \end{pmatrix}. \tag{4}$$

The number of narrowing operations is equal to $m(m-1)$, which means that this is the $m(m-1)$ rank tensor overlapping the operation [1] on itself (i.e. on the tensor Tp''). The protocol security state may be described in the binary space, basing on the so called tokens $tk_{i,j}$. In this case, for each attribute in all modules, we may design the threshold of security acceptation $th_{i,j}$. These limits are presented in the following way:

$$Tth^{(2)} = \begin{pmatrix} th_{1,1} & th_{1,2} & \cdots & th_{1,m} \\ th_{2,1} & th_{2,2} & \cdots & th_{2,m} \\ \vdots & \vdots & & \\ th_{n,1} & th_{n,2} & \cdots & th_{n,m} \end{pmatrix}, \tag{5}$$

where $th_{i,j} \in [0,1]$ - security threshold for the i-th attribute in the j-th module.

The transition from probability to binary space is realized by threshold condition verification and leads to the creation of the token structure:

$$
Ttk^{(2)} = \begin{pmatrix} at_{1,1} \geq th_{1,1} & at_{1,1} \geq th_{1,1} & \dots & at_{1,1} \geq th_{1,1} \\ at_{2,1} \geq th_{1,1} & at_{2,1} \geq th_{2,1} & \dots & at_{2,1} \geq th_{2,1} \\ \vdots & \vdots & & \vdots \\ at_{n,1} \geq th_{n,1} & at_{n,1} \geq th_{n,1} & \dots & at_{n,1} \geq th_{n,1} \end{pmatrix}
$$

$$
= \begin{pmatrix} th_{1,1} & th_{1,2} & \dots & th_{1,m} \\ th_{2,1} & th_{2,2} & \dots & th_{2,m} \\ \vdots & \vdots & & \vdots \\ th_{n,1} & th_{n,2} & \dots & th_{n,m} \end{pmatrix}, \tag{6}
$$

$tk_{i,j} \in 0,1$ - binary security token adequate to the i-th attribute in the j-th module.

In the process of the analysis, the aims of prognosis are also regarded. In this case, the information about the pace of attribute changes in time is essential. Therefore, the tensor of time modification Tdt, with dimensions $n \cdot m$, is introduced as follows:

$$
Tdt^{(2)} = \begin{pmatrix} \frac{\partial at_{1,1}}{\partial t} & \frac{\partial at_{1,1}}{\partial t} & \dots & \frac{\partial at_{1,1}}{\partial t} \\ \frac{\partial at_{2,1}}{\partial t} & \frac{\partial at_{2,2}}{\partial t} & \dots & \frac{\partial at_{2,m}}{\partial t} \\ \vdots & \vdots & & \vdots \\ \frac{\partial at_{n,1}}{\partial t} & \frac{\partial at_{n,2}}{\partial t} & \dots & \frac{\partial at_{n,m}}{\partial t} \end{pmatrix}. \tag{7}
$$

Additionally, we also regard the mutual attribute value correction. This is represented by the tensor Tda with valence 3, and with dimensions $m \cdot n(n-1)$.

$$
Tp^{(3)} = tp_{i,j} = \begin{pmatrix} \frac{\partial at_{i,j}}{\partial at_{k,j}} \\ \frac{\partial at_{i,j}}{\partial at_{k+1,j}} \\ \vdots \\ \frac{\partial at_{i,j}}{\partial at_{k+n-1,j}} \end{pmatrix}_{i,j}, \quad i \in \{1,n\}, \quad j \in \{1,m\}, \quad k \neq i. \tag{8}
$$

Atomic components - granules are presented in the following way:

$$
tda_{i,j,k} = \frac{\frac{\partial at_{i,j}}{\partial t}}{\frac{\partial at_{k,j}}{\partial t}} = \frac{\partial_{i,j}}{\partial_{k,j}}, \tag{9}
$$

where: i, k - attribute codes, j - module codes.

With respect to the possibility of appearance of zero in the denominator (when the k-th attribute does not change its value), we modify the above expression and obtain:

$$
tda_{i,j,k} = \begin{cases} \frac{\partial_{i,j}}{\partial_{k,j}} & when \quad \frac{\partial_{k,j}}{\partial t} \neq 0 \\ 1 & when \quad \frac{\partial_{k,j}}{\partial t} = 0 \end{cases}. \tag{10}
$$

The security assessment of an oncoming threat, with reference to the base component (the atomic part of security structure), is estimated as follows:

$$Satt_{i,j} = \prod_{k=1}^{n} tda_{i,j,k} \, . \tag{11}$$

The first stage of generalization leads to the estimation of communication module security, but in this case the greater level of attribute variance appoints the greater threat scale. Therefore, module security is assessed as an inverse of the component value product:

$$Sm_j = 1 \bigg/ \left(\prod_{i=1}^{n} \prod_{k=1}^{n} tda_{i,j,k} \right). \tag{12}$$

The next generalization stage is connected with the estimation of security of the current part of a protocol run. This security assessment is expressed in the following way:

$$Sr = 1 \bigg/ \left(\prod_{j=1}^{m} \prod_{i=1}^{n} \prod_{k=1}^{n} tda_{i,j,k} \right). \tag{13}$$

3   Description of Security State with the Help of Probabilistic - Timed Automaton Nodes

The security state is assigned to communication run factors, such as: user, message, protocol, key, nonce, secret, etc. Adequate probabilistic - timed automaton is described in [10]. Generally, a communication run consists of interleaving protocols. Protocol is created as a sequence of operations built on the basis of users (sender, receiver, intruder), shared keys, nonces, secrets. The structure and operation (action) components are arguments exploited by rules [3] to extend the set of security parameters. The same kind of parameters is directly included in protocol operations:

$$set\_sec\,1 = sp \subseteq OP,$$
$$set\_sec\,2 = r\,(sp \subseteq OP)\,,$$

$$set\_sec = set\_sec\,1 \cup set\_sec\,2, \tag{14}$$

The task of a user consists in the selection of the set of security attributes, which will be the components of nodes of probability time automaton (PTA). Components can be evaluated in two ways: as real and binary values. Real values express the probability of security parameters and binary values express the acceptable level of parameters. According to [2] the relation between the PTA and the Petri net is mutually convertible. Therefore, it was decided to introduce a notation of tokens which will be adequate to node security components in the binary form.

Security attribute types infer from protocol logic (BAN or PCL) formalisms [3,9] which are partly described by the character of communication dealings in the following way:

$A \leftrightarrow^K B$ - users A and B communicate via the shared key K,

$\rightarrow^K A$ - user A has K as its public key,

$A \Leftrightarrow^Y B$ - users A and B share Y as a secret,

$\{X\}_K$ - the message X is encrypted by key K,

$\{X\}_K{}^A$ - the message X is encrypted by key K by user A,

$<X>_Y$ - the message X with an attached secret Y,

$A| \equiv X$ - user A believes the message X,

$A \rhd X$ - user A sees the message X,

$A \lhd X$ - user A sends the message X once,

$A| \Rightarrow X$ - user A has jurisdiction over X,

$\#(X)$ - the message is fresh

At this point, we cite [4] one rule from the BAN logic as example: Authentication rule – type I:

if $(A| \equiv ((A \leftrightarrow^K B), A \rhd X_K)$ then $(A| \equiv (B \lhd X)$.

The rule can be interpreted as follows: if A and B shared key K and A sees the message X, then A believes that this message is from B.

In the similar way we can define rules of nonce, vision, jurisdiction, freshness etc. [11]. Due to the determined character of the attribute number la and their binary form, the number of the security state (level) is strictly defined and equal to l^{2a}. The created node, in the investigation process (accompanying the realization of the communication run), saves its structure but changes its values of attributes (and consequently the security state). The security level can only decrease. In the proposed parallel system, different, independently converted, security nodes are created for selected security elements (the so called main security factors). Generally, this situation is presented as in Fig. 1.

4   Parallel Process of Correction Communication Security Attributes

The corrections of attributes can be realized simultaneously. The new recognized action is used with the help of communication logic rules to activate the set of attributes. At this moment adequate processor units start to correct the clock and value of attributes. After the correction, the medicated attribute is sent to different processors analyzing communication security level in accordance with particular main factors. The estimation of acceleration inferring from parallelization can be defined as $acc = la * (1 + lmf)/2$, where lmf - the number of main factors. The number of main factors is the sum of number of selected protocols, messages, keys, users, nonce's. The main security factor(s) is (are) declared for the current action. An action usually influences one or several attributes.

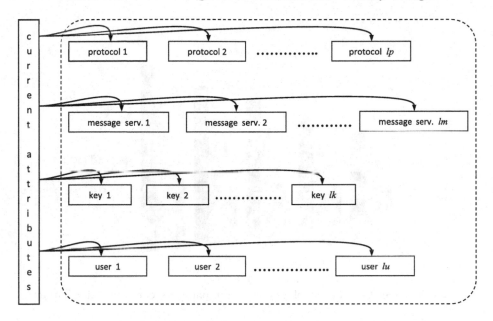

Fig. 1. Security modules (security nod structures for different main factors) are built on the basis of chosen attribute sets. Each module (bold frame) consists of a different set of attributes.

Analyzing security situation in network several (their number is e.g. equal lac) processors can serve set of communication actions. Therefore, the acceleration parameter will be estimated as interval in following way:

$$accn = [la * (1 + lmf)/2; lac * la * (1 + lmf)/2]. \qquad (15)$$

The upper bound of acceleration is achieved when the sets of attributes, evoked by actions, are mutually independent:
$set_{at}(i) \cup set_{at}(j) = \emptyset$, where: i, j - numbers of actions [12].

The stages of this algorithm are as follows:

1. action input,
2. the recognition of the attribute corrected by the action,
3. the recognition of the type of correction,
4. correction realization*,
5. go to the 3-rd point until the last attribute,
6. the recognition of the main factor activated by the action,
7. token structure creation for the main factor**,
8. security state estimation for the main factor**,
9. go to the 6-th point until the last factor,
10. auxiliary analysis (threaten state prognosis creation, the distribution of probabilities of transitions to the next stages)**,
11. go to the 1-st point until the last action***.

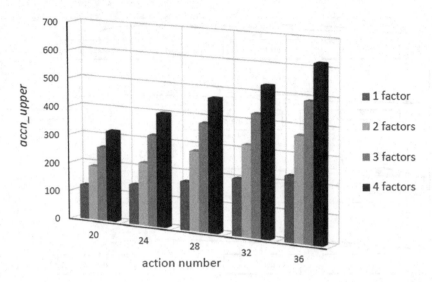

Fig. 2. The upper bound of acceleration $accn$ in parallel security checking variant.

There are three stages of parallelization: *- simultaneous corrections of attributes, **- simultaneous main security factor analysis, ***- the simultaneous serving of actions.

The type of action influences is practically regarded by two forms of algorithm attribute corrections:

$mc = \{0,1\}$ - correction by multiplication by a given updating coefficient MCC in case logic and heuristic rules influence, $mc = 1$ - the activation this form of attribute correction, $mc = 0$ - the rejection this form of correction.

$ec = \{0,1\}$ - correction by exchanging to current level (represented by current coefficient value of ECC) in case of lifetime or users(intruders) influences.

Therefore, it is possible to use simultaneously two form of correction for single attribute. So, if $ec = 1$ then attribute value does not have to be increased:

$$at_{t=k+1}(i) \xrightarrow{mc=0,\ ec=0} at_{t=k}(i),$$

$$at_{t=k+1}(i) \xrightarrow{mc=1,\ ec=0} at_{t=k}(i) * MCC,$$

$$at_{t=k+1}(i) \xrightarrow{mc=0,\ ec=1} ECC,$$

$$at_{t=k+1}(i) \xrightarrow{mc=1,\ ec=1} \min\{at_{t=k}(i) * MCC, ECC\}.$$

The experiments have approved that heuristic rules that influence in specific cases (for example in multi usage of the same nonce) are more effective when correction is realized in the following way:

$$at_{t=k+1}(i) \xrightarrow{mc=1,\ ec=0} at_{t=k}(i) * (1 - MCC),$$

or

$$at_{t=k+1}(i) \xrightarrow{mc=1,\ ec=0} at_{t=k}(i) * (1 - at_{t=k}(i)).$$

The actual value of ECC, in case of lifetime type of influence, will be counted by formula:

$$ECC = 1 - et^{t_j - lt_i}, \tag{16}$$

where:
t_i - the time of attribute activation,
lt_i - the attribute lifetime.

In reality, the time activity is transformed into probability attribute value, in accordance with the given attribute lifetime.

The actual value of ECC in case of additional users (intruders) type of influence will be counted by formula:

$$ECC = if \ (nus < nht) \ then \ ECC = 1, \tag{17}$$

else

$$ECC = e^{nht - nus},$$

where:
nus - the number of users (in the environment of main security factor),
nht - the number of honest users.

In reality, the time activity is transformed into probability attribute value, according with the given number of honest users.

5 Conclusions

The security investigation is acquiring an increasing importance with the growing network communication. Therefore, the problem of dynamic security estimation is increasingly grasping the interest of the data mining community [6]. The parallel approach guarantees not only increase the possibility to accelerate the reaction (from 100 to 600 times) (Fig. 2) to impending threats, but also permits us to treat chosen main security factors independently and to simultaneously provide security attribute corrections as the result of the effect of influence on the same protocol actions. The parameter integration approach is proposed also in [5] but there it serves a different purpose. The PTA and Petri net structures are the convenient forms to realize state transition procedures according to both particular secure attributes and their compositions. Timed, structural parameters and logic rules are regarded. New specific approach based on searching communication threads in parallel strategy permit us to simultaneously protect group of users against communication intruders activities.

References

1. Abraham, R., Marsden, J.E., Rating, T.: Manifolds, Tensor Analysis and Application, 2nd edn. Springer, New York (1988)
2. Beauquier, D.: On probabilistic timed automata. Theoret. Comput. Sci. **292**, 65–84 (2003)

3. Burrows, M., Abadi, M., Needham, R.: A logic of authentication. ACM Trans. Comput. Syst. **8**(1), 18–36 (1990). doi:10.1145/77648.77649

4. Gjøsteen, K.: A new security proof for Damgård's ElGamal. In: Pointcheval, D. (ed.) CT-RSA 2006. LNCS, vol. 3860, pp. 150–158. Springer, Heidelberg (2006)

5. Kwiatkowska, M., Norman, G., Segala, R., Sproston, J.: Automatic verification of real-time systems with discrete probability distribution. Theoret. Comput. Sci. **282**, 101–150 (2002)

6. Li, X., Xiong, Y., Ma, J., Wang, W.: An efficient and security dynamic identity based authentication protocol for multi-server architecture using smart cards. J. Netw. Comput. Appl. **35**(2), 763–769 (2012)

7. Lindell, Y., Pinkas, B.: A proof of security of Yaos protocol for two-party computation. J. Cryptol. **22**(2), 161–188 (2009)

8. Luu, A., Sun, J., Liu, Y., Dong, J., Li, X., Quan, T.: SeVe: automatic tool for verification of security protocols. Front. Comput. Sci. **6**(1), 57–75 (2012)

9. McIver, A., Morgan, C.: Compositional refinement in agent-based security protocols. Formal Aspects Comput. **23**(6), 711–737 (2011)

10. Piech, H., Grodzki, G.: Probability timed automata for investigating communication processes. Appl. Math. Comput. Sci. **25**(2), 403–414 (2015)

11. Piech, H., Grodzki, G.: The system conception of investigation of the communication security level in networks. In: Abramowicz, W. (ed.) BIS Workshops 2013. LNBIP, vol. 160, pp. 148–159. Springer, Heidelberg (2013)

12. Tudruj, M., Masko, L.: Toward massively parallel computation based on dynamic clusters with communication on the fly. In: IS on Parallel and Distributed Computing, Lille, France, pp. 155–162 (2005)

Extending Enterprise Architectures for Adopting the Internet of Things – Lessons Learned from the smartPORT Projects in Hamburg

Ingrid Schirmer[1], Paul Drews[2(✉)], Sebastian Saxe[3], Ulrich Baldauf[3], and Jöran Tesse[1]

[1] Department of Informatics, University of Hamburg, Hamburg, Germany
[2] IEG, Leuphana University of Lüneburg, Lüneburg, Germany
paul.drews@leuphana.de
[3] Hamburg Port Authority, Hamburg, Germany

Abstract. In many industries, companies are currently testing and adopting internet of things (IoT) technology. By adopting IoT, they seek to improve efficiency or to develop and offer new services. In current projects, a variety of IoT systems is used and gets interconnected with existing or newly developed application systems. Due to the integration of IoT and the related cloud systems, existing enterprise architecture (EA) models have to be extended. By drawing on the example of the Hamburg smartPORT initiative, we analyze the consequences of IoT projects on the enterprise architecture. As a result, we present an EA meta-model extension, which includes (1) sensor, physical object, smart brick, and fog system types, (2) a smart brick management database and (3) data streams, cloud systems and service applications. Furthermore, we discuss implications regarding a to-be architecture.

Keywords: Enterprise architecture · Internet of things · smartPORT · Smart brick

1 Introduction

The adoption of internet of things (IoT) technologies, is one of the so called IT mega trends that are assumed to have impact on diverse industries [7]. For 2020, Gartner expects a total number of 25 billion installed IoT units [6]. Today, a broad range of sensors, actuators, smart labels and other systems is available. While the manufacturing industry, for example, considers these technologies as one of the major drivers for change today [3], other industries are still undertaking innovation projects to identify use cases that are relevant for their business.

In the logistics industry, companies began early with testing and adopting smart technologies, e.g. for monitoring the cold chain [1]. In other fields, like container logistics, smart technologies are still not used at large scale because of missing standardization [15]. Nevertheless, some actors in the logistics industry continue with testing and piloting IoT systems.

At the harbor of Hamburg, the port authority (HPA – Hamburg Port Authority) started to test and evaluate IoT systems as a part of its "smartPORT" initiative [10]. This initiative comprises projects that strive for implementing smart technologies like sensors

© Springer International Publishing Switzerland 2016
W. Abramowicz et al. (Eds.): BIS 2016, LNBIP 255, pp. 169–180, 2016.
DOI: 10.1007/978-3-319-39426-8_14

into roads, bridges and railway points. These IoT systems generate data, which can be aggregated and used for improving traffic management or for predicting the failure of important physical components (predictive maintenance).

As the new IoT systems and the related processes are getting part of the enterprise architecture (EA), the HPA's EA model has to be modified and the meta-model needs to be analyzed whether it is still appropriate. Though a technical oriented reference architecture for IoT architectures has been published, the impact of IoT systems on EA models, EA meta-models and EA management is a rather new topic that is only briefly covered in the literature so far.

Hence, the research question for this paper is: How can the IoT-related changes of HPA's enterprise architecture be adequately captured in its EA model and EA meta-model? In order to address this research question, we followed an action design research based approach in cooperation with the HPA.

The remaining paper is structured as follows: In Sect. 2, we briefly summarize the related research regarding EA and IoT. Section 3 provides an overview of the smart-PORT initiative. Section 4 describes the methodological approach we employed. In Sect. 5, we present the results including the meta-model. The paper closes with a conclusion and an outlook in Sect. 6.

2   Related Literature: EA and IoT

During the last 30 years, the modelling of enterprise architecture shifted from a niche topic to a well-researched field [14, 18]. Today, many companies use models and tools that are based on enterprise architecture approaches for improving the alignment between business and IT. For managing the enterprise architecture, appropriate models need to be build, kept up-to-date and should be taken as a basis for decision making regarding issues that affect the relation between business and IT. As change in organizations today is also driven by the impact of IT innovations, EA approaches and models have to adopt to new challenges like cloud computing, mobile computing, social networks or the increasing interconnectedness with external partners and customers [5]. Hence, the EA models and meta-models need to be modified accordingly. Meta-models play an important role, as they ensure "semantic rigor, interoperability and traceability" [16]. While existing meta-models in the literature as well as those defined by frameworks like TOGAF [19] provide a high level of abstraction, new technologies still might impose the need for adapting them [16].

As one of the IT megatrends, IoT technologies are expected to be one of the forces that will lead to changes of IT infrastructures, processes and business models. By using a large number of small sensors and computers, the physical world and the "information world" are getting directly interlinked. Companies from diverse industries are seeking to implement IoT devices [8]. Often, they start with identifying use cases for IoT systems that match the company's needs. Though the implementation of a large number of new devices and the changing processes are expected to impose severe changes to the enterprise architecture, research on the relation between EA and IoT is scarce so far.

From the EA research perspective, Zimmermann et al. recently published a paper on "digital enterprise architecture" that seeks to identify relevant IoT architectural objects [20]. The paper only provides vague suggestions for how IoT objects might be considered in an EA meta-model. The second related field, IoT research, covers a broad range of applications in diverse industries like smart retail, smart transportation and smart city [8]. The most relevant source from the IoT field to the topic of this paper is the final publication of the IoT architecture project "enabling things to talk" [2]. In this book, the authors present a detailed model for IoT architectures (ARM – architectural reference model) as well as related processes. A basic part of the ARM is the domain model, which captures "the main concepts and the relationships that are relevant for IoT stakeholders" [2, p. 118]. The authors propose an "augmented entity" that combines a physical with a virtual component. From an EA perspective, the ARM mainly focusses on technical artifacts and their properties. Relations between the business and the IT perspective are only covered very briefly in the ARM.

3   Context: The Hamburg smartPORT Initiative

In 2012, the Hamburg Port Authority (HPA) launched the smartPORT initiative. This initiative consists of two sub-initiatives: smartPORT Logistics (SPL) and smartPORT Energy (SPE) [9–12]. While the SPE projects aim at reaching sustainability goals, the SPL projects strive for optimizing the logistic processes in the harbor mainly by adopting smart technologies. The SPL projects are carried out together with several partners from industry. Most of the projects had an innovative and pilot character. The results of these projects were presented at the IAPH 2015 (29th World Ports Conference) in Hamburg [13]. The SPL projects strive for increasing "the efficiency of the port as an important link in the supply chain" [10]. As the space in the harbor of Hamburg is limited, the HPA needs to manage the existing infrastructure in an efficient manner. The goals are to establish an "intelligent infrastructure" and to optimize "the flow of information to manage trade flows efficiently" [10]. In several sub-projects, the SPL projects explore the use of smart technologies for better managing the traffic on the roads (smart road, port road management system, depiction of the traffic situation, parking space management), rail and water as well as for predictive maintenance scenarios (intelligent railway point, smart maintenance for bridges) [10].

Our research project at the HPA started, after the results of the exploratory IoT projects were presented at the IAPH. In this phase, the HPA had to evaluate and further integrate the projects. They also had to decide, which projects will be transformed to a productive mode and which partners and vendors will be chosen for this mode. Furthermore, information about the projects had to be spread within the organization. We assume that such a situation is typical for innovative and exploratory IoT projects. During the exploratory phase, the focus lies on demonstrating the feasibility of adopting a certain IoT system. A complete alignment with the existing EA might be skipped due to limited budgets and unknown outcome. Proceeding in such a way is well known in IS research and captured by the term "bricolage" [4].

In this paper, we look at this consolidation phase mainly from an enterprise architecture viewpoint as it can contribute by: (1) analyzing the projects' results in a uniform way, (2) identifying relationships among different projects, (3) providing tools for performing these tasks, (4) developing models and visualizations for supporting decisions, (5) defining and ensuring compliance with architectural guidelines, (6) determining a to-be architecture for roll-out and transfer to operations.

4   Methodological Approach: Action Design Research

This research project was carried out as an action design research (ADR) project according to Sein et al. [17]. It started with the problem of the HPA to integrate the results of the IoT projects in its existing enterprise architecture model and tool during the consolidation phase (principle 1: practice-inspired research). We categorized this as an IT-dominant BIE (building, intervention and evaluation) type of ADR project. However, during the project we learned that the HPA wants to use the EA model and visualizations based on this model for communicative purposes with the organization as a part of the digital transformation process. This would lead to a second, organization-dominant BIE task, which will become more relevant in the future.

Our research team consisted of two senior researchers and several graduate students. The research team was interested in exploring whether EA meta-models – as a fundamental type of artifacts for EA research – need to be changed and adopted if companies integrate IoT systems into their EA (principle 2: theory-ingrained artifact).

A first analysis of the existing literature and discussions with the EA tool vendor lead to the conclusion that – regarding both problem formulations – scarce information is available in publications as well as in practice. These findings encouraged us to start a cooperative project for developing the required EA model and meta-model. During the ADR project, mixed teams (researchers and practitioners) conducted 17 informal interviews (with project managers, internal and external experts and other stakeholders within the HPA), analyzed documents about the projects and the existing EA as well as the EA tool and the EA model. For discussing the intermediate results and for planning next steps, regular workshops were established as well as less frequent review meetings. During these meetings, researchers and practitioners evaluated the applicability, coherence and correctness of the related artifacts (models and meta-model) (principle 4: mutually influential roles).

In several iterations, the EA models for each IoT project were evaluated and refined and the EA meta-model was co-developed and evaluated (principle 3: iterative reciprocal shaping and principle 5: authentic and concurrent evaluation). Based on the insights from the interviews and the document analysis, the project team decided, which information about the IoT projects should be captured, held and managed in the EA model. The team's decisions were also driven by the goal of using concepts for the EA model that are easy to understand and to communicate while also being comprehensive and precise. The intermediate results were regularly evaluated in the above mentioned meetings. They were also discussed at a practice-oriented conference to gain further general feedback from practitioners from different domains.

During the cooperation so far, some decisions were made that may be understood as design principles. For example, sensor types were modelled instead of sensors, the concept of smart bricks guides the understanding of sensors attached to infrastructural elements, the distinction of raw data, information streams and information services makes transparent how sensor data get transformed and enriched to useful business-oriented information. Furthermore, design principles for a generalized (cross-project) to-be architecture were developed (principle 6: guided emergence). Since the cooperation will be extended, the evolving of design principles and their ongoing shaping by organizational use and different stakeholder perspectives is ensured. Due to the cross-project comparison, we also achieved a first step of generalization from the individual projects. Additionally, we also take first steps for generalizing from the HPA case at the end of this paper (principle 7: generalized outcomes).

Out of the four ADR stages, we passed through an extended and reflected problem formulation (stage 1), an ongoing intertwined BIE (stage 2) and reflection and learning (stage 3) and started with the formalization of learning (stage 4).

5   Results: Capturing IoT in Extended EA Meta-Models

In this section, we present the results we developed in this project regarding the extension of the EA meta-model. We start by reasoning our suggestions of how to model sensors and physical objects as "smart bricks" in Sect. 5.1. In the following sections, we describe changes regarding data streams and cloud systems (5.2) and summarize the suggested changes in a combined meta-model (5.3). This is followed by a recommendation for the to-be architecture (5.4).

5.1   Smart Bricks: Introduction, Level of Abstraction and Database

In the smartPORT projects, hundreds of sensors were installed on several real infrastructure elements at the harbor. Like in other IoT initiatives, the number of sensors is likely to increase dramatically in the future, if the explorative projects will be transferred into a productive mode. Hence, the first question is, whether the EA as a strategic tool is the appropriate place for administrating each instance of the operative sensors. For maintenance reasons and evaluation tasks within the IoT systems, each installed and activated sensor needs to be known, described and its description needs to be kept up-to-date. A company seeking to use IoT systems has to decide, which tool is appropriate for storing and managing this information. In our case, we learned that a geographic information system (GIS) is used to briefly document and visualize these sensor instances. During our analysis, we considered this as a similar situation compared to the use of a configuration management database (CMDB) that stores detailed information on IT infrastructure components. Hence, we decided to keep a similar separation of concerns and to only model sensor types and not instances within the EA model.

Since sensors are only one part of smart infrastructure elements, we introduced a combined element, the smart brick. A smart brick consists of a brick and a sensor. A brick describes a physical element of the harbor's infrastructure (such as streets, bridges, quay walls, berths, etc.). As the past and current main task of the HPA is to maintain the harbor's infrastructure, we chose the word brick as it fits to the organization's identity. As each brick is usually constituted out of a hierarchy of (sub-)bricks, a brick is made smart by adding at least one sensor to it or to one of its sub-bricks. In this way, the concept can be used to classify the "smartness" of a smart brick or states of its IoT enrichment.

We illustrate the concept of a smart brick by drawing upon the example of a smart bridge (see Fig. 1). The bridge is enhanced with sensors for predictive maintenance purposes. As illustrated, a typical bridge might be composed of track beams and steel beams. Track beams will become vibration-sensitive by attaching vibration sensors, steel beams strain-sensitive by strain gauges or tilt-sensitive by tilt gauges. Hence, some smart bridges might be only vibration sensitive whereas others are tilt sensitive, strain sensitive or it may be monitored by a combination of different sensor types.

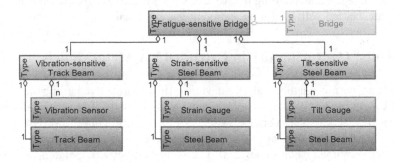

Fig. 1. Smart bridge as an example for a smart brick

The corresponding EA meta-model for smart brick types expresses their hierarchical (tree) structure by a recursive relationship (See Fig. 2). It is a "simple" concept in which the smart brick types mainly follow the tree structure of their corresponding brick types. Each smart brick type is related to exactly one brick type (at the appropriate tree level). However, there might be more smart brick types based on one brick type due to different related sensor types, see e.g. the strain-sensitive steel beam type and the tilt-sensitive steel beam type above – both are based on a steel beam brick type. Hence, we have a 1:n instead of a 1:1 relationship between brick and smart brick types. Further integrity constraints exist for this meta-model: The set of sensor types on each smart brick node is the same as the sensor types of all smart brick nodes of the related subtree.

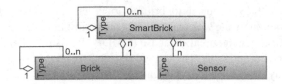

Fig. 2. EA meta-model extension for smart brick types

The meta-model allows to derive models with detailed or less detailed granularity. For a detailed model, we would propose to model sensor types on leaf smart brick nodes only and to allow exactly one related sensor type. The sensor types on the non-leaf nodes would be calculated in this case. Alternatively, the sensor types on the root node can be completely omitted like in the above fatigue smart bridge type example.

Attributes on sensor types comprise information about sensor ID types. They further allow distinguishing sensor types regarding their vendors and versions. As this distinction is important for maintenance and strategic decisions, we differentiate sensor type instantiations in a concrete architecture from those stemming from different sensor type vendors and versions. Attributes on brick types will indicate their purpose, location, material and status of maintenance.

In correspondence to the decisions described above, we propose the implementation of a "smart brick management database" (SBMDB) for documenting smart brick instances. The instances will be classified by the smart brick types, which form the link between the two documentation repositories (EA tool and SBMDB). Apart from maintenance purposes, the SBMDB can play a major role in the resolution of identifiers in information streams needed within the cloud system layer (see below). In the future, sensor integration might follow automated subscription patterns and models [20]. But in our case, the diversity of sensors, intermediate systems and vendors will inhibit such an approach for the upcoming years.

The extended EA meta-model with smart brick types can be integrated with a corresponding SBMDB for answering concerns that were mentioned by stakeholders in the interviews such as: Which are fatigue-sensitive bridges that are not yet vibration sensitive? Which sensor types can be attached to steel under water? Where are instances of smart brick types located? Which sensor types are provided by vendor A and in which smart bricks of which type are they installed?

We challenged our modelling approach by applying it to other smartPORT projects. In a predictive maintenance project related to marker posts in road construction areas, the smart bricks (marker posts) can be moved. The changing location can be captured with our meta-model by an attribute. This movement would be handled by the SBMDB. Furthermore, we ensured that the concept of smart bricks can also easily be adopted for capturing actuators in the future. Within a project aimed at reducing power consumption by implementing a usage-dependent light control, a key aspect to be modelled were dimmable lights. These actuators were modelled analogous to smart bricks by replacing the sensor part with an actuator. This extension captures the required details and thus fits in well with the existing model.

5.2 Data Streams, Cloud Systems and Service Applications

The diverse smart bricks are the source of various continuous data streams, which are typical for IoT architectures and applications. Data is being processed and transformed in different types of systems for which we introduce different layers. We distinguish fog systems, cloud systems and service application systems. Fog systems (such as section controllers for induction loop sensors) aggregate and transform raw data. Similar to sensors, these systems are only modeled by types due to the high number of instances.

IoT systems today often come with accompanying cloud systems for integrating, analyzing and managing the sensor instances and the data they provide. The data streams are consumed by service applications like logistics systems, consumer related apps or monitoring services. Accordingly, we distinguish different types of data streams between these layers. Raw data streams flow between smart bricks and fog systems, information streams between fog and cloud systems and information services from cloud systems to service application systems.

This is exemplified by the project EVE (a German acronym for "effective depiction of the traffic situation in the Port of Hamburg"), which provides traffic status and forecast information. This information is used to provide several information services like internal monitoring services, public monitoring services and navigation and logistics services. The project extends (and partially replaces) an existing system that estimates the current traffic situation (traffic analysis system) in four ways: (1) the new system utilizes further types of sensors (Bluetooth, video cameras, floating car data (FCD) from an automobile association), (2) it provides a sophisticated traffic simulation component for better forecasts, (3) it uses data storage in the cloud (unlike typical application systems at the HPA), and (4) it receives from and provides data for a public cloud, the mobile data marketplace (MDM).

The smart bricks in this project are smart roads, which are further subdivided into different road segments. Each sensor type requires different ways of defining road segments. Some of the smart bricks need fog systems whereas others do not require fog systems and deliver their information streams directly to the cloud systems.

Apart from common attributes for applications such as vendor, version and operator, the new data processing systems need additional attributes as they are operating in the cloud. These attributes include the cloud type (public, hybrid or private), data-related attributes (such as location of data repositories, data storage format, time and volume, scalability, archiving regulations) as well as infrastructure and management-related attributes (like administrative responsibilities and location, e.g. for data privacy regulations).

In the content component of the different data streams, the semantic enrichment of the stepwise processed information becomes apparent. E.g. in case of the induction loops (1) the raw data describes inductivity in Henry, (2) the information stream exhibits speed by vehicle class and the time vehicles were above the induction loop, whereas (3) the information service provides travel density and travel time per segment. This data can be used by a service application system for delivering individually estimated arrival times.

Further support for optimizing the traffic flow combines services of the EVE project with those of other projects, e.g. from smart parking. Here, we want to point out that the proposed way of modeling smart bricks also fits to the scenario where different information streams are generated out of the same raw data. In these two projects, we have a dissimilar use of induction loops. While induction loops always measure inductivity, the measured data is interpreted differently depending on its later usage. For traffic situation estimation, the attached fog systems aggregate data such as vehicle speed and the number of passing cars. On smart parking lots, the data is aggregated to "vehicle footprints", uniquely identifying each truck on the parking lot. These fog systems and the attached information streams can be easily distinguished by attaching either a road segment brick type or a parking lot brick type. Thus, the smart brick logic provides intuitive means of differentiating the way that one type of sensor is used on a use case level (Fig. 3).

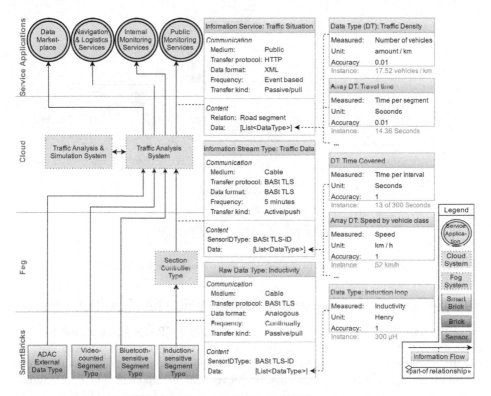

Fig. 3. EA model slice for the EVE project

5.3   An Extended Meta-Model for the IoT-Related EA at the HPA

We summarize the required extensions in a combined meta-model. This model comprises the IoT specific layers with smart bricks and fog systems (each on a type level) and their relations to layers of cloud systems and service applications (each on an instance level). The meta-model classifies the related interfaces between elements of each layer in raw data, information streams and information services. Figure 4 also briefly indicates an EA tool based on the meta-model together with a link to the SBMDB. This combination of tools contributes to a future concern-directed IoT information platform, which can be used to answer concerns based on an integration of instance-level smart brick data and sensor data (SBMD – smart brick management data) from the SBMDB and the architectural relations from the EA model based on the extended meta-model (see Sect. 5.1). The models of both systems are connected by sharing the same smart brick and sensor types.

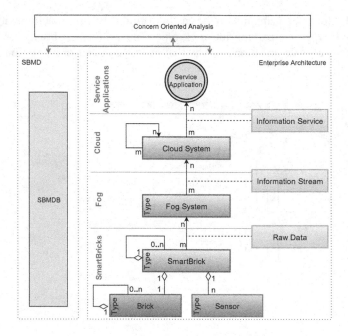

Fig. 4. Extended meta-model for the IoT-related EA at the HPA

5.4   Towards a to-be Architecture for the IoT-Related EA at the HPA

After documenting and analyzing impact of the smartPORT projects on the EA, we recognized that each IoT data processing system typically consists of three vendor-specific components: a data acquisition and conversion component, a data fusion and calculation component and an information provisioning component. The data acquisition and conversion component resolves identifiers of sensors stemming from the information streams in order to relate them to smart bricks. In the EVE project, identifiers are substituted by road segments (logical location) and geographical references (physical location). In the two parallel traffic analysis systems, this is done separately based on different segment types and different repositories relating the sensor instances and the respective segment instances. As a consequence, each change in the sensor farm requires changes in both sensor lists.

For the to-be architecture as shown in Fig. 5, we suggest a vendor-independent component. This component provides acquisition (Acqu), filtering (Filt) and conversion (Conv) functionality and a unique smart brick identity resolution management component. The latter is exactly the SBMDB we introduced in Sect. 5.1. In Fig. 5, we use icons combining letters for sensor types (F for FCD, V for Video, B for Bluetooth and I for Induction loops) for indicating their smart brick type (here: two lines for road segments). The outlined to-be architecture should be considered for defining future contracts with vendors for developing the productive cloud systems. This again underlines the role of EAM in the interrelated consolidation tasks mentioned in Sect. 3.

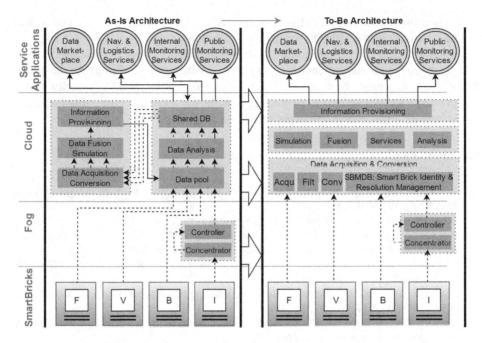

Fig. 5. As-is and to-be architecture for the IoT-related EA: suggested change on the cloud layer

6   Conclusion and Outlook

Based on the insights gained from a real case, we modeled concrete architectures for different explorative IoT-based projects. During this process, we investigated possible IoT-related extensions for the EA meta-model. The extensions mainly consist of new layers that are used for modelling smart brick types and fog system types and a cloud system and a service application layer. Furthermore, we emphasize the information streams that connect the different layers and distinguish raw data, information streams and information services.

For the HPA, the unified models for several explorative IoT-based projects contributed to the ongoing consolidation phase. The project-focused models form a valuable basis for better understanding interrelationships among the projects and for improving strategic decision making for the future development and roll-out of IoT. The developed meta-model introduces standardized naming conventions and concepts that are understandable for several stakeholders within and beyond the organization (including e.g. system vendors). From a research perspective, we developed an IoT extension for EA meta-models that is so far rooted in a cross-case analysis of several different IoT applications in one organization. Determining whether the suggested extensions are applicable for other port authorities or in related domains like smart city initiatives has not been subject of our study and requires further research. Yet, during our research process, we identified several basic issues that we consider relevant for other IoT related EA endeavors: (1) the level of abstraction (instances/types), (2) business-oriented conceptualizations of IoT system and (3) the introduction of a SBMDB together with a concern oriented analysis platform.

Ongoing research is focusing on IoT-based application services, business models and inter-organizational processes (ecosystem architecture). Additionally, the meta-model extensions might be integrated into EA frameworks like TOGAF. Security, safety and privacy issues should to be covered appropriately and visualizations of IoT-specific EA information and analysis patterns should be developed.

References

1. Abad, E., et al.: RFID smart tag for traceability and cold chain monitoring of foods. J. Food Eng. **93**(4), 394–499 (2009)
2. Bassi, A., et al.: Enabling Things to Talk. Springer, New York (2013)
3. Bitkom e.V., VDMA e.V, ZVEI e.V.: Implementation Strategy Industry 4.0 – Report on the results of the Industrie 4.0 Platform. Technical report, Bitkom (2016)
4. Ciborra, C.U.: From thinking to tinkering. In: Ciborra, C.U., Jelassi, T. (eds.) Strategic Information Systems. Wiley, Chichester (1994)
5. Drews, P., Schirmer, I.: From enterprise architecture to business ecosystem architecture. In: IEEE EDOCW 2014, pp. 13–22 (2014)
6. Gartner: Gartner Says 4.9 Billion Connected "Things" Will Be in Use in 2015. http://www.gartner.com/newsroom/id/2905717
7. Gartner: Gartner's Hype Cycles for 2015: Five Megatrends Shifting the Computing Landscape. Gartner
8. Gubbi, J., Buyya, R., Marusic, S., Palaniswami, M.: Internet of things (IoT): a vision, architectural elements, and future directions. Future Gener. Comp. Syst. **29**(7), 1645–1660 (2013)
9. HPA: Energy Cooperation Port of Hamburg. http://www.hamburg-port-authority.de/de/presse/broschueren-und-publikationen/Documents/broschuere_smartportenergy_web.pdf
10. HPA: Port of Hamburg - digital Gateway to the World. http://www.hamburg-port-authority.de/de/presse/broschueren-und-publikationen/Documents/140401_HPA_Broschuere_spl_web.pdf
11. HPA: Container schneller an Bord. http://www.hamburg-port-authority.de/de/presse/pressearchiv/Seiten/Pressemitteilung-26-09-2012.aspx
12. HPA: HPA legt ersten Nachhaltigkeitsbericht vor. http://www.hamburg-port-authority.de/de/presse/pressearchiv/Seiten/Pressemitteilung-02-09-2013.aspx
13. IAPH: IAPH 2015. https://www.iaph2015.org/
14. Kappelman, L., McGinnis, T., Pettite, A., Sidorova, A.: Enterprise architecture: charting the territory for academic research. In: Proceedings of AMCIS 2008, paper 162 (2008)
15. Montreuil, B.: Toward a physical internet: meeting the global logistics sustainability grand challenge. Logistics Res. **3**(2), 71–87 (2011)
16. Saat, J., Franke, U., Lagerström, R., Ekstedt, M.: Enterprise architecture meta models for IT/business alignment situations. In: IEEE EDOC 2010, pp. 14–23 (2010)
17. Sein, M.K., et al.: Action design research. MIS Q. **35**(1), 37–56 (2011)
18. Simon, D., Fischbach, K., Schoder, D.: An exploration of enterprise architecture research. Commun. AIS **32**(1), 1–72 (2013)
19. The Open Group: TOGAF Version 9.1, U.S., The Open Group (2011)
20. Zimmermann, A., et al.: Digital enterprise architecture – transformation for the internet of things. In: IEEE EDOCW 2015, pp. 130–138 (2015)

Towards the Omni-Channel: Beacon-Based Services in Retail

Anja Thamm[1(✉)], Jürgen Anke[2], Sebastian Haugk[1], and Dubravko Radic[1]

[1] Universität Leipzig, Leipzig, Germany
thamm@studserv.uni-leipzig.de,
{sebastian.haugk,radic}@wifa.uni-leipzig.de
[2] Hochschule Für Telekommunikation Leipzig, Leipzig, Germany
anke@hft-leipzig.de

Abstract. The integration of online and offline channels is a key challenge for retailers pursuing an omni-channel strategy to improve consumer experience. The prevalence of smartphones offers an opportunity to connect the physical and digital world. Bluetooth Low Energy beacons are small devices, which send out a signal that can be detected by consumer's smartphones to enable location-based services. However, there are very few documented cases of beacon usage in Germany, whereas they seem to have a much higher adoption in the US. In this paper, we investigate the challenges associated with the use of beacons in retail. Using a survey, we aim to understand the attitude towards beacons-based services from a sample of consumers in Germany.

Keywords: Bluetooth beacons · Omni-channel · Retail · Internet of things

1   Introduction

The struggle between retail and internet retailers is still in full swing but in the course of time, retailers have developed strategies to improve their competitive position. Bricks-and-mortar retailers adapt their business models and generate options for selling stationery and on the internet. Due to this fact, multi-channel, cross-channel and omni-channel retailers emerged. The difference between these concepts is the mix between online and offline sales channels. In multi-channel, both channels exist, but they cannot be changed by the customer during a purchasing transaction. In the cross-channel concept, customers can change channel within one transaction, e.g. buy online and pick up in a store. However, in both variants, the online and offline channel are technically and organizationally separated. Omni-channel connects both channels on the basis of a central infrastructure and looks at the customer journey in its entirety [1]. By using their smartphone customers already swap between online and offline channels, e.g. to check the online prices or to obtain product information [2]. With the help of beacons, the channels can be integrated by providing customers the information they seem to need at specific locations during their stay in a store [3] and thus improve the overall consumer experience in omni-channel scenarios [6].

© Springer International Publishing Switzerland 2016
W. Abramowicz et al. (Eds.): BIS 2016, LNBIP 255, pp. 181–192, 2016.
DOI: 10.1007/978-3-319-39426-8_15

Beacons are small devices which are detected by smartphones or other devices in proximity and indicate the presence of a user within the area of the beacon. With this information, retailers are able to offer many different services like sending out advertisements, coupons, product information or navigation support [7]. Successful examples can be found in the US, e.g. in Video Game stores [4], malls and shopping centers (Simon Property Group) [5], retail stores, stadiums, airlines, and airports [8]. In contrast to US-based companies, German retailing companies seem to be more reluctant to utilize beacons for innovative services [9]. In this paper, we investigate whether beacons are a technology that could provide benefit to German retailers.

We approach this question in the following way: First, we introduce the beacons technology to identify possible applications as well as technical requirements and the potential impact of beacon-based services on the retailer's business model. To identify the interest in specific applications, as well as the willingness to use beacon-based services in general, we conducted a survey with German consumers. Using the results of the survey, we propose actions for retailer to improve utilization of beacons. The paper closes with a conclusion and an outlook on further research questions.

2 Fundamentals of Beacon-Based Interactions

2.1 Bluetooth Low Energy Beacons

Beacons are a new approach to tag physical locations [10]. They work similar to light-houses by broadcasting signals (advertising messages) in short intervals (typically less than a second). Beacons are inexpensive and can be deployed without additional infrastructure. The communication takes place on the technical basis of Bluetooth Low Energy (BLE). It allows for very low energy consumption; hence, a coin battery can power it for up to two years. As the communication is unidirectional (beacons cannot receive data), provide a location to phones while the users remain anonymous to the beacon [11]. Depending on the signal strength, an app can determine the distance to the beacon within the three proximity levels "immediate" (a few centimeters), "near" (a few meters) and "far" (more than 10 meters) or "unknown", if the distance could not be determined. If more than three beacons are in range, the position of the receiver can be determined using trilateration, e.g. to support indoor navigation [12, 13]. Chawathe proposes a method to determine the placement of beacons for such scenarios [14].

While the underlying Bluetooth communication protocol is standardized, the message formats supported by beacons vary by manufacturer. One example is the "iBeacon" offered by Apple, which sends out messages sent containing the following data: (1) a 16-byte UUID (universally unique identifier), which describes a large collection of related beacons, (2) a major and (3) minor number (2 bytes each), which are used to further subdivide the location [15]. Smartphones and other Bluetooth-enabled devices can receive these advertising messages, connect to the beacon and process received information using dedicated apps.

Apps can combine the received beacon-message with context-based information from a cloud-server to provide location-based services to user, e.g. showing location related information. The provisioning of information to a customer or visitor can therefore be tailored to the respective context consisting of time and location – a concept also known as "proximity marketing" [16]. To provide such beacon-based services, a system consisting of the elements depicted in Fig. 1 is required. The role of these elements can briefly be described as follows:

- *Beacons:* They need to be placed at all locations where an interaction is desired, e.g. at the entrance, at information desks or at certain products.
- *Bluetooth Low Energy-enabled devices:* They have to support Bluetooth 4.0 which includes devices with iOS 7 or higher and Android 4.3 or higher.
- *Smartphone App:* These native apps have to be enabled for beacon communication. They can retrieve personalized information from the Internet or report the presence of the user to the venue operator.
- *Beacon management system:* Cloud-based services to register beacons, assigning them to locations, updating their UUID as well as version numbers, and attaching data (for Eddystone compatible beacons)
- *Content Provider:* A cloud-based backend system, which delivers product details, navigation support, or other information to the user's smartphone based on the provided context, e.g. location, proximity, time, and user.
- *Local interaction:* Information displays for both customers (e.g. personal greeting) and local staff (e.g. about currently present customers).

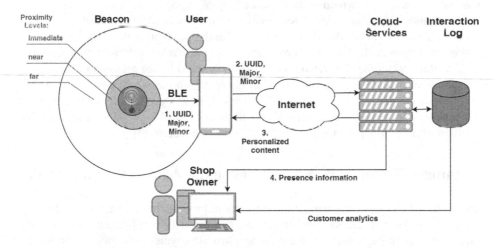

Fig. 1. System architecture for beacons-based services in retail

2.2 Context Creation for Personalized Interaction

It is obvious that more precise context information is the prerequisite for highly personalized user interaction. However, how much context information is needed, depends on the concrete service design.

The range of beacons is limited to approximately 50 meters and even less in case of obstructions through walls, furniture and people. However, this short range is beneficial, as it enables more precise positioning of users [14]. Depending on the setup of beacons and the user's behavior, the usage context can be more precisely derived, which is subsequently used for tailored content provision. For beacon-based interaction, the simplest form is to get information about a certain item (such as product in a store). The following table shows examples of context information required to create the desired form of interaction. It highlights that beacons only provide part of the context information needed for a beacon-based service (Table 1).

Table 1. Required context depending on interaction form

Interaction	Required context
Item information	Item (at location)
Special offers for new customers	Location, running campaign (at certain time)
Personalized greeting at location	Location, user name
Personalized coupon	Location, user name, preferences/interests
Navigation support	Location, destination

To create this context, various parameters have to be determined. Location is obviously provided by the beacon itself. Time can be important context information to provide time-related information such as special offers, alternative routes, bus delays etc. For user-related information, the user needs to provide his or her identity. This can be achieved via an app, which requires some kind of login.

Even more important is the determination of user interests. In case of navigation services, the user is most likely willing to provide the destination through the app. For marketing purposes this is more challenging, as not many customers will actively disclose their preferences or needs. One approach could be the analysis of the user's purchase history to identify cross- or up-selling opportunities which could be part of a personalized coupon. Another option could be real-time analysis of time spend at various departments of a department store. This could be used as an expression of interest but will likely lead to mistrust from users due to privacy concerns [3].

3   Impact of Beacon-Based Services on the Retailer Business Model

Beacon-based services carry a great potential for improving the satisfaction and loyalty of the customers of retailers by moving retailers towards the omni-channel [5, 17]. The major areas for strategic improvement for are: Mobile payment and mobile couponing [18]. While the majority of the customers already meets the technical requirements for using beacon-based services and barriers to entry are reasonably low for the customer (mainly app download and account registration), the number of customers using their smartphones while shopping is still growing [2, 19]. There are also other trends [20] in shopping behavior (e.g. growing numbers of mobile/digital coupons [21, 22]), growing interest in mobile payment technologies [18, 21] and technology trends in general (e.g. big data [21]), which enable the application of beacon-based services in the retail sector.

With regard to a company's success, their business model must not be neglected as this perspective presents a connection between strategic and operational orientation [23]. Thus the term business model describes "an integrative concept that combines all the relevant fields of the company in a meaningful way" [24]. In the following it will be described how beacons can affect various parts of the business model of retailers by referencing the Business Model Framework of Fraunhofer MOEZ. This approach contains all relevant key components and is based on five main business model dimensions [24]: (1) value proposition (product and/service offering, pricing model), (2) value communication (communication channels, story), (3) value creation (core competencies, key resources, value networks), (4) value delivery (distribution channels, target market segments), (5) value capture (revenue model, cost structure, profit allocation [24, 25].

While obviously affecting the cost structure of the store owner (e.g. costs of app development, purchase of beacons, maintenance costs etc.), beacons can also enhance the service experience of the customer [26] (e.g. indoor-navigation, automated checkout, electronic receipt/guarantee, real-time location-based advertising/discounts) and by that become part of the value proposition of the retailer itself. Other obvious parts of the business model, which hold potential for innovation in this context, are channels and customer relationships [27, 28]. How are retailers reaching their customers? Which channels are the most cost-effective and how are they integrated with the rest of the business model [24]? By turning the smartphone of the customer into a distribution channel, that supports fast and effective couponing, discounts, (location-based) advertising, upselling and payment processes, beacons carry great potential for innovating the shopping experience in general [29].

These enhancements can also be seen as mere innovation of channels and customer relationships of the retailer, but they are a new part of the value proposition of the retailer because they are enhancements of the customer experience [26]. Especially fast payment processes or automated checkout services hold potential for increasing customer satisfaction and loyalty. Additionally, there is potential for improving loyalty programs (mobile loyalty) further by keeping track of what a regular customer is interested in, what he buys, when and how long he is usually shopping at the store and by communicating directly with the customers through beacons and apps. Also, beacons serve as an enabler for loyalty programs, because retailers can target customers (e.g. with advertising, discounts) based on their location and past shopping behavior [30, 31].

Another part of the business model that could be affected by beacons is the revenue stream [30]. By using beacon-based services as part of a larger digital strategy, we argue that sufficient value can be created for participants and retailers will be able to monetize on these services (e.g. transaction-based).

Adopting the technology will also lead to changes on the supply side of the business model. Key partners and key resources will be affected, because retailers will need to cooperate strategically with manufacturers of beacons and app developers (key partners) as beacons and apps become part of their key resources for delivering on their value propositions. Other possible key partners are financial services companies for payment services or external companies, which offer loyalty programs (e.g. Payback in Germany, which is a service of American Express).

The key activities of retailers are also affected by these changes. Applying beacons in retail stores will yield significant amounts of data, which retailers in turn can use for optimizing their store design and layout [29], shift planning or various analytical purposes like store traffic analysis or product range optimization [7]. Last but not least, customer segments are also affected by adopting beacons in retailing [32]. Retailers try to compete with e-commerce companies like Amazon but suffer a competitive disadvantage from the convenience of online shops and a slightly outdated image of department stores. We argue that beacons and apps might be able to help gaining a more modern, up-to-date image, which attracts younger customers, who are most likely to be the early adopters of the technology and its services [19].

4   Survey

To investigate people's mindset towards the use of beacons, a cross-sectional empirical study was conducted. This instrument was chosen to reach as many survey participants as possible in a short time. The study was posted on Facebook to acquire young people who might be more interested in technology. The survey's purpose was to explore the interest of end customers in beacon-based services and the willingness to use them.

4.1   Research Design

To this end, we designed a questionnaire with the following five sections:

1. In the *introductory section*, participants were asked, whether they (1) own a smartphone, (2) know Bluetooth, (3) know Bluetooth Low Energy, (4) know beacons, (5) have used beacons before. This helps to understand, whether consumers are aware of beacons.
2. In section two, they were given a *short introduction of the beacon technology*, as a preparation for the remaining questions.
3. In section three, participants were asked to *assess the usefulness of typical beacon-based applications*, which were based on already existing scenarios and pilot projects. They were grouped into four categories:
    *   *General situations*, such as navigation on airports, coupons in stores, information on exhibits in museums etc.
    *   *Specific retail store types*, e.g. supermarkets, outdoor stores, furniture shops
    *   *Applications in a super-market*, e.g. personal welcome message, navigation to products on the shopping list, information on products (fruits, vegetables, meat, coffee), special offers, and electronic payment at the checkout.
    *   *Applications in a stadium*, e.g. offers for seat category upgrade, navigation to wardrobe/restrooms, special offers for drinks and snacks.
4. In section four, participants were asked whether they were *willing to use beacon-based services*, and if not, for what reasons. Participants were also asked to make an assessment whether beacon-based services will be successful in Germany.
5. The last section contained *socio-demographics*, e.g. age, gender and occupation.

4.2   Data Collection

The survey was conducted between 21st and 29th June 2015. A total number of 99 participants were recruited using social networks. The survey was conducted using an online as questionnaire. All statements of the 99 participants have been made correctly and completely.

The gender distribution was well balanced with 58 % female and 42 % male participants. 81 % of the participants were between the age of 17 to 34, 14 % between 35 and 49, and 5 % 50 to 69 years old. More than half (54 %) were students and apprentices. 43 % were employees, civil servants or self-employed. 5 % of the participants were unemployed.

4.3 Results

Introductory questions: The first questions did start with the technological context to check, whether people are already using the technology that is required to use beacons services. It turned out, that optimal conditions are given, as 93 % of the participants own a smartphone, 58 % know and use Bluetooth and 30 % know Bluetooth and its purpose but do not use it. Participants are unfamiliar with Bluetooth Low Energy and the beacon-based technology. Only 7 % do know Bluetooth Low Energy, 4 % know beacons and 3 % did use this technology already. 93 % did not know beacons before taking part in the survey.

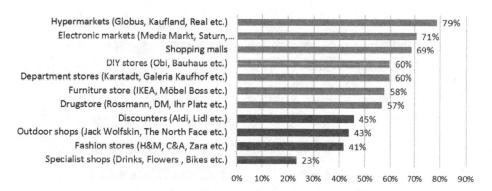

Fig. 2. Commercial types where usage of beacons is imaginable by respondents (N = 99)

Assessment of potential scenarios and application areas: After the introduction of beacon technology, imaginable situations for the use of the beacons were presented and had to be rated. Most participants (79 %) perceive the usage of beacons for navigation in airports, inner cities, shopping centers and town hall as "mostly useful" and "very useful". 69 % of all respondents perceive the usage of beacons next to posters, monuments, buildings and artworks as "mostly useful" and "very useful". As Fig. 2 shows, people seem to prefer using the advantages of beacons in large sites. Nearly three quarters of all participants can imagine using beacons in Hypermarkets (79 %), Electronic markets (71 %) and Shopping malls (69 %) (Fig. 3).

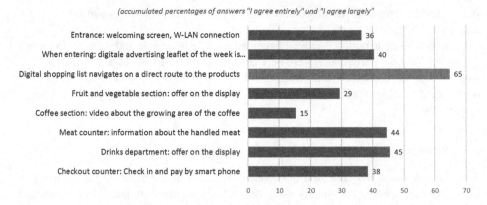

Fig. 3. Potential beacon application in supermarket (N = 99)

Looking at the applications in a stadium (Fig. 4), respondents prefer getting informed about event details (68 %) and being navigated at the end of the event.

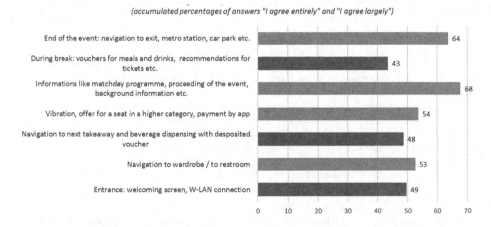

Fig. 4. Possible applications of beacons in stadiums (N = 99)

Willingness to use and reservations: 56 % of the participants can imagine using beacon-based services, 22 % are undecided. Only 22 % state that they are not willing to use beacons in future. The main reasons for declining the beacons technology are the installation of too many apps/too many notifications on every beacon event (73 %) and the fear of misuse of data (64 %). The percentage of positive and neutral assessments shows that there is a large acceptance for the beacons among the participants of the survey. Regarding to the success of the beacons-based technology, participants are undecided. 42 % believe beacons will have success, and 43 % are not sure. Only 14 % of the participants do not think that beacons will succeed.

4.4 Recommended Actions for Retailers

The technical prerequisites (Smartphone, Bluetooth) for using beacon-based services are largely met by the users, but unfortunately beacons are not yet well known. As the interest in this technology is driven by users, beacons should be advertised more extensively, to provide familiarity with it and increase the acceptance. As the survey results showed, most people are interested in the technology and can see themselves using it in the future. As (indoor-)navigation (and distribution of additional information) is one of the most promising areas of beacon-based services for most potential users, applications of the technology in large stores or other public buildings (e.g. malls, airports) or other places of interest [33] (e.g. historical sights, landmarks, museums) seem most promising to increase user numbers and acceptance of the technology on a wider scale [28].

The major drawbacks of beacon-based services in retail are privacy concerns of the customers and the public, which could be addressed through marketing communication and quality seals or guarantees for improving the acceptance of the technology of the customers. This is the main challenge, which needs to be addressed by retailers. Although a customer "has to opt-in or give permission for a company to track his location, when he downloads a company's app, companies need to walk a fine line between offering useful information and being invasive" [34]. Companies should only keep a minimum of the data which are required and delete those which are not needed for the business. In order to be credible, the companies should be audited by an independent third party [35].

Another drawback, which could arise, might be too much complexity on the side of the customer. Customers might struggle to adopt the technology if they have to install a new app for every single retailer or department store. Integration between retailers might be one solution to this drawback. App installation has to be easy and understandable for users to adopt the new technology [29]. Finally, it could be argued that customer preferences might not be sufficiently understood from purely technical systems. Furthermore, it should be communicated which kind of data are used and in which way the application communicates with the customer [29].

5 Conclusions and Outlook

Overall, beacon-based services are a promising approach, if they are part of a more comprehensive digital marketing strategy [36]. Currently, a growing share of consumers is using their smartphones as part of their shopping activity anyway, be it for additional product information, pricing or customer reviews. Employing beacons at useful locations to provide context-specific information, might create a more seamless and convenient shopping experience. From a technical point of view, the chances of widespread adoption are good, as Bluetooth support and Internet connection are common in today's smart phones. The major question is, if consumers are willing to share their information with retailers for a better shopping experience in terms of convenience and savings. If the overall channel strategy is well-designed, the customer can be catered for very individually to increase sales and customer loyalty at low cost due to high degree of automation in the process. At the same time, the analysis of data created from beacon-interactions provides potential for shop owner to better understand its customer's behavior and preferences. This knowledge allows adapting personnel capacity, opening hours, product range and other factors of the store accordingly [7].

A number of issues related to the employment of beacons in retail are still open, as the technology is still not widely established. Future research should concentrate on the actual acceptance of such services by consumers and the related critical success factors, e.g. via pilot installations. Additionally, to the best of our knowledge, the effect on retailer's sales and margins have not been investigated yet. To achieve that, framework of both analytical and operational benefits should be elaborated and used as basis for empirical studies.

References

1. Härtfelder, J., Winkelmann, A.: Opportunities and challenges for local retailing in an environment dominated by mobile internet devices –literature review and gap analysis. In: Nissen, V., Stelzer, D., Straßburger, S. et al. (eds.) Multikonferenz Wirtschaftsinformatik (MKWI) 2016
2. Taylor, G.: More Than 90 % Of Consumers Use Smartphones While Shopping In Stores (2015). http://www.retailtouchpoints.com/topics/mobile/more-than-90-of-consumers-use-smartphones-while-shopping-in-stores. Accessed 23 Mar 2016
3. Moody, M.: Analysis of promising beacon technology for consumers. Elon J. Undergraduate Res. Commun. 6(1), 59–68 (2015)
4. Berthiaume, D.: Beacons at your command: GameStop uses beacons to put customers in charge of store promotions (2015). http://www.chainstoreage.com/article/beacons-your-command#. Accessed 18 Mar 2016
5. Korber, S.: How retailers are using your phone to tell you: Buy (2015). http://www.cnbc.com/2015/05/26/retails-newest-brick-and-mortar-bet.html. Accessed 22 Mar 2016
6. Martin, C.: How beacons are changing the shopping experience. Harvard Bus. Rev. (2014)
7. Dudhane, N.A., Pitambare, S.T.: Location based and contextual services using Bluetooth beacons: new way to enhance customer experience. Lect. Notes Inf. Theory 3(1), 31–34 (2015)
8. Maycotte, H.O.: Beacon Technology: The Where, What, Who, How and Why (2015). http://www.forbes.com/sites/homaycotte/2015/09/01/beacon-technology-the-what-who-how-why-and-where/#7c132ec84fc1. Accessed 22 Mar 2016
9. Bach, C.: Warum deutsche Händler die Beacon-Technologie (nicht) brauchen (2015). http://locationinsider.de/warum-deutsche-haendler-die-beacon-technologie-nicht-brauchen/. Accessed 23 Mar 2016
10. Takalo-Mattila, J., Kiljander, J., Soininen, J.: Advertising semantically described physical items with Bluetooth low energy beacons. In: 2013 2nd Mediterranean Conference on Embedded Computing (MECO), pp. 211–214 (2013)
11. Dahlgren, E., Mahmood, H.: Evaluation of indoor positioning based on Bluetooth smart technology. Master thesis, Chalmers University (2014)
12. Fujihara, A., Yanagizawa, T.: Proposing an extended iBeacon system for indoor route guidance. In: 2015 International Conference on Intelligent Networking and Collaborative Systems (INCOS), pp. 31–37 (2015)
13. Yun, C., So, J.: A Bluetooth beacon-based indoor localization and navigation system. Adv. Sci. Lett. 21(3), 372–375 (2015). doi:10.1166/asl.2015.5796

14. Chawathe, S.S.: Beacon placement for indoor localization using Bluetooth. In: 11th International IEEE Conference on Intelligent Transportation Systems (ITSC), pp. 980–985 (2008)
15. Apple Inc.: Getting Started with iBeacon (2014)
16. Boidman, M.: Proximity-based communications: adding beacons yields more intelligent digital signage in retail and out-of-home (2015)
17. Deloitte, A.B.: Omni-channel retail: a deloitte point of view (2015)
18. Crowe, M., Rysman, M., Stavins, J.: Mobile payments at the retail point of sale in the United States: prospects for adoption. Rev. Netw. Econ. **9**(4) (2010). doi:10.2202/1446-9022.1236
19. Boniversum Demographic data on smartphone usage for mobile shopping in Germany from 2013 to 2015. http://www.statista.com/statistics/454370/mobile-shoppers-age-and-gender-distribution-germany/. Accessed 23 Mar 2016
20. Vend University: Retail Trends and Predictions 2016: 12 Retail trends and predictions to watch for (2016). https://www.vendhq.com/university/retail-trends-and-predictions-2016. Accessed 23 Mar 2016
21. Paul, A.K., Hogan, S.K.: Deloitte's 2014 Annual Holiday Survey: Making a List, Clicking it Twice. Deloitte University Press (2014)
22. Paul, A.K., Hogan, S.K.: On the Couch: Understanding Consumer Shopping Behavior. Deloitte University Press (2015)
23. Abdelkafi, N., Salameh, N.A.: Geschäftsmodellmuster im Dienstleistungssektor- Dargestellt am Beispiel der Internationalisierung deutscher Berufsbildungsdienstleister. In: Schallmo, D.R.A. (ed.) Kompendium Geschäftsmodellinnovation, pp. 385–415. Springer, Wiesbaden, Germany (2014)
24. Abdelkafi, N.: Open Business Models for the Greater Good – A Case Study from the Higher Education Context. Die Unternehmung, pp. 299–317 (2012). doi: 10.5771/0042-059X-2012-3-299
25. Rayna, T., Striukova, L.: The impact of 3D printing technologies on business model innovation. In: Benghozi, P.-J., Krob, D., Lonjon, A., Panetto, H. (eds.) Digital Enterprise Design & Management. AISC, vol. 261, pp. 119–132. Springer, Heidelberg (2014)
26. SITA IT Review: Beacon Technology - Improving The Customer Service Experience (2014)
27. van Belleghem, S.: 5 Ways Beacons are Changing Relationships with Customers (2015). http://digitalmarketingmagazine.co.uk/customer-experience/5-ways-beacons-are-changing-relationships-with-customers/1986. Accessed 19 Mar 2016
28. Nesamoney, D.: Personalized Digital Advertising: How Data and Technology Are Transforming How We Market. Pearson Education, Old Tappan (2015)
29. Böpple, O., Glende, S., Schauber, C.: Innovative Einkaufserlebnisse mit Beacon-Technologie gestalten. In: Linnhoff-Popien, C., Zaddach, M., Grahl, A. (eds.) Marktplätze im Umbruch: Digitale Strategien für Services im Mobilen Internet, pp. 299–307. Springer, Berlin (2015)
30. Smith, C.: How beacons — small, low-cost gadgets — will influence billions in US retail sales (2015). http://www.businessinsider.de/beacons-will-impact-billions-in-retail-sales-2015-2? r=US&IR=T. Accessed 23 Mar 2016
31. Swirl Networks: Swirl Releases Results of Retail Store Beacon Marketing Campaigns|Swirl (2014). http://www.swirl.com/swirl-releases-results-retail-store-beacon-marketing-campaigns/. Accessed 23 Mar 2016
32. Hamka, F.: Smartphone's customer segmentation and targeting: defining market segment for different type of mobile service providers. Master thesis, TU Delft (2012)
33. Fleck, M., Frid, M., Kindberg, T., et al.: From informing to remembering: ubiquitous systems in interactive museums. IEEE Pervasive Comput. **1**(2), 13–21 (2002). doi:10.1109/MPRV. 2002.1012333

34. Ehrens, T.: Privacy concerns arise with proximity-based beacon technology (2014). http://searchsalesforce.techtarget.com/feature/Privacy-concerns-arise-with-proximity-based-beacon-technology. Accessed 23 Mar 2016
35. Smith, A.: Location-based apps present opportunities - and data challenges (2014). http://searchcrm.techtarget.com/feature/Location-based-apps-present-opportunities-and-data-challenges. Accessed 23 Mar 2016
36. Burdick, G.A.: In-store beacons mature, make shopping personal. ECONTENT **38**(4), 6–8 (2015)

Process Management

Batch Processing Across Multiple Business Processes Based on Object Life Cycles

Luise Pufahl$^{(\boxtimes)}$ and Mathias Weske

Hasso Plattner Institute at the University of Potsdam, Potsdam, Germany
{Luise.Pufahl,Mathias.Weske}@hpi.de

Abstract. Batch processing is a means to synchronize the execution of multiple process instances for certain activities to improve process performance. Current batch processing concepts for business processes focus only on single process models whereas in practice large process model repositories exist with repeating activities. In this paper, we introduce a concept to specify batch processing requirements in centrally given object life cycles, which describe allowed data manipulations in order to identify candidates for batch processing during run-time across multiple processes and propose them to the user. We evaluate the applicability of this concept by implementation for an open source BPM platform.

Keywords: BPM · Batch processing · Process repository · Object life cycle

1 Introduction

Business process management (BPM) allows organizations to specify, execute, monitor, and improve their business operations [1] based on process models stored in process model repositories [2] and enacted in business process management systems (BPMS) [3,4]. In current BPMS, process instances – the concrete executions of business processes – run independently. However, the synchronized execution of process instances can improve process performance. For instance, in healthcare, it is more time-efficient to first collect a set of blood samples taken from patients to deliver them to the laboratory instead of sending a nurse for each separately. In e-commerce and logistics, it is more cost-efficient to consolidate packages sent to the same recipient instead of handling each separately. In this work, a concept to synchronize process instances of multiple process models in their execution of certain activities is presented.

A recently introduced means for synchronized execution of process instances is batch processing [5–7]. Batch processing in business processes allows for a process which usually acts on a single item, to bundle the execution of a group of process instances for certain activities in order to improve performance. These approaches allow various improvements such as data-enabled instance grouping, optimized batch assignment or a rule-based batch activation. Nevertheless, current approaches are limited to process instances of a single process model only.

© Springer International Publishing Switzerland 2016
W. Abramowicz et al. (Eds.): BIS 2016, LNBIP 255, pp. 195–208, 2016.
DOI: 10.1007/978-3-319-39426-8_16

However, in real world process model repositories, activities may be reused in multiple process models, e.g., in finance, sending notifications to the customer is required in multiple processes, such as account opening, credit card issue, and loan approval. Although customer notifications are created in different processes, they can be batched for sending only one letter to the customer instead of several ones for saving costs.

Based on these considerations, this paper introduces a concept for extending existing BPMS to allow *batch processing across process model boundaries* for synchronizing process instances of different process models. Our work aims at the following three objectives O1 through O3: At design time, the batch processing requirements have to be specified. Our concept will provide (O1) a centralized specification being valid for multiple process models. It will be independent from process models such that addition, deletion, and changes of process models in the repository have no influence on it. Thereby, we want to use object life cycles [8,9] which are centrally given and accompany process models. An object life cycle describes allowed data manipulations by process activities for one class of data objects. Batch processing can be conducted optionally or compulsory. The latter case has the risk of undesirable waiting times which has to be controlled. As several process models are included, an optional batch processing solution (O2) is targeted in which only instances are paused in their execution and executed as batch where a matching partner can be identified based on run-time information. However, neither run-time information nor design-time specifications might cover all information required to decide whether multiple activity instances shall be batched. We aim at (O3) a solution where the user can optionally decide about batch assignment – whether all, some or no process instance is executed as batch.

The remainder of this paper is structured as follows. Section 2 introduces the running example used for concept discussions throughout the paper. Section 3 provides theoretical foundation based on which Sect. 4 introduces the new concept of batch processing by considering object life cycles as a two-fold approach. First, we discuss the basics for single data state transitions followed by the generalization to multiple, connected data state transitions in the second part. Section 5 discusses our prototypical implementation. Section 6 is devoted to related work and Sect. 7 concludes the paper.

2   Scenario: Waste Management Organization

Our example organization is active in the waste management domain and operates a process model repository containing the process models with respect to the business operations. These process models are not only used for documentation purposes, but also they are executed in a BPMS. Figure 1 represents two processes of this repository on which we discuss the motivation of our work and the requirements.

Figure 1a shows the process of creating and sending an invoice to a customer either electronically or by a traditional postal service. If an *Invoice* is required, it first gets *prepared*. After preparation, an electronic invoice is directly sent as

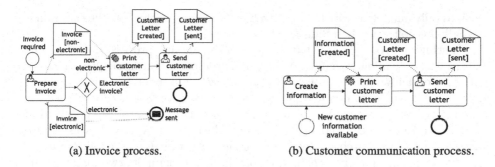

(a) Invoice process.   (b) Customer communication process.

Fig. 1. Two example processes of the repository of waste management company.

represented by the *Message sent* end event. In case of a non-electronic invoice, the *Invoice* is *printed* resulting in a *Customer Letter* that is then *sent* to the customer. The *Customer communication process* in Fig. 1b is started when some information shall be provided to the customer. For example, a customer shall be informed that, in future, the waste container is picked-up on Tuesday instead of Monday. This process consists of a sequence of first *creating* the *Information*, then *printing* it resulting in a *Customer Letter*, and finally *sending* the *Customer Letter* to the customer.

From these two process models, the object life cycle given in Fig. 2 can be derived for objects of type *Customer Letter* [10]. The actions used for transitioning from one state to another

Fig. 2. Object life cycle of *Customer Letter*.

directly map to activities in the corresponding process models, e.g., to reach state *sent* from state *created*, activity *Send customer letter* must be executed.

Executing these processes may lead to a situation where the employee John has four open tasks as presented in the task list in Fig. 3. We can observe that four instances of activity *Send customer letter* were assigned to John; three of them require a sending to customer *Boyson* in London – one for the invoice process and two for the customer communication process – and one of them requires a sending to customer *Thomsan* in Madrid through

Task list	
Send customer letter (Invoice process)	Boyson, London
Send customer letter (Invoice process)	Thomsan, Madrid
Send customer letter (Customer communication process)	Boyson, London
Send customer letter (Customer communication process)	Boyson, London

Fig. 3. Task list of employee John showing recently assigned activity instances.

the traditional postal service. Although handled by two different processes, the three customer letters to be sent to Boyson could be enveloped and sent as one mail. Consolidating the handling of these three customer letters has the advantages that (1) costs for sending mails can be reduced and (2) the customer receives only one letter which he has to handle instead of three at the same time. We call this collective handling *batch processing*.

Existing approaches for batch processing in the process domain, e.g., [6,7,11,12], only allow specification of batch processing in a single process model but do not allow batch processing across process model boundaries. These approaches have in common that the requirements for batch processing, e.g., which activity instances are executed under which conditions, are specified at some central point to avoid update issues. Most of the approaches except [7] have a compulsory batch processing approach and wait for an explicit number of activity instances to be executed as batch. If, for example, two instances are required to start batch execution, the activity instance of *Send customer letter* for customer *Thomsan, Madrid* (see Fig. 3) can only be started after another letter is required to be sent to him. This means that the invoice is sent later than intended and required leading to a delay in payment. In the given case, the costs saving for sending only one letter does not compensate the costs for waiting for the second letter. Thus, we conclude that scenarios exist where an optional batch processing approach is more suitable especially if more than one process model is involved. However, batch processing is not always useful. The second information letter to customer Boyson (see Fig. 3) could be an invitation to a customer event which is supposed to be sent separately and in a "golden" envelope. Thus, this letter shall not be sent together with the other two letters. That means that the system calculates potential batches, proposes these candidates to the user, but the user finally decides which identified activity instances can be handled as a batch and which not. To summarize, we identified three objectives – already introduced in Sect. 1 – which we will tackle while introducing a concept for batch processing over multiple process models: (O1) specify batch information at one central point to avoid update issues, (O2) optional batch processing based on run-time information, and (O3) allow user-based decisions for creating actual batches.

3   Foundation

In this paper, we combine process and data modeling to derive batch processing information from the data side and steer process execution accordingly. We next introduce formalisms for both sides which are then used for concept discussion in Sect. 4. First, we give a generic process model definition and require it to be syntactically correct with respect to the used modeling notation. Formally, we define a process model as follows.

Definition 1 (Process Model). A *process model* $m = (N, D, \mathfrak{C}, \mathfrak{F}, type)$ consists of a finite non-empty set $N \subseteq A \cup E \cup G$ of control flow nodes being activities A, events E, and gateways G (A, E, and G are pairwise disjoint) and a finite non-empty set D of data nodes (N and D are disjoint). $\mathfrak{C} \subseteq N \times N$ is the control flow and $\mathfrak{F} \subseteq (A \times D) \cup (D \times A)$ is the data flow relation specifying input/output data dependencies of activities. Function $type : G \rightarrow \{AND, XOR\}$ gives each gateway a type.   ◇

We refer to a data node $d \in D$ being read by an activity $a \in A$, i.e., $(d, a) \in \mathfrak{F}$, as *input data node* and to a data node d being written by an activity a, i.e., $(a, d) \in \mathfrak{F}$, as *output data node*. Figure 1a shows a process model in BPMN notation [13] with one start event, two alternative end events, three activities, one gateway, and multiple data nodes read and written by activities. Each data node has a name, e.g., *Customer Letter*, and a specific data state, e.g., *created* or *sent*. Data nodes sharing the same name refer to the same data class; here: *Customer Letter*. A data class describes the structure of data nodes (objects) in terms of attributes and possible data states. A data object is the run-time representation of a data node consisting of a value for each attribute. A data node (object) maps to exactly one data class. A data state denotes a situation of interest for the execution of the business process, e.g., *Customer Letter* is *created*. The logical and temporal order of data states is represented by means of object life cycles.

Definition 2 (Data Class). A *data class* $c = (J, olc)$ consists of a finite set J of data attributes and an object life cycle *olc*.   ⋄

Definition 3 (Object Life Cycle). An *object life cycle olc* $= (S, s_i, S_F, \mathfrak{T}, \Sigma, \zeta)$ consists of a finite non-empty set S of data states, an initial data state $s_i \in S$, a non-empty set $S_F \subseteq S$ of final data states, and finite set of actions representing the manipulations on data objects (S and Σ are disjoint). $\mathfrak{T} \subseteq S \times S \setminus \{s_i\}$ is the data state transition relation describing the logical and temporal dependencies between data states. Function $\zeta : \mathfrak{T} \to \Sigma$ assigns an action to each data state transition.   ⋄

Definition 4 (Data State). Let V be a universe of data attribute values. Then, the *data state* $s_o : J_c \to V$ is a function which assigns each attribute $j \in J_c$ a value $v \in V$ that holds in the current state of data object o.   ⋄

Definition 5 (Data Object). During its life time, a *data object* $o = (T_S, c, pid)$ traverses a sequence $T_S = \langle s_1, s_2, \ldots, s_k \rangle$ of data states. The data class c describes its structure and the identifier *pid* refers to the process instance processing this object.   ⋄

At any point in time, each data attribute of an object can get assigned a value. If it is not defined, the value is set to $\perp$. Regarding the interplay of the process and the data side, we assume process models and object life cycles to be consistent; i.e., all data manipulations induced by some activity in the process model are covered by data state transitions in the corresponding object life cycles (cf. notion of object life cycle conformance [14,15]). Additionally, an activity utilizes at least one single data state transition of an object of some data class such that function $\epsilon : A \to 2^U$ with $U \subseteq \bigcup \mathfrak{T}_{olc}$ (all transitions of all object life cycles) returns a set of transitions for a given activity. For reducing complexity of discussion, we apply the connection between process models and object life cycles through consistent label matching, i.e., the same activity respective action label refers to the same action. Alternatively, in practice, similarity matching techniques could also handle the connection.

Executions of process models are represented by process instances with each instance belonging to exactly one process model m. At any point in time, a process instance has a current *process instance state* consisting of a set of enabled activity instances – the run-time representation of an activity – and a set of current states of processed data objects.

Each activity instance follows an activity life cycle during its execution. We define the activity life cycle as state-transition-diagram such that $init \overset{enable}{\longrightarrow} ready \overset{start}{\longrightarrow} running \overset{terminate}{\longrightarrow} terminated$ and $ready \overset{disable}{\longrightarrow} disabled \overset{re\text{-}enable}{\longrightarrow} ready$. The link between an activity instance and the task performer of an activity is realized in a BPMS by a work item. A work item represents the to-be-executed activity instance. If an activity instance is enabled (i.e., in state *ready*), work items are provided to selected task performers. The work item which is selected by one of the task performer is executed while the others are withdrawn [1].

In our approach, we group activity instances based on their data characteristic. For example, only letters addressed to the same customer are combined. Therefore, we will rely on a *grouping characteristic* that allows to group multiple process instances into batch clusters – a batch cluster is a container for compatible process instances. Two process instances are compatible if their data view is equal. Instance grouping based on data characteristics was first introduced in [6] using the concept of *data views*. In the scope of this paper, we ease this concept such that a *data view DV* is a projection on the values of logically combined data attributes contained by a single data class as specified by the *data view definition DVD*, a list of fully qualified data attributes.

4   Batch Specification in Object Life Cycles

In this section, we introduce the concept to specify parameters for batch execution in object life cycles to allow the grouping of instances of compatible activities from various process models into one batch based on run-time data information. We distribute the discussion into two parts. In Sect. 4.1, we introduce the design of batch transitions in object life cycles and the corresponding execution semantics. In Sect. 4.2, we present the concept of connected batch transitions. It allows to batch activity instances providing the same business result, but having different data inputs.

4.1   Part 1 – Design and Execution

Design. One object life cycle per data class is centrally defined in an organization. Following our foundation, a state transition can occur in different process models. They are utilized to specify batch information at a central point (O1). A new type of state transition: the *batch transition* is introduced. A batch transition carries information to create and subsequently execute a batch consisting of multiple activity instances.

Definition 6 (Batch Transition). Let $t \in \mathfrak{T}_{olc}$ be a transition of an object life cycle *olc*. Then, function $\beta : \mathfrak{T}_{olc} \to \{true, false\}$ returns true, if t is a *batch* *transition*, and returns false otherwise. $\mathfrak{T}'_{olc} \subseteq \mathfrak{T}_{olc}$ denotes the set of all batch transitions of *olc*. Each batch transition gets assigned a data view definition $dvd \in DVD$ as grouping characteristic by function $\gamma : \mathfrak{T}' \to DVD$. A batch transition *t* can be jointly processed for multiple activity instances referring to *t*, if their data view *DV* match.

Figure 4 shows an example object life cycle for data class *Customer Letter* from our waste management scenario (cf. Sect. 2). It contains a batch transition – the one referring to action *Send customer letter* –, with a specific grouping characteristic. We

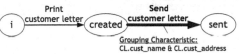

Fig. 4. Object life cycle of data class *Customer Letter* with batch transition *Send customer letter*.

visualize batch transitions by bold edges and an corresponding textual annotation for the grouping characteristic referencing a data view definition.

All activities that reference a batch transition are candidates for batch processing. In our example, instances of activities referencing the batch transition *Send customer letter* are candidates for batch processing. Given a data view definition consisting of data attributes *cust_name & cust_adress*, we ensure that only activity instances for which customer letter objects are addressed to the same customer may be executed as a batch.

An activity of a process model can have several output data nodes referring to different data classes. Thus, several object life cycle transitions may exist for an activity. Following the concept of case objects from business artifacts [16], one class of objects drives the execution and thus, we require only one of them being a batch transition.

Execution. Based on the design-time batch specifications, we present the algorithm on processing batches of activity instances first focusing on user interaction activities and later discussing the generalization. The algorithm consists of five subsequent steps. The steps for the running example are visualized in Fig. 5.

(1) Check whether an activity instance is candidate for batch processing. During process execution, activity instances can be in states *initialized, ready, running, terminated*, and *disabled* (cf. Sect. 3). If an activity instance is *ready* for execution, all matching object life cycle transitions *U* of the corresponding activity $a \in A$ are retrieved by function $\epsilon : A \to 2^U$ introduced in the foundation section. The resulting set *U* of transitions is checked for existence of a batch transition, i.e., each transition $t \in U$ is checked whether $\beta(t) = true$. As mentioned above, we require that β returns *true* for at most one transition per activity. If a batch transition $\hat{t}$ was found, the activity instance is a candidate for batch processing and the corresponding data object $\hat{o}$ is retrieved. As each data object has a reference *pid* to its process instance in which it is processed, the data object $\hat{o}$ can be obtained from the data storage with the *process instance id* of the just enabled activity instance and the *data class* containing the object life cycle of the batch transition $\hat{t}$. For example, activity instance *1-21* is identified as batch candidate. Since, the identified batch transition belongs to the *Customer Letter*

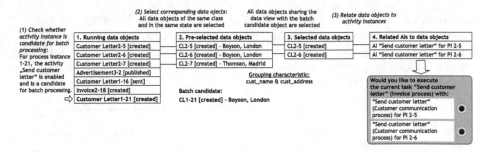

Fig. 5. Illustration of algorithm to process activity instances of multiple process models in a batch.

object life cycle, the *Customer Letter* object in state *created* and with id *1-21* is retrieved (see last row of most-left table in Fig. 5).

(2) Select corresponding data objects based on data view. Next, all data objects sharing the data class and data state with the in (1) identified data object $\widehat{o}$ are pre-selected. In Fig. 5, the most-left table shows all currently processed data objects O of our waste management organization. The *Customer Letter* in state *created* referring to the process instance with id *1-21* to be processed by the current activity instance is the candidate for batch processing and is shown in bold font. The second table of Fig. 5 visualizes the projection O' on the set of data objects O such that all objects refer to data class *Customer Letter* and are in state *created*. In total, three data objects are referred in the given example. From this set, all objects sharing the data view with $\widehat{o}$ are finally selected. In our example, the grouping characteristic consists of data attributes *cust_name* and *cust_address* of data class *Customer Letter*. For the current processed data object $\widehat{o}$, we find the values *Boyson* and *London*. All data objects $O'' \subseteq O'$ having identical values are selected as shown by the third table in Fig. 5. If the data view definition is empty or not defined, then $O'' = O'$ holds.

(3) Relate data objects to activity instances. Batch processing works on activity instance level. Thus, next, the activity instances of the selected data objects O'' must be determined. As given in Definition 5, each object contains a reference *pid* to its corresponding process instance. For activity instance determination, the auxiliary function $\eta : O \rightarrow AI$ is used where AI denotes the set of all currently existing activity instances. Subsequently, the corresponding related activity instance is retrieved that is currently in state *ready* and that, in turn, refers to an activity a reading a data node d, where the state matches the one from $\widehat{o}$ $((d, a) \in \mathfrak{F})$. The most-right table in Fig. 5 shows the resulting activity instances for the data objects O''. The identified activity instances may refer to different process models. In our example, two identified activity instances refer to the *customer communication process* while the batch candidate refers to the *invoice process*.

(4) Provide batch processing candidates to the user. The identified activity instances are transferred from state *ready* into state *disabled*; a disabled activ-

ity instance is temporarily deactivated in its processing [17]. Then, the user is asked which of the identified activity instances shall be processed in a batch with the batch candidate. In our example, activity instances for activity *Send customer letter* of the process instances *2-5* and *2-6* may be grouped in a batch with the activity instance of *1-21*. The gray-colored box on the bottom-right in Fig. 5 visualizes this. One of the responsible task performers decides whether all additionally identified activity instances, a subset of them, or none are joined into one *work item batch*. A *work item batch* is an aggregated work item of all selected activity instances allowing joint visualization and execution in one step. The task performer is then responsible for executing the *work item batch* if at least some activity instances are joined. Otherwise, the batch candidate is handled as a common single activity instance. Finally, those activity instances that were not selected for batch processing are transferred back into state *ready*.

(5) Execution of a work item batch. All activity instances selected by the task performer and the batch candidate are jointly realizing the *work item batch* are added to a batch cluster. A batch cluster is a container collecting activity instances that can be executed together and may pass multiple states during its lifetime [18] First, a batch cluster is *initialized*. In this state, it aggregates the activity execution information of all contained activity instances into a single work item – the *work item batch*. Afterward, the batch cluster is *ready* for execution but can still be extended. Further activity instances

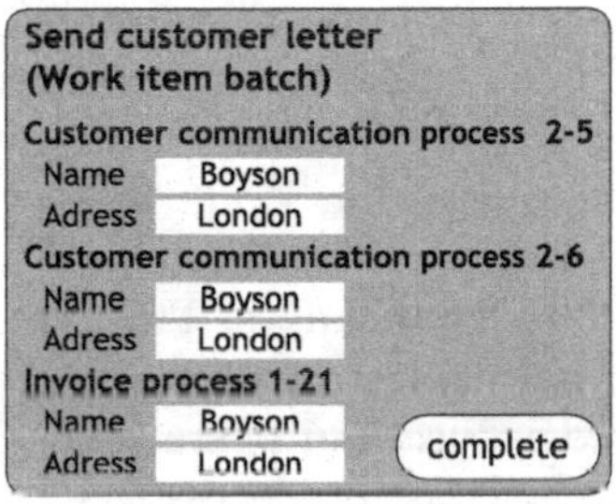

Fig. 6. Resulting *work item batch* from Fig. 5 if all activity instances are selected for batch processing.

can be added to the batch cluster by expanding the work item batch accordingly. Additions are allowed until the task performer starts its execution, i.e., the batch cluster transitions into state *running*. Figure 6 visualizes the *work item batch* for the example given in Fig. 5 under the assumption all identified activity instances were selected. Upon completion of the *work item batch*, the output data is stored by the batch cluster for each activity instance individually to easily distribute the results, all activity instances are terminated. Subsequently, the batch cluster terminates as well and the succeeding activity instances are handled again separately .

In BPMS, activities are distinguished into user interaction activities, system activities and manual activities [1]. Handling of user interaction activities is discussed above. Targeting, system activities, step (4) needs to be automated by automatically adding all batch candidates to the batch. Alternatively, a responsible role can be specified to handle step (4) interfering with automated system activity execution. Step (5) remains conceptually unchanged. Instead of providing the *work item batch* to the user as form, the batch cluster aggregates the input information and provides it to the information system processing the system activity. Targeting manual activities and assuming the BPMS shows them to the user as form, the approach introduced above remains unchanged.

4.2   Part 2 – Connected Batch Transitions

Design. In real-world business processes, we observed that multiple tasks provide the same result, e.g., a letter is sent by mail, given various input data represented by different states, e.g., the letter to be sent can be either signed or unsigned or it can be approved, rejected, or unchecked. The algorithm presented in Sect. 4.1 cannot handle these observations properly. Only letters being in the same data state before sending can be executed as batch. However, despite of different input states, the customer letters can still be grouped as batch and sent within one envelope if they share the addressee. In this section, the concept of *connected batch transitions* is presented to cope with this challenge. Assume, the customer communication process in Fig. 1b requires the *Customer Letter* to be signed before it is sent, a corresponding activity is added between activities *Print customer letter* and *Send customer letter*. The data flow is adapted accordingly.

The object life cycle is changed as well to match these data flow changes. The additional data state *signed* is added as successor of state *created* and as predecessor of state *sent*. The corresponding data state transitions get associated to the actions *Sign customer letter* and *Send customer letter* respectively. Figure 7 visual-

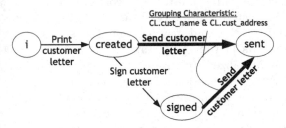

Fig. 7. Object life cycle of customer letter with the new state *signed* and connected batch transitions.

izes the extended object life cycle for data class *Customer Letter*.

Allowing batch processing of activity instances referring to different batch transitions, a connection between them is required. The property of *Connected batch transitions* is introduced describing batch transitions that have different input states and might have different labels but have the same output state. Visually, this is represented by associations between corresponding batch transitions. We assume that the object life cycle designer decides based on the business context which transitions can be connected. Referring to consistency, all connected batch transitions must be assigned to the same grouping characteristic. Toolingwise, this can be handled by specifying the grouping characteristic only once and applying it to all.

Execution. Based on above design adaptations, the execution semantics need to be adapted as well. In step (2), after checking whether the currently enabled activity instance is a candidate for batch processing, the set O' of pre-selected data objects is expanded. All data objects sharing the same data class and being in one of the data states that are the source state of some connected batch transition are added. Thereby, alignment of data objects to processes must be considered since the same data state may be utilized differently in different process models (cf. state *created* of data class *Customer Letter*). In above example, this includes all data objects of class *Customer Letter* being in state *created* – if they refer to the *invoice process* – or *signed* – if they refer to the *customer communication process*. The remaining steps are applied as described in Sect. 4.1.

5   Implementation

For our prototypical implementation, we utilized the Camunda engine in Version 7.1 to show that our approach discussed in Sect. 4 can be implemented in a standard BPMS without direct support for data objects. The implementation and a screen-cast are available at http://bpt.hpi.uni-potsdam.de/Public/BatchProcessing.

The Camunda engine does not handle data nodes annotated in process models as all current standard BPMS [19]. Thus, we first extended the parser to consider annotated data nodes and associations. Further, we added a parser to retrieve the information from the object life cycles which are represented in an extended format of the jBPT library [20]. Second, we extended the engine itself. This is exemplary explained on the user tasks. When a user task gets enabled, its related data state transitions are checked whether one of them is a batch transition. If yes, all currently enabled user tasks are retrieved and checked based on their input and output nodes whether they refer to the same batch transition. For all positively identified user tasks, the attribute values of the grouping characteristic are compared with the batch candidate's values. If some matching user tasks were found, they are provided to the task performer in a *Choose-Tasks*-form presenting the batch candidate and all selected ones. A work item batch is created, if one or more user tasks are selected by the user. In such work item batch, all form variables are shown in one view and can be terminated by one click.

Our implementation shows that work item batches can be created and executed and several work item batches can run in parallel. Further, it shows that if the *Choose-Tasks*-form is not yet executed, new user tasks can be added dynamically, and that an existing batch activity can be combined with a new user task.

6   Related Work

In recent years, several approaches and techniques were introduced in the novel area of batch processing in activity-centric processes. They introduce means to configure simple options like the maximum number of jointly executed instances [5,11,12] to more complex ones proposing a constraint-based [7] or rule-based [6,18] activation. From these, [7] (i.e., for optional batch processing) and [6] (i.e., for an optimized batch enablement) are the only works explicitly considering run-time data for batch processing. However, they do not allow –

likewise all other mentioned approaches – to jointly execute instances of *different* process models.

The already existing approaches could be adapted to support batch specification across process model boundaries. However, this would result in challenging tool support in terms of visualization, specification, and maintenance, because multiple diagrams must be handled in parallel. Alternatively, BPMN [13] provides the concept of a global task which is referenced in multiple process models as call activity. Still, all process designers have to ensure that the call activity is used in the corresponding process models again resulting in a multi-model handling. As we aimed at providing a broader concept that allows batch processing of activity instances performing similar business steps with the same business goal, we did not use the call activity for our approach.

The increasing interest in the development of process models for execution has shifted the focus from control flow to data flow perspective leading to integrated scenarios providing control as well as data flow views. One step in this regard is object-centric processes [16,21,22] that connect data classes with the control flow of process models by specifying object life cycles. Batch processing in terms of jointly processing data objects is of interest for object-centric processes as well [22]. Conversely to our approach, data objects have to be selected manually by the user and attribute values entered in the form are assigned to all processed data objects instead of separating them per object instance.

7   Conclusion

This paper presents a concept to process a group of activity instances of different processes as batch in order to improve process performance in large process collections with repeating activities. For instance, in the bank sector, in case of a bank account opening several documents can be sent as one mail. The requirements for batch processing are centrally defined in object life cycles (O1) which describe the allowed data manipulations of data objects used in the business processes. The presented concept is an optional batch processing approach (O2) being started as soon as matching partners could be identified based on run-time information, the currently existing data objects. Further, identified batch processing candidates are proposed to a task performer (O3) who finally decides which of them form a batch to ensure a correct batch assignment in a multi-process setting. We consider our approach light-weight, since it only uses few additional concepts (most prominently the batch transitions) well embedded in recent research works. Our implementation showed that the approach is capable to have several batches running in parallel and to extend batches dynamically as long as they are not executed.

In this paper, activity instances are not delayed to avoid unnecessary waiting times. Further cost saving potentials by grouping future activity instances also into a batch are not considered yet. Considering this, existing concepts tackling this challenge, e.g., activation rules [6], can easily be integrated into our approach. Additionally, we also aim at investigating to group a set of subsequent data state transitions into one batch transition similar to the concept of batch regions consisting of a set of connected activities. Further, we intend to provide a method to quantify the awaited cost savings.

Acknowledgments. We thank Andreas Meyer for his helpful input and Stephan Haarmann for his support in extending the Camunda engine.

References

1. Weske, M.: Business Process Management: Concepts, Languages, Architectures, 2nd edn. Springer, Heidelberg (2012)
2. Yan, Z., Dijkman, R.M., Grefen, P.W.P.J.: Business process model repositories - framework and survey. Inf. Softw. Technol. **54**(4), 380–395 (2012)
3. Camunda: camunda BPM Platform. https://www.camunda.org/
4. Lanz, A., Reichert, M., Dadam, P.: Robust and flexible error handling in the AristaFlow BPM suite. In: Soffer, P., Proper, E. (eds.) CAiSE Forum 2010. LNBIP, vol. 72, pp. 174–189. Springer, Heidelberg (2011)
5. Sadiq, S., Orlowska, M., Sadiq, W., Schulz, K.: When workflows will not deliver: the case of contradicting work practice. In: BIS, pp. 69–84 (2005)
6. Pufahl, L., Meyer, A., Weske, M.: Batch regions: process instance synchronization based on data. In: EDOC, pp. 150–159. IEEE (2014)
7. Natschl"ager, C., B"ogl, A., Geist, V.: Optimizing resource utilization by combining running business process instances. In: Toumani, F., et al. (eds.) ICSOC 2014. LNCS, vol. 8954, pp. 120–126. Springer, Heidelberg (2015)
8. Rumbaugh, J., Blaha, M., Premerlani, W., Eddy, F., Lorensen, W.E.: Object-Oriented Modeling and Design. Prentice-Hall, Englewood Cliffs (1991)
9. Kappel, G., Schrefl, M.: Object/behavior diagrams. In: ICDE, pp. 530–539. IEEE (1991)
10. Meyer, A., Weske, M.: Activity-centric and artifact-centric process model roundtrip. In: Lohmann, N., Song, M., Wohed, P. (eds.) BPM 2013 Workshops. LNBIP, vol. 171, pp. 167–181. Springer, Heidelberg (2014)
11. van der Aalst, W.M.P., Barthelmess, P., Ellis, C.A., Wainer, J.: Proclets: a framework for lightweight interacting workflow processes. Int. J. Coop. Inf. Syst. **10**(4), 443–481 (2001)
12. Liu, J., Hu, J.: Dynamic batch processing in workflows: model and implementation. Future Gener. Comput. Syst. **23**(3), 338–347 (2007)
13. OMG: Business Process Model and Notation (BPMN), Version 2.0, January 2011
14. Meyer, A., Weske, M.: Weak conformance between process models and synchronized object life cycles. In: Franch, X., Ghose, A.K., Lewis, G.A., Bhiri, S. (eds.) ICSOC 2014. LNCS, vol. 8831, pp. 359–367. Springer, Heidelberg (2014)
15. Küster, J.M., Ryndina, K., Gall, H.C.: Generation of business process models for object life cycle compliance. In: Alonso, G., Dadam, P., Rosemann, M. (eds.) BPM 2007. LNCS, vol. 4714, pp. 165–181. Springer, Heidelberg (2007)
16. Cohn, D., Hull, R.: Business artifacts: a data-centric approach to modeling business operations and processes. IEEE Data Eng. Bull. **32**(3), 3–9 (2009)
17. Weske, M., Húndling, J., Kuropka, D., Schuschel, H.: Objektorientierter Entwurf eines flexiblen Workflow-Management-Systems (Object-oriented design of flexible WfMS). Inform. - Forsch. und Entwickl. **13**(4), 179–195 (1998)

18. Pufahl, L., Weske, M.: Batch activities in process modeling and execution. In: Basu, S., Pautasso, C., Zhang, L., Fu, X. (eds.) ICSOC 2013. LNCS, vol. 8274, pp. 283–297. Springer, Heidelberg (2013)
19. Meyer, A., Pufahl, L., Fahland, D., Weske, M.: Modeling and enacting complex data dependencies in business processes. In: Daniel, F., Wang, J., Weber, B. (eds.) BPM 2013. LNCS, vol. 8094, pp. 171–186. Springer, Heidelberg (2013)
20. Polyvyanyy, A., Weidlich, M.: Towards a compendium of process technologies: the jBPT library for process model analysis. In: CAiSE Forum, CEUR, pp. 106–113 (2013)
21. Yongchareon, S., Liu, C., Zhao, X.: A framework for behavior-consistent specialization of artifact-centric business processes. In: Barros, A., Gal, A., Kindler, E. (eds.) BPM 2012. LNCS, vol. 7481, pp. 285–301. Springer, Heidelberg (2012)
22. Künzle, V., Reichert, M.: PHILharmonicFlows: towards a framework for object-aware process management. J. Softw. Maint. **23**(4), 205–244 (2011)

Towards a Methodology for Industrie 4.0 Transformation

Isabel Bücker[1(✉)], Mario Hermann[1], Tobias Pentek[2], and Boris Otto[1]

[1] TU Dortmund University, Dortmund, Germany
{isabel.buecker,mario.hermann,boris.otto}@tu-dortmund.de
[2] CDQ AG, St. Gallen, Switzerland
tobias.pentek@cdq.ch

Abstract. Implications of market and environmental changes have always influenced the industrial world. Combined with new technologies, the current changes are summarized under the term Industrie 4.0. Since the first announcement, Industrie 4.0 is one of the most discussed topics in research and industry. However, for companies in the industrial sector, it is a challenge to assess the implications of Industrie 4.0 for their organizations, and to decide whether and how to respond. Therefore, a methodology to transform an organization towards Industrie 4.0 is required. This paper provides a metamodel for the transformation of organizations towards Industrie 4.0 as well as the first technique of this method, a framework, to structure the implications of Industrie 4.0 and to identify Industrie 4.0 action fields as a first step towards Industrie 4.0 transformation. Furthermore, it provides an outlook how to implement the identified action fields systematically in existing process change management.

Keywords: Industrie 4.0 · Process change management · Organizational change · Framework · Method engineering

1 Introduction

INDUSTRIE 4.0 is an increasingly discussed topic in the German-speaking area [1]. Since 2011 as the German Federal Government announced Industrie 4.0 as one of the key initiatives of its high-tech strategy [2], numerous academic publications, conferences and practical articles have focused on that topic [3]. Since the first announcement of the term Industrie 4.0, researchers and practitioners are confronted equally with keywords like Cyber-Physical-Systems or the Internet of Things. Both, practitioners and researchers, are trying to build up a common understanding and appreciation of what this term could mean for their activities [4]. Although the "Industrie 4.0 Working Group" [2] describes the idea's basic technologies, its vision and selected scenarios for Industrie 4.0, it is still not clear how the industry will be affected of these impacts [4–6]. As change has become a permanent characteristic in modern business, companies have to improve their ability to respond rapidly and dynamically to market forces and competition [7]. As companies are confronted with Industrie 4.0, they have to address the question of how to organizationally prepare for this change, a methodology to implement changes well-structured and systematically is requested [8]. To face the industrial future and to understand how it will look like, it will require an appropriate framework [2]. This paper

© Springer International Publishing Switzerland 2016
W. Abramowicz et al. (Eds.): BIS 2016, LNBIP 255, pp. 209–221, 2016.
DOI: 10.1007/978-3-319-39426-8_17

aims at closing the gap in research, how the upcoming challenges connected to Industrie 4.0 can be structured and how the transformation process can be methodically supported. By providing a framework based on Industrie 4.0 design principles and an approach to structure an organization, the paper allows practitioners to understand how these challenges influences their enterprise. By using the framework, practitioners can identify action fields for an Industrie 4.0 transformation of their activities. Because the implementation of the change is as important as what the change is [9], the authors categorize their framework in existing process change management models and provide a methodology to implement the identified action fields.

The paper is structured as follows: Sect. 2 introduces the challenges and impacts of Industrie 4.0 for the industry, seizes related work and points out the research gap. Section 3 outlines the applied research process and research methods consisting of a qualitative literature review and interview. Section 4 elucidate the methodology to implement identified potentials caused by Industrie 4.0 action fields. In Sect. 5, the components of the framework are illustrated. Additionally the authors describe the construction of the framework. In conclusion, Sect. 6 gives an outlook and points out further need for investigation.

2   Industrie 4.0

Three revolutions, which have fundamentally changed the industrial world, have taken place until now [1]. The introduction of mechanical production facilities starting in the second half of the 18th century is considered as the first industrial revolution. The second industrial revolution was started in the 1870s with the electrification and the division of labor (i.e. Taylorism). Around the 1970s, the advanced electronics and information technologies developed further the automation of production processes and led to the third industrial revolution, also called "the digital revolution" [1, 6].

2.1   Impacts of Industrie 4.0 on Organizations

Currently market and environment changes like the increase of variant diversity, the increase of individualization, globalization or the decrease of product lifecycles are forcing companies to reactions in order to stay competitive [10]. Meanwhile, technologies like smartphones or apps, which are already frequently used in private life, find their way to industrial practice and have implications like the increase of automatization and mechanization or digitalization and connectivity [10]. In consequence, companies have to deal with global complexities and these discontinuous technology changes to secure their competitive situation [7]. Furthermore, companies should be able to identify future implications and implement process changes to secure continuous adaptions [8]. Moreover, researchers proclaim that the vision of Industrie 4.0 describes a new kind of industrial world including an increasing internal and intercorporate connectivity. This confronts companies with new challenges for all dimensions of an enterprise, especially humans (e.g. qualification), technology (e.g. highly connected systems) and organization (e.g. self-organization) [4].

2.2 Related Work

With the target of providing guidance on the description, standardization, implementation and improvement of Industrie 4.0 scenarios, the German-centered Plattform Industrie 4.0 as well as the US-based Industrial Internet Consortium (IIC) have published a reference architecture.

The Plattform Industrie 4.0's Reference Architectural Model Industrie 4.0 (RAMI 4.0) describes Industrie 4.0 technologies with the help of three perspectives: from a *hierarchy level*, a *life cycle and value stream*, and a *layer* angel. The hierarchy level represents the functionalities of factories (i.e. product, field device, control device, station, work center, enterprise, and connected world). The life cycle and value stream point of view distinguishes between types and instances of products and facilities. The layer view describes machine properties (i.e. business, functional, information, communication, integration, and asset). Based on the integration of the three views, RAMI 4.0 provides a basis for mapping and classifying objects and, thus, describing and implementing Industrie 4.0 scenarios [11].

The IIC's Industrial Internet Reference Architecture (IIRA) describes conventions, principles and practices for industrial internet systems. The architectural framework comprises four viewpoints: *business, usage, functional*, and *implementation*. The business viewpoint addresses the system's stakeholders and their respective vision, values, and objectives for applying industrial internet scenarios. The usage viewpoint represents the sequences of activities of human or machine users required to provide the necessary functionalities of the system. The functional viewpoint centers around the functional components, "their interrelation and structure, the interfaces and interactions between them, and the relation and interactions of the system with external elements in the environment" [12]. Finally, the implementation viewpoint addresses the required technologies for implementing the functional components [12].

Both introduced reference architectures have a technical focus on Industrie 4.0 and do not provide any guidance on Industrie 4.0 specific processes. Consequently, these architectures support the technical implementation but lack a broader understanding of Industrie 4.0 and its impact on companies and therefore they are not suitable for the approach of the paper.

3   Research Process and Research Method

The research process aimed at identifying a method to classify Industrie 4.0 impacts on organizations. The authors decided to use qualitative research methods, because they are accounted as being useful for several areas of research connected with organizational change situations, "including theory development, theory testing, construct validation, and the uncovering of new, emerging phenomena" [13]. The research process contained several research steps, which are briefly described in the following.

As a first step, the authors scoured several publication databases (Scopus, ECONIS, ScienceDirect, ACM Digital Library, AIS Electronic Library, IEEE Xplore) for approaches how to structure an organization holistically and implement the impacts of major changes in market and environment. These databases have been searched for items like "organizational design", "enterprise configuration" or "socio-technical system". As already considered in Subsect. 2.1, Industrie 4.0 affects all domains of a company [4]. Based on this perception, the search focused on holistic concepts to describe an organization. It could be determined that in order to gain a complete understanding of an organization it requires to see an organization as a system of interacting components which are people, tasks, technology and structure [14]. Using the literature review, the Human-Technology-Organization (HTO) method has been identified by the authors to fulfil the requirements of a method to represent all components of an organization.

When analyzing the term Industrie 4.0, it can be determined that there is a wide range of existing definitions with different focuses. Accordingly, it is difficult to grasp a common understanding of Industrie 4.0 and to incorporate it into a framework. One approach to systemize and to describe a phenomenon like Industrie 4.0 is to derive design principles from existing work on the topic [15]. So as a second step, the authors decided based on existing preparatory work to clarify the term of Industrie 4.0 by using Industrie 4.0 design principles [6]. The authors searched for different approaches to describe Industrie 4.0, but could not identify a comparable concept. Consequently, they decided to use the design principles of Hermann et al. [6].

As a third step, the authors chose a qualitative research interview because it is most appropriate when the research field "focuses on the meaning of particular phenomena to the participants" [16]. For this paper, the qualitative research interview has been conducted to evaluate the framework as a method to structure the impacts of Industrie 4.0 for companies. Following the recommendations of Dicicco-Bloom [17], in a sub step the tentative draft of the framework has been presented to three different focus groups, each composed of nine participants of research and industry. The group consists in each case of two department managers, one division manager, three operative workers, one university professor with industrial information management background and two doctoral students, which graduate at the faculty of mechanical engineering and investigate the transforming process of organizations towards Industrie 4.0. The industry representatives had various functional backgrounds (from operative to strategic) in order to represent a company as holistic as possible. They have been chosen as representatives for one operative process. The authors started the interview by asking questions about how the impacts of Industrie 4.0 could concern the organization, e.g. what impacts they see connecting to Industrie 4.0 for their business or how decentralized decision making could influence the organization structure, which led to an open discussion about the holistic viewpoint of the framework. The participants of the interviews accepted the framework as a holistic method to structure the impacts of Industrie 4.0 but requested a methodology to implement the identified action fields. Based on this feedback, the authors developed a metamodel to describe the whole transformation process, which is outlined in Sect. 4, and evaluated it by interviewing the same workshop groups for a second time. Several adjuvant annotations of the first and second discussions have been considered by the authors and evaluated in a third interview round.

4   A Metamodel for Industrie 4.0 Transformation

"Business process improvement has been a perennial concern of companies ever since the industrial revolution began in the late eighteenth century" [7]. Because changes have always been a part of industrial development, the aim of process change management is to "provide a systematic approach for the consistent development of process models" [18]. In literature, there exists a number of models to guide and instruct the implementation of changes [9], exemplary Kotter's strategic eight-step model for transforming organizations [19], Jick's tactical ten-step model for implementing change [20], and General Electric (GE)'s seven step change acceleration process model consider this topic [9, 21]. Even though the defined steps of these approaches have diverging focal points and differ in the level of detail, they have in common that – first of all the – expected changes have to be defined [9]. Afterwards a vision has to be created based on these identified changes. Furthermore, the identified vision has to be implemented continuously [9]. Inspired by this approach, the authors aligned their Industrie 4.0 framework in the first level of managing change by structuring Industrie 4.0 impacts and providing a framework to section them. Furthermore, to manage change and to proceed in a structured way an overall concept is necessary [22]. For this purpose, the authors developed a method to structure the transformation process, which is based on the method engineering. Since only a limited number of scientific publications address the engineering process of methods [23], in this context, Gutzwiller [24] analyzed numerous publications on method engineering and derived five universal elements:

Activities: Activities have the objective to achieve one or more defined goals. They can be divided into sub-activities. By bringing the activities into a temporal and proper logical sequence, a procedure model can be provided.

Actors: When executing the activities, humans or committees have different roles, which comprise a combination of different activities.

Deliverables: Activities develop new deliverables or revise existing deliverables. Therefore, they may use previous activities as an input.

Metamodel: The metamodel of the method includes the individual components of the method, their dependencies as well as their semantics.

Technique: Techniques support the execution of the activities by giving guidance on the development of deliverables.

As Gutzwiller [24] recommended, the authors developed a metamodel of the method, visualized in Fig. 1, which contains the activities and the supporting techniques of the transformation process. The contents and dependencies of the metamodel have been developed in interaction with the annotations of the interviewed practitioners. For the metamodel, the paper uses Unified Modeling Language (UML) modelling to visualize the coherences of the different actions. Because the use of class diagrams is recommended in order to structure components of a system, the authors decided for this model notation [25]. Hereafter, the different activities of the metamodel are briefly described below:

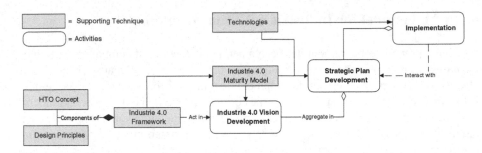

Fig. 1. Metamodel of the Industrie 4.0 transformation process

Development of an Industrie 4.0 scenario: The authors recommend referring to existing change management processes [9] to develop an Industrie 4.0 vision in workshops with practitioners, e.g. for an operative process. During the workshop it needs to be determined which areas within the framework of Industrie 4.0 are relevant. Therefore, the Industrie 4.0 framework supports to structure the impacts of Industrie 4.0. By using the framework, it can be determined how processes could look like in future and furthermore, which capabilities the organization, the technology and the humans have to provide and which potentials resultant. The deliverable of this activity is the Industrie 4.0 vision.

Development of a strategic plan: The next activity is to construct a strategic plan, as the deliverable of this step, for implementing the vision. This step should be supported by the levels of Industrie 4.0 in the maturity model and a technology tool, which provides specific technologies compatible to specific situations, e.g. autonomous driver assistance systems for the transportation of goods. The usage of an Industrie 4.0 maturity model is adjuvant; because maturity models are appropriate to asses, where the organization stands today in terms of Industrie 4.0 by providing steps in which the company can classify itself [26]. Thereby, the highest maturity level can be derived from the frameworks' action fields and provides thus ideas to shape the Industrie 4.0 vision.

Implementation: In the last step, the strategic plan should be implemented. The authors recommend to use an action plan and to interact continuously with the strategic plan to ensure that all ideas of the Industrie 4.0 vision are taken appropriately into consideration. The actors of the transformation method are the practitioners. The role of the actors has to be defined individually in each case depending on the process. Also depending on the process is the definition of concrete projects, e.g. the pilot testing of technologies.

5   Development of the Industrie 4.0 Framework

The Industrie 4.0 framework constitutes two main perspectives that need to be considered when analyzing the impact of Industrie 4.0 on organizations. The first perspective is the organizational perspective. Its purpose is to holistically analyze a company without disregarding important aspects. For the conceptual frame of this dimension, the authors chose the HTO concept, which will be described in Subsect. 5.1. The second dimension of the framework is Industrie 4.0 specific. For this dimension, the authors applied the four Industrie 4.0 design principles, which are explained in Subsect. 5.2.

5.1 Conceptual Frame of the HTO Concept

The conceptual idea of the HTO concept is that organizations can be described and analyzed by the interactions between the subsystems human (H), technology (T), and organization (O) [27–29]. This idea originates from the sociotechnical system theory [28], which implies that organizations consist of social and technical subsystems that are associated with each other via the task of the overall system [28, 30]. Sociotechnical system interact with their environment. They receive inputs from their environment and provide outputs for their environment as well [28, 31].

In addition to the classic sociotechnical system theory, the HTO concept includes humans as a separate subsystem. Within the HTO concept, humans are at the same level as technology and organization. This emphasizes the importance of humans [27]. Case studies have proven the approach of the HTO concept and showed that the performance of the overall system depends on how well the interactions between the three subsystems are [27]. Thus, the subsystems can be seen as enablers, which need certain capabilities to enhance the interactions between the subsystems. This enables potentials that may increase the productivity of the overall system. Consequently, when analyzing an organization all three subsystems need to be taken into account individually as well as their interaction with each other [28]. Figure 2 visualize these interactions between the subsystems. Because of the importance of the interactions, each interaction is described hereinafter.

Fig. 2. The HTO concept with its interactions (own representation based on [28])

Human–Technology Interaction: The task of the overall system determines the most appropriate differentiation of labor between humans and machines. This affects the degree of automation as well as how humans and machines need to interact with each other [28].

Human–Organization Interaction: The degree of automation specifies the kind of tasks humans needs to perform and therefore sets his correlating role within the organization [32].

Organization–Technology Interaction: It is important that the available technology meets the requirements coming from the organization and conduces to the strategic goals. Furthermore, it has to be ensured that the capabilities of the technology are used as beneficial as possible [28].

5.2   Industrie 4.0 Design Principles

The authors decided based on existing preparatory work and a qualitative literature review to clarify the term of Industrie 4.0 by using Industrie 4.0 design principles [6]. The four Industrie 4.0 design principles are described below.

Interconnection: In the Industrie 4.0 era, all members of an organization are connected with each other. Not only objects (i.e. machines, products, and devices) are connected over the Internet of Things (IoT) [33], furthermore, people are connected over the Internet of People (IoP) as well [6, 34].

Information transparency: The increasing number of interconnected objects and people [10] makes it possible to collect data coming from the physical world in real time [33]. By linking the data with digitalized models, a virtual copy of the physical world is created [6, 35].

Decentralized decisions: Since all objects and people in Industrie 4.0 scenarios can access all relevant data, they are empowered to make informed decisions on their own that contribute to the overall goal of the organization [36]. Objects and people perform their tasks as autonomous as possible. Only in case of exceptions, interferences, or conflicting goals, tasks are delegated to a higher level [6, 37].

Technical assistance: Caused by the increasing complexity of organizations, humans need to be assisted in their work activities. Virtual assistance systems provide humans the necessary information to make an informed decision and are equipped with context-sensitive user interfaces [38]. Additionally, physical assistance systems support humans by conducting a range of tasks that are unpleasant, too exhausting or unsafe for their human co-workers [6, 39, 40].

5.3   Design of the Industrie 4.0 Framework

As described above, the HTO concept can be utilized to represent the organizational dimension of the framework. By providing a frame to analyze an organization, the HTO concept helps to structure the Industrie 4.0 framework. These two areas are directly influenced by Industrie 4.0. As illustrated in Fig. 3, this results in new Industrie 4.0 capabilities and potentials.

Since the subsystems and their interactions influence each other, they cannot be analyzed separately [28]. As a result, the influence of the Industrie 4.0 design principles needs to be coherently analyzed on the subsystems as well as their interactions. This results in a matrix with 24 fields as shown in Fig. 4.

To describe the usage of the framework's methodological structure, the influence of the Industrie 4.0 design principle "decentralized decisions" on the subsystems human and technology as well as their interaction will be described exemplarily and are summarized in Table 1.

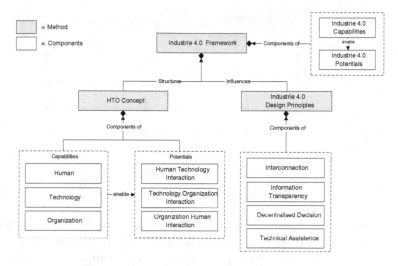

Fig. 3. Metamodel of the Industrie 4.0 framework

Table 1. Impact of "Decentralized decisions" on human, human-technology-interaction, and technology

	Human	Human-technology-interaction	Technology
Decentralized decisions	Control of complex manufacturing systems, responsibility of system operation, and strategic decision maker [38]	More efficient processes through a high degree of automation [5]	Autonomous decisions-making [2]
	Resolving of disturbance in case of failure [41]	Focus of employees on creative and value-added activities [2]	Cooperation with each other to reach common goals [33]

In Industrie 4.0, businesses will establish global networks where cyber-physical systems CPS (i.e. machinery, warehousing systems and production facilities) make decental decisions and cooperate self-regulated with each other [2]. This will lead to a further increase in automation [10], which will transform the role of employees significantly [38]. Hereinafter, the main aspects are described. The aspects are build up on the results of the interviews.

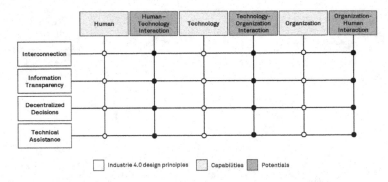

Fig. 4. Industrie 4.0 framework

Human: The increasing level of automation will shift the role of employees from an operator of machines towards a strategic decision-maker and a flexible problem solver [38]. This will transform employees' jobs and require new competence profiles [2]. In future, flexible and highly qualified employees will be necessary in order to resolve disturbance within the complex manufacturing systems in case of failure [41]. Therefore, it will be necessary to implement appropriate training strategies and to organize work in a way that enables lifelong learning of employees [2].

Technology: It will be necessary to equip manufacturing technologies and IT systems of existing facilities with CPS capabilities [2]. From a technical point of view, these capabilities are enabled by processing elements, network connections, sensors, and actuators [42]. These components allow CPS to make autonomous decisions in response to different situations [2] and to cooperate with each other to reach common goals [33].

Human–Technology-Interaction: The described capabilities of humans and technology will lead to a higher level of automation and will enable more efficient processes [5]. Thus, employees can be released from executing routine tasks and further focus on creative and value-added activities [2].

6   Conclusion and Outlook

The introduced methodology provides a conceptual overview of how the transformation of organizations towards Industrie 4.0 can be organized. For practitioners this constitutes a structure for approaching Industrie 4.0 within their companies. In further research, the additional techniques of the method need to be developed and verified within case studies. The presented framework is the first step within this method. It can be utilized to gain a vision of how organizations and processes will look like in the future. This foresight is an essential basis for the Business Process Change Management [9]. The framework allows researchers to organize and to classify the existing knowledge about Industrie 4.0 and to structure the impacts on organizations. However, within this work only the impact of the design principle "decentralized decisions" has been investigated. In order to further elaborate the framework, the other three design principles have to be examined as well to gain a holistic approach. By developing the content of the framework, 24 Industrie 4.0 action fields will be created that describe how Industrie 4.0 will influence the different areas of organizations. By combining these 24 action fields an Industrie 4.0 vision can be derived. As both the components of the HTO concept as well as the Industrie 4.0 design principles have interdependencies it further needs to be investigated how the 24 action fields are interconnected. Researchers are welcome to use the framework to structure their Industrie 4.0 knowledge and to investigate whether there might be further research gaps within the Industrie 4.0 action fields.

References

1. Drath, R., Horch, A.: Industrie 4.0: hit or hype? IEEE Ind. Electron. Mag. **8**(2), 56–58 (2014)
2. Kagermann, H., Wahlster W., Helbig, J. (eds.): Recommendations for Implementing the Strategic Initiative Industrie 4.0: Final Report of the Industrie 4.0 Working Group. Frankfurt (2013)
3. Bauernhansl, T., Ten Hompel, M., Vogel-Heuser, B. (eds.): Industrie 4.0 in Produktion, Automatisierung und Logistik. Springer, Wiesbaden (2014)
4. Bischoff, J. (ed.): Erschließung der Potenziale der Anwendung Industrie 4.0 für den Mittelstand (2015)
5. Bauer, W., Schlund, S., Marrenbach, D., Ganschar, O.: Industrie 4.0 – Volkswirtschaftliches Potenzial für Deutschland, Berlin (2014)
6. Hermann, M., Pentek, T., Otto, B.: Design principles for Industrie 4.0 scenarios. In: Proceedings of the 49th Annual Hawaii International Conference on System Sciences (HICSS), pp. 3928–3937. Computer Society Press (2016)
7. Harmon, P.: Business Process Change, 3rd Edition Provides a Balanced View of the Field of Business Process Change. Morgan Kaufmann Publishers, San Francisco (2014)
8. Binner, H.: Industrie 4.0 bestimmt die Arbeitswelt der Zukunft. Elektrotech. Informationstechnik **131**(7), 230–236 (2014)
9. Mento, A.J., Jones, R.M., Dirndorfer, W.: A change management process: grounded in both theory and practice. J. Change Manage. **3**(1), 45–59 (2002). Henry Stewart Publications
10. Lasi, H., Kemper, H.-G.: Industry 4.0. Bus. Inf. Syst. Eng. Int. J. Wirtschaftsinformatik **6**(4), 239–242 (2014)
11. ZVEI - German Electrical and Electronic Manufacturers' Association, Industrie 4.0 (eds.): Reference architecture model Industrie 4.0 (RAMI4.0) (2015). http://www.zvei.org/Downloads/Automation/5305%20Publikation%20GMA%20Status%20Report%20ZVEI%20Reference%20Architecture%20Model.pdf
12. Industrial Internet Consortium (IIC), Industrial Internet Reference Architecture: Technical report (2015)
13. Garcia, D., Gluesing, J.C.: Qualitative research methods in international organizational change research. J. Organ. Change Manage. **26**(2), 423–444 (2013)
14. Aquinas, P.G.: Organizational Behavior: Concepts, Realities, Applications and Challenges. Excel Books India, New Delhi (2006)
15. Gregor, S.: Building theory in the sciences of the artificial. In: Vaishanvi, V. (ed.) Proceedings of the 4th International Conference on Design Science Research in Information Systems and Technology, pp. 1–18. Malvern, PA (2009)
16. King, N.: The qualitative research interview. In: Cassel, C., Symon, G. (eds.) Qualitative Methods in Organizational Research: A Practical Guide (1994)
17. Dicicco-Bloom, B., Crabtree, B.F.: The qualitative research interview. Med. Educ. **40**, 314–321 (2006)

18. Gerth, C.: Business Process Models - Change Management. Springer, Heidelberg (2013)
19. Kotter, J.P.: Why transformation efforts fail. Harvard Bus. Rev. **74**(2) (1995). (Reprint No. 95204)
20. Jick, T.: Note on the Recipients of Change. Harvard Business School Press, Boston (1991). Note 9-491-039
21. Garvin, D.: Learning in Action: A Guide to Putting the Learning Organisation to Work. Harvard Business School Press, Boston (2000)
22. Cameron, E., Green, M.: Making Sense of Change Management. Kogan Page, London (2009)
23. Höning, F.: Methodenkern des Business Engineering - Metamodell, Vorgehensmodell, Techniken, Ergebnisdokumente und Rollen, St. Gallen, Universität, Institut für Wirtschaftsinformatik. Dissertation (2009)
24. Gutzwiller, T.: Das CC RIM-Referenzmodell für den Entwurf von betrieblichen, transaktionsorientierten Informationssystemen. Physica, Heidelberg (1994)
25. De Bruin, T., Freeze, R., Kulkarni, U., Rosemann, M.: Understanding the main phases of developing a maturity assessment model. In: ACIS 2015 Proceedings. Paper 109 (2005)
26. Rupp, C., Queins, S.: UML 2 glasklar. Praxiswissen für die UML-Modellierung. Hanser Verlag, München (2012)
27. Karltun, A., Karltun, J., Eklund J., Berglund, M.: HTO-a complementary ergonomics perspective. In: Human Factors in Organizational Design and Management–XI Nordic Ergonomics Society Annual Conference–46 (2014)
28. Ulich, E.: Arbeitssysteme als Soziotechnische Systeme – eine Erinnerung. Journal Psychologie des Alltagshandelns **6**, 4–12 (2013). Innsbruck university press, Innsbruck
29. Bullinger, H.-J., Spath, D., Warnecke, H.-J., Westkämper, E. (eds.): Handbuch Unternehmensorganisation, Strategien, Planung, Umsetzung. Springer, Heidelberg (2009)
30. Fox, W.M.: Sociotechnical system principles and guidelines: past and present. J. Appl. Behav. Sci. **31**(1), 91–105 (1995)
31. Grote, G., Künzler, C.: Sicherheitskultur, Arbeitsorganisation und Technikeinsatz. Vdf Hochschulverlag AG, Zürich (1996)
32. Blumberg, M.: Towards a new theory of job design. In: Karwowski, W., Parsaei, H.R., Wilhelm, M.R. (eds.) Ergonomics of Hybrid Automated Systems I, pp. 53–59. Elsevier, Amsterdam (1988)
33. Giusto, D., Iera, A., Morabito, G., Atzori, L. (eds.): The Internet of Things. Springer, New York (2010)
34. Vilarinho, T., Farshchian, B.A., Floch, J., Mathisen, B.M.: A communication framework for the internet of people and things based on the concept of activity feeds in social computing. In: 9th International Conference on Intelligent Environments (2013)
35. Kagermann, H.: Change through digitization: value creation in the age of Industry 4.0. In: Albach, H., Meffert, H., Pinkwart, A., Reichwald, R. (eds.) Management of Permanent Change, pp. 23–45. Springer, New York (2015)
36. Malone, T.W.: Is 'empowerment' just a fad? Control, decision-making, and information technology. BT Technol. J. **17**(4), 141–144 (1999)
37. Ten Hompel, M., Otto, B.: Technik für die wandlungsfähige Logistik: Industrie 4.0. In: 23. Deutscher Materialfluss-Kongress (2014)
38. Gorecky, D., Schmitt, M., Loskyll, M., Zühlke, D.: Human-machine-interaction in the Industry 4.0 era. In: 12th IEEE International Conference on Industrial Informatics (INDIN), pp. 289–294 (2014)
39. Awais, M., Henrich, D.: Human-robot interaction in an unknown human intention scenario. In: 11th International Conference on Frontiers of Information Technology, pp. 89–94 (2013)
40. Kiesler, S., Hinds, P.: Human-robot interaction. Hum. Comput. Interact. **19**(1 & 2), 1–8 (2004)

41. Hirsch-Kreinsen, H.: Wandel von Produktionsarbeit: Industrie 4.0. Soziologisches Arbeitspapier 38/2014 (2014)
42. Rajkumar, R., Lee, I., Sha L., Stankovic, J.: Cyber-physical systems: the next computing revolution. In: Design Automation Conference 2010, Anaheim, California, USA, pp. 731–736 (2010)

A Generic Process Data Warehouse Schema for BPMN Workflows

Thomas Benker[✉]

Chair of Business Information Systems – Systems Engineering
Otto-Friedrich-University of Bamberg, An der Weberei 5, 9047 Bamberg, Germany
thomas.benker@uni-bamberg.de

Abstract. Companies in dynamic environments have to react to certain market events. Reactions can be short-term and influence the behavior of running process instances or they can be mid-term or long-term and cause the redesign of the process. In both situations, insights into the process flow are necessary and provided by Process Data Warehouse Systems. This paper proposes to derive the data warehouse structures from the meta model of the BPMN (Business Process Model and Notation), the actual de-facto standard of workflow languages. The resulting data structure is generic in order to be portable between application domains and to be stable in case of changing workflows.

Keywords: Process data warehouse system · Data warehouse schema · Business Process Model and Notation · Workflow

1 Introduction

Data Warehouse Systems (DWH systems) are established in most companies. They are used to support strategical and tactical decisions at managerial level by providing relevant multidimensional information. The multidimensional data model enables the aggregation and analysis of quantitative measures (e.g., number of sales) along qualitative dimensions (e.g., region, customer group or time). The schema design is aligned to certain areas of business decisions and analysis. The measures are provided aggregated and abstracted from concrete business transactions or processes.

Companies in highly competitive and dynamic markets often have to react appropriately and with an adequate latency to changing conditions [1]. Process Data Warehouse Systems (PDWH systems), a specialization of subject-oriented DWH systems, are appropriate to optimize business processes and operations on a daily or intraday basis [2]. They can be used to support the identification of changing conditions as well as the design of adequate reactions by providing insights into the processes. The difference to the subject-oriented DWH concept is shown in Fig. 1. A PDWH system focuses on a certain process type while subject-oriented DWH systems are providing data abstracted from process type information. This is also true of the data instances. A PDWH system has a lower aggregation level and explicitly provides data that is related to a process and its behavior. Processes are executed by Workflow Management Systems (WfMS). These systems also keep track of the execution and enable some runtime monitoring and

© Springer International Publishing Switzerland 2016
W. Abramowicz et al. (Eds.): BIS 2016, LNBIP 255, pp. 222–234, 2016.
DOI: 10.1007/978-3-319-39426-8_18

restricted analytical functions at a technical level. But they are restricted in case of business level analysis in order to support the following PDWH scenarios: The behavioral (operational) scenario describes short-term reactions that influence the behavior of running process instances. The structural scenario describes the mid-term or long-term redesign of the process structure based on data of finished instances [3]. The process relationship is the common basis as well as the multidimensional data structuring. E.g., [4, 5] motivate multidimensional structures for the structural scenario. For the behavioral scenario the multidimensional data model is proposed to analyze process events in a historical context [6, 7]. Of course, the data structure has to be extendable with scenario-specific information.

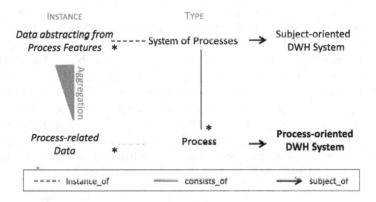

Fig. 1. Distinction of subject- and process-oriented DWH systems

The term *process* is rather unspecific in the context of PDWH systems. The term *workflow* is more precise and better suited. A workflow is understood as a special kind of process that is designed to be executed by humans or WfMS. Basically, it is described by activities and their relationship [8]. Workflow languages are close to the executing systems and for this appropriate as basis to derive multidimensional structures. The Business Process Model and Notation (BPMN) [9] is one of the dominant workflow modeling languages [8]. The OMG (Object Management Group) standard is widespread and accepted by modelers and tool developers [10]. But the BPMN has not been considered so far in multidimensional process data structures.

The goal of this paper is to present a multidimensional data structure in order to realize the following requirements: First, the data structure should support decision making in the behavioral and the structural PDWH scenario. Second, it should support the workflow language BPMN. And finally, the multidimensional structure should be flexible in order to react on dynamically changing workflow schemas. The strategy to realize these requirements is to derive the multidimensional data structure based on the BPMN meta model. Of course, it has to be enhanced with business information. Further, the data structure will be specified to be generic. This means that it is independent of the workflow schema. First, this enables the application in heterogeneous business domains (e.g. sales or human resources). Second, the redesign of a workflow schema due to changing market conditions does not imply the redesign of the DWH schema.

The benefit is that existing reports are repeatable and results are reproducible. This is a key feature of DWH systems (e.g. in case of compliance checking). The redesign of DWH structures often has a negative influence on repeatability and reproducibility.

To introduce the generic PDWH schema, the paper is structured as follows: First, the relevant basics of the BPMN are introduced. Section 3 is a discussion of related work. Section 4 presents the concept of the generic PDWH schema that is applied and demonstrated in Sect. 5 on a real-world case study. Finally, Sect. 6 summarizes and reflects the concept and gives an outlook on future work.

2   The Business Process Model and Notation

PDWH systems focus on process behavior and interaction between processes. The BPMN provides *process* (workflows) and *collaboration diagrams* for this purpose [11]. The other diagram types are not relevant for the research problem. The BPMN also defines conformance classes in order to determine the conformance between modeling tools and the specification. To control the complexity of the BPMN in this work, the focus of the presented concept is reduced to the elements of the *descriptive process modeling conformance* class. This class defines the basic concepts for process and collaboration modeling. Section 6 shortly explains the extension of the concept to support the *full process modeling conformance*. The meta model for the conformance class is based on the BPMN specification [9] and shown in Fig. 2.

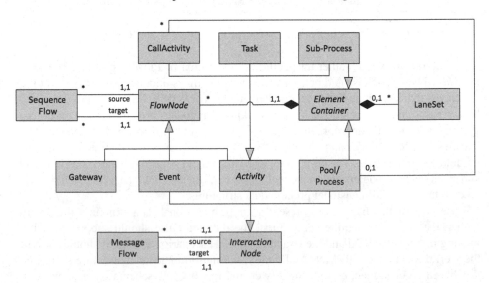

Fig. 2. The relevant part of the BPMN meta model (representation based on [9])

Flow Nodes [9, 11]: Flow Nodes are the building blocks of BPMN process models. *Activities* are used to model working steps. A *Task* represents an atomic working step that can be described more closely as *user, service, send* or *receive task*. A *Sub-Process* is a non-atomic working step and encapsulates a process itself. A *Call Activity* is used to model a callable and reusable process. *Gateways* control the flow of the process. An *exclusive* gateway splits/joins alternating sequence flows. *Parallel* gateways create/synchronize parallel sequence flows. *Events* of a process are differentiated by their position (*start, intermediate, end*) and their trigger. Relevant events are the *empty start/end event*, the *message start/end event*, the *timer start event* and the *terminate end event*. Sub-types of *Task, Event* and *Gateway* are not shown in Fig. 2.

Connecting Objects [9, 11]: The *Sequence Flow* is used to model the process behavior. A Sequence Flow element always runs from a start to a target *Flow Node*. The interaction between processes is modeled by *Message Flows* with optionally annotated message descriptions.

Lanes [9, 11]: A process is contained within a *Pool* that represents a participant of a collaboration. A Pool can be hierarchical structured by *Lanes*. Typically, they are used to model roles or responsibilities within a process.

The *Data Objects* of the BPMN specification are left out of this work. The BPMN semantics of the term *Data* are not precise enough and because of this not operable. The element *Data* is not only used to annotate input/output data. Even more, it is used to assign physical objects and products. Further, the BPMN does not provide the possibility to specify data structures like UML Class Diagrams or Entity Relationship Models. But such data structures could be important to identify business dimensions. Because of this, Sect. 5.2 demonstrates the derivation of business dimensions based on object-oriented operational structures.

3   State of the Art

A number of publications are presenting multidimensional data structures for workflow schemas. A first group of publications base their concepts on informal [12] or proprietary [4, 13] workflow specifications. The authors of [14] present a generic multidimensional schema. The used workflow specification is kept abstract. The proprietary workflow specifications are often incomplete, restricted and only for theoretical usage. A second group of concepts is defined for certain application domains. Multidimensional data structures for surgery workflows are presented in [5]. In [15] parts of multidimensional data structures for service and sales processes are introduced. Because of their domain-specific contexts, the portability to other application domains is restricted. The authors of [16] base their concept on the workflow specification of the WfMC (Workflow Management Coalition). The multidimensional data structures are not specified in general terms. They are designed to support assumed queries. The data structure of [17] abstracts from application domain and process language. It is designed for the identification of information requirements based on generated process data. Due to the abstraction from language specifics it is limited for tracking and analyzing relevant processes data at runtime.

The presented approaches support heterogeneous workflow concepts as dimensions (Table 1). Only few approaches support the analysis of the hierarchical structures or of the sequence flow of the workflows. Data usage is mostly considered rudimentary or only domain-specific. The dimensions *Time, Actor* and *Instance* (Assignment of node instances to workflow instance) are supported in all proposals. Yet, the BPMN has not been considered in related work. Compared to the BPMN, the used workflow specifications and multidimensional structures are restricted. E.g., the interaction between participants/processes is not subject of any of the concepts. Furthermore, the related multidimensional structures are not defined based on a formal specification and, for this, could not be checked for completeness. A further feature of the BPMN is its clearly defined and human readable graphical syntax. This is important for process redesign in order to localize workflow elements that, for example, have been identified to be inefficient. The related publications do not discuss workflow specifications with an appropriate graphical syntax compared to the BPMN.

Table 1. Supported dimensions in related concepts

	[4]	[5]	[12]	[13]	[14]	[15]	[16]	[17]
Instance	X	X	X	X	X	X	X	X
Hierarchy		X	X				X	X
Sequence		X			X			X
State		X	X	X	X		X	X
Time	X	X	X	X	X	X	X	X
Actor	X	X	X	X	X	X	X	X
Organization	X		X	X		X	X	X
Data		dom	X	X	dom	dom		X
Events		X	X					

X → *feature is realized*; dom → *domain-specific realization*

4   The Generic Process Data Warehouse Schema

The PDWH schema is derived and justified based on the meta model of the BPMN (Fig. 2). The data instances of a PDWH schema should be provided not aggregated as they are available at the source system (workflow engine) [18, 19] in order to flexibly realize ad-hoc queries. Nodes (Task, Gateway, Event) are the low-level elements of the BPMN and, for this, the lowest level for dimensions of the multidimensional structure. Figure 3 shows the resulting dimensions that are derived based on the element *Flow Node* of the meta model. For representing the multidimensional schema, the *Semantic Data Warehouse Model* (SDWM) [20] is used. A dimension is shown with all of its hierarchy levels and the aggregation relationships between these levels. The highest level (*Total*) means a full aggregation of the dimension hierarchy. Normally, there is a n:1-aggregation relationship between a hierarchy level and the next higher one. As graphical representation crow's feet are used at the n-side. The SDWM assumes that a hierarchy level is implicitly described by ID and a semantic attribute.

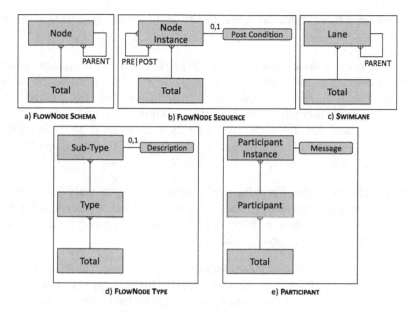

Fig. 3. BPMN-specific dimensions

Scope of the PDWH schema is a *BPMN process* (workflow). It is embedded within a *collaboration diagram* to show the interaction with other participants. A multidimensional data structure covers the process of one participant. The other participants are regarded as black boxes. Their processes can be subject of separated PDWH schemas.

FlowNode Schema: The dimension *FlowNode Schema* (Fig. 3a) is used to make the workflow and nodes analyzable at schema level. At the lowest level of this dimension the semantic of Node is its name within the workflow schema. Nodes may be hierarchically structured (*Parent*-relationship). E.g., a Task may be part of a Sub-Process and the Sub-Process may also be part of a higher Sub-Process itself. So the dimension is unbalanced at the instance level. For the *FlowNode Schema* dimension, the *Total* level means the aggregation up to the whole workflow.

FlowNode Sequence: The behavior of a certain workflow instance is tracked and analyzable through the dimension *FlowNode Sequence* (Fig. 3b). A *Node Instance* can have zero or multiple pre- and successors (pre/post-relationship). Multiple successors are, for example, the result of parallel gateways. This enables the aggregation along the behavioral path of a workflow instance. An aggregation over the whole instance is modeled by the *Total* dimension level. The *Post Condition* is a dimensional attribute that is especially useful to understand gateway behavior.

Swimlane: The dimension *Swimlane* (Fig. 3c) is derived from the meta element *LaneSet*. It enables the analysis of organizational structures within a participants workflow schema. In a collaboration diagram, a node is assigned to a *Lane*. *Swimlanes* are hierarchically structured with an a priori unknown depth. This is considered by the parent relationship.

FlowNode Type: The dimension *FlowNode Type* (Fig. 3d) is a generic dimension to consider the node types and enables the aggregation along the BPMN-type hierarchy (generalization of *Flow Node* for *Event*, *Activity*, and *Gateway* in Fig. 2). The instance level of the dimension is common for all BPMN workflows. The aggregation of the *Sub-Types Exclusive* and *Parallel* Gateway to the *Type Gateway* is an example for the instance level. The attribute *description* could be used (0,1-relationship) to provide further information (e.g., *message* as trigger of a *Start Event*).

Participant: The dimension *Participant* (Fig. 3e) is necessary to analyze the interaction within a collaboration between the instance of a workflow and of related participants. It is closer described by the message exchange. The dimension enables the aggregation of participant instances to types and the *Total* level.

To enable the aggregation of all workflow instances, a dimension *Process Instance* is needed that is not derived from the meta model. This also applies to the obligatory dimension *Time* of DWH schemas, which enables the analysis of workflow executions on the time axis.

Finally, measures have to be identified for the multidimensional schema. Following the argumentation of BOEHNLEIN [20], measures of a DWH system are defined to assess certain objectives. The objectives for the structural DWH scenario (process design) that could be assessed in a workflow schema are (i) short process runtime, (ii) low process costs, and (iii) high process quality [21]. The measures *time* and *cost* are appropriate to assess objective (i) and (ii). For the assessment of the process quality a number of publications propose static procedures [22]. A PDWH system could be used additionally to analyze the true behavior at runtime. The analysis of normal and exceptional behavior may be an example. So, a counting measure (*number* of executions) is necessary to track the executed activities. Based on these measures, further case-specific measures could be derived. Measures for the behavioral scenario are application-domain dependent and cannot be specified in general terms.

5   Application of the Process Data Warehouse Schema

Section 5 demonstrates the application of the generic multidimensional data structure on an extract of a real-world case study (Sect. 5.1). Section 5.2 shows the extension of the data structure by business dimensions to support the structural DWH scenario.

5.1   Application of the Generic Data Structure

The generic PDWH schema is demonstrated by application to the sales process of an e-car (electronic car) rental company. The sales process is specified according to the case study presented in [23]. Figure 4 shows an extract of the BPMN sales workflow.

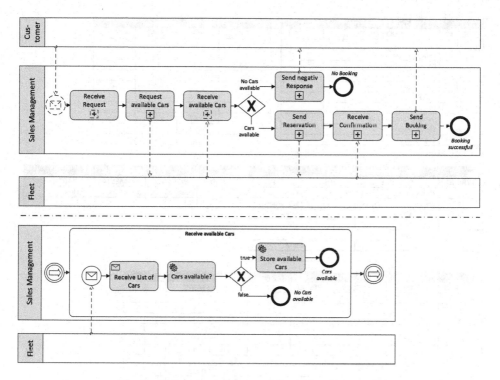

Fig. 4. Sales management workflow for e-Car rental

The upper part of Fig. 4 shows the overall process starting with the receipt of a customer request and ending with successful or non-successful booking. The availability of e-Cars is the crucial success factor. The process is specified within a collaboration diagram and interacts with the black box participants *Customer* and *Fleet Management System*. The main activities are modeled as embedded and collapsed Sub-Processes. The lower part of Fig. 4 shows the insight of the Sub-Process *Receive available Cars*. The *Fleet* participant sends a list of available cars. Only if cars are available, they are persisted locally and the Sub-Process ends up in the final state *Cars available*. Otherwise, the final state is *No Cars available*.

Figure 5 illustrates the application at instance level using tables according to the star schema paradigm of relational DWH systems. The *Facts* table combines the dimensions by foreign key relationships and has the attributes for the measures. The example shows the measure n (number of executions). The dimensions for process instance (*process*) and time (*timestamp*) are not shown explicitly. They are only considered by the attributes *instance* and *timestamp*.

FlowNode Schema		
NODE-ID	NODE-DESC	PARENT
[...]	[...]	[...]
2	Request available Cars	-
3	Receive abailable Cars	-
4	Decision Cars available	-
301	Event Receive List of Cars	3
302	Receive List of Cars	3
303	Cars available?	3
304	Decision Cars available	3
305	Store available Cars	3
306	Event Cars available	3
[...]	[...]	[...]

FlowNode Sequence			
INSTANCE	PRE	POST	P.COND.
[...]	[...]	[...]	[...]
2-1	[...]	3-1	-
3-1	2-1	4-1	-
4-1	3-1	[...]	[...]
301-1	-	302-1	-
302-1	301-1	303-1	-
303-1	302-1	304-1	-
304-1	303-1	305-1	true
305-1	304-1	[...]	[...]
[...]	[...]	[...]	[...]

FlowNode Type			
TYPE-ID	SUBTYPE	DESC.	TYPE
T1	start_event	message	event
T2	embedded_sub_process	-	sub_process
T3	exclusive_gateway	-	gateway
T4	receive_task	-	task
T5	service_task	-	task
T6	send_task	-	task
[...]	[...]	[...]	[...]

Participant			
P-ID	INSTANCE	P-MESSAGE	PARTICIPANT
P1	Fleet-App	List of Cars	Fleet
P2	[...]	[...]	[...]

Facts						
NODE-ID	INSTANCE	PROCESS	TYPE-ID	P-ID	TIMESTAMP	N
[...]	[...]	[...]	[...]	[...]	[...]	1
2	2-1	1	T2	-	2015-12-13T13:15:20	1
3	3-1	1	T2	-	2015-12-13T13:15:50	1
4	4-1	1	T3	-	2015-12-13T13:18:45	1
301	301-1	1	T1	P1	2015-12-13T13:15:50	1
302	302-1	1	T4	-	2015-12-13T13:15:52	1
303	303-1	1	T5	-	2015-12-13T13:16:40	1
304	304-1	1	T3	-	2015-12-13T13:17:50	1
305	305-1	1	T5	-	2015-12-13T13:17:57	1
[...]	[...]	[...]	[...]	[...]	[...]	1

Fig. 5. The process data warehouse schema of the sales management workflow

A node instance (data set of the *Facts* table) is identified by the attribute *instance* and is assigned to a process instance by the attribute *process* (e.g., instance = 301-1). The table *FlowNode-Type* shows for instance 301-1 (type-id = T1) that it is of the Sub-Type *message start event* and could be aggregated to the Type *event*. For the same node instance the interaction with the participant instance *Fleet-App* (p-id = 1) is associated in table *Participant*. The message exchange is described as *List of Cars*. Node instance 301-1 is further defined by the schema element 301. The table *FlowNode Schema* shows the semantic description of the event *Receive List of Cars* as well as its integration in the workflow hierarchy. It is a child of node 3, the Sub-Process *Receive available Cars*. The behavior of process instance 1 is shown in table *FlowNode Sequence*. E.g., node instance 304-1 is an exclusive gateway. The decision of this gateway in process instance 1 is true, meaning cars are available and node instance 305-1 is executed next.

It is also possible to manage loops in a workflow schema. Each time a node is accessed in a workflow instance, a new node instance is created and put in the *Facts* table. The timestamp and the counter within the node instance description enable to keep track of the execution sequence of the node instances within the loop. The presented data structure is, due to its generic character, able to deal with changes within the workflow schema. Changes in workflow hierarchy and flow node semantics are realized by new instances within the dimension *FlowNode Schema*. Changes in the sequence of the workflow schema are tracked with the workflow instances in the dimension *Flow-Node Sequence*.

5.2 Domain-Specific Extension of the Multidimensional Data Structure

For comprehensive analysis it is necessary to add business-specific dimensions and measures. Customer or sales dimensions would be useful for the presented case study. For example, the sales process could be executed by call center or web shop. The sales dimension would allow to compare performance measures between these types. A customer dimension could show differences in the sales process between young and older customers. Of course, these dimensions could not be derived from the meta model of the BPMN. As mentioned in Sect. 3, the BPMN *Data* semantics are not useful for the derivation of dimensions. According to [20], we suggest to identify business dimensions based on object-oriented diagrams. As mentioned yet, data structures are not provided by the BPMN and have to be specified additionally. The author of [20] gives hints in order to reveal business-specific dimensions in an object-oriented data schema. On the left side Fig. 6 shows an example of the object-oriented structures of the case study. As stated, a *Sales Channel* could be an *e-Shop* or a *CallCenter*. A *CallCenter* is the composition of *CallCenter Employees*.

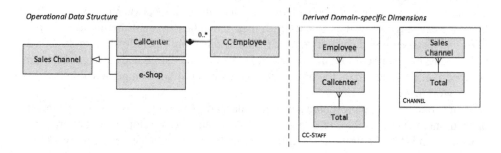

Fig. 6. Domain-specific dimensions of the case study

The right side of Fig. 6 shows possible business dimensions. The generalization of *CallCenter* and *e-Shop* to Sales Channel results in a dimension *Channel*. E.g., it enables to differentiate performance measures of activities/workflows by the sales channel type (*e-Shop* vs. *CallCenter*). The dimension CC-Staff results from the composition relationship of *CallCenter* and *Employee*. Measures could be analyzed at the grain of single employees or for different engaged call center providers.

Generalizations or compositions in object-oriented data structures are just hints for the modeler to identify dimension structures. Of course, domain knowledge and experience of the modeler are important for this task.

6 Summary, Reflection and Future Work

This paper presents a generic Data Warehouse Schema to support process-related decision making. *Generic* means that it is suitable for BPMN processes independent of its application context and stable in case of changing workflow schemas.

As mentioned initially, the work is restricted to the *descriptive modeling conformance* class. Only one additional attribute of the *Facts* table (*association*) is necessary to support complex BPMN elements of the *full process modeling conformance* as *Compensations*. A number of elements are just supported by the presented data structure, e.g. further sub-types of gateways (*complex, event-based*) and events (*intermediate*). Other elements are understood as alternating representations of supported modeling scenarios (e.g., looping activities).

According to [24], process warehousing approaches can be reflected by their support of the process modeling perspectives of [25]: *functional, behavioral, organizational,* and *informational*; subsequently, the presented approach is also reflected based on these perspectives. The initial classification of the BPMN in [26] is used as basis. The *functional* perspective is realized by the dimension *FlowNode Schema*, showing the hierarchical structure and the semantical names of the Tasks. The dimension *FlowNode Sequence* and the dimension *FlowNode Type* build up the *behavioral* perspective. Using Pools and Lanes to specify responsibilities enables the *organizational* perspective. The differentiation between *Service Tasks* and *User Tasks* is also understood as an organizational aspect, as well as the consideration of the interaction between participants. The message exchange as well as Events with their semantics are contributing to the *informational* perspective. The derivation of dimensions from operational data structures (UML class diagram) complements the BPMN-based dimensions and the *organizational* perspective. To summarize, the presented concept can provide comprehensible insights into a workflow.

For future work, the presented schema is planned to be integrated in a holistic development method that enables two features: (i) the integrated development of operational workflow and PDWH structures; (ii) the interpretation of multi-dimensional workflow data in business semantics. The idea is to interpret conceptual process models as business views on more technically oriented workflow schemas.

References

1. Dorn, C., Burkhart, T., Werth, D., Dustdar, S.: Self-adjusting recommendations for people-driven ad-hoc processes. In: Hull, R., Mendling, J., Tai, S. (eds.) BPM 2010. LNCS, vol. 6336, pp. 327–342. Springer, Heidelberg (2010)
2. Panian, Z.: Delivering actionable business intelligence through service-oriented architecture. In: Cordeiro, J., Filipe, J. (eds.) Proceedings of the 10th International Conference on Enterprise Information Systems, pp. 191–194 (2008)
3. Benker, T.: Konzeption einer Komponentenarchitektur für prozessorientierte OLTP- and OLAP-Anwendungssysteme. In: Bamberger Beiträge zur Wirtschaftsinformatik und Angewandten Informatik, vol. 97. Bamberg (2015)

4. Eder, J., Olivotto, G.E., Gruber, W.: A data warehouse for workflow logs. In: Han, Y., Tai, S., Wikarski, D. (eds.) EDCIS 2002. LNCS, vol. 2480, pp. 1–15. Springer, Heidelberg (2002)
5. Mansmann, S., Neumuth, T., Scholl, M.H.: OLAP technology for business process intelligence: challenges and solutions. In: Song, I.-Y., Eder, J., Nguyen, T.M. (eds.) DaWaK 2007. LNCS, vol. 4654, pp. 111–122. Springer, Heidelberg (2007)
6. Eckerson, W.: Best practices in operational BI: converging analytical and operational processes. TDWI Best Practices Report, TDWI (2007)
7. Russom, P.: Operational data warehousing. The integration of operational applications and data warehouses. TDWI Best Practices Report, TDWI (2011)
8. Pütz, C., Sinz, E.J.: Model-driven derivation of BPMN workflow schemata from SOM business process models. Enterp. Model. Inf. Syst. Architect. 5, 57–72 (2010)
9. OMG: Business process model and notation (BPMN). Version 2.0. http://www.omg.org/spec/BPMN/2.0/PDF
10. Silver, B.: BPMN Method and Style: A Structured Approach for Business Process Modeling and Implementation Using BPMN 2. Cody-Cassidy Press, Aptos (2011)
11. Weske, M.: Business Process Management: Concepts, Languages Architectures. Springer, Berlin (2012)
12. Pau, K., Si, Y., Dumas, M.: Data warehouse model for audit trail analysis in workflows. In: Cheung, S.C., Li, Y., Chao, K.-M., Younas, M., Chung, J.-Y. (eds.) Proceedings of IEEE International Conference on e-Business Engineering. IEEE (2007)
13. Grigori, D., Casati, F., Castellanos, M., Dayal, U., Sayal, M., Shan, M.-C.: Business process intelligence. Comput. Ind. 53, 321–343 (2004)
14. Casati, F., Castellanos, M., Dayal, U., Salazar, N.: A generic solution for warehousing business process data. In: Koch, C. et al. (eds.) Proceedings of the 33rd International Conference on Very Large Data Bases, pp. 1128–1137. ACM, New York, NY, USA (2007)
15. Kueng, P., Wettstein, T., List, B.: A holistic process performance analysis through a performance data warehouse. In: Proceedings of the American Conference on Information Systems, pp. 349–356 (2001)
16. List, B., Schiefer, J., Tjoa, A., Quirchmayr, G.: Multidimensional business process analysis with the process warehouse. In: Abramowicz, W., Zurada, J. (eds.) Knowledge Discovery for Business Information Systems, pp. 211–227. Springer, Boston (2000)
17. Sun, X.: Ein szenario- und prototypingbasiertes Konzept zur Informationsbedarfsanalyse für Business-Process-Intelligence-Systeme. Entwicklung und Evaluation. Eul, Lohmar (2014)
18. Baars, H., Sun, X.: Wo sind die Klippen im Prozess? Business process intelligence - das unterschätzte Potenzial. BI Spektrum 8, 10–12 (2013)
19. Inmon, W.H.: Operational and informational reporting. http://www.information-management.com/issues/20000701/2349-1.html
20. Böhnlein, M.: Konstruktion semantischer Data-Warehouse-Schemata. Dt. Univ.-Verl, Wiesbaden (2001)
21. Fischermanns, G.: Praxishandbuch Prozeßmanagement. Schmidt, Gießen (2013)
22. Overhage, S., Birkmeier, D.Q., Schlauderer, S.: Quality marks, metrics, and measurement procedures for business process models. Bus. Inf. Syst. Eng. 4, 229–246 (2012)
23. Leunig, B., Wagner, D., Ferstl, O.K.: E-Car-Szenario - Hochflexible Geschäftsprozesse in der Logistik als Integrationsszenario für den Forschungsverbund forFLEX. In: Sinz, E.J., Bartmann, D., Bodendorf, F., Ferstl, O.K. (eds.) Dienstorientierte IT-Systeme für hochflexible Geschäftsprozesse, pp. 15–38. University of Bamberg Press, Bamberg (2011)

24. Shahzad, K., Johannesson, P.: An evaluation of process warehousing approaches for business process analysis. In: Barjis, J., Kinghorn, J., Ramaswamy, S. (eds.) Proceedings of the International Workshop on Enterprises and Organizational Modeling and Simulation, pp. 1–14. ACM, New York, NY, USA (2009)
25. Curtis, B., Kellner, M., Over, J.: Process modeling. Comm. ACM **35**, 75–90 (1992)
26. Korherr, B.: Business Process Modelling - Languages, Goals and Variabilities (2007)

Business and Enterprise Modeling

Discovering Decision Models from Event Logs

Ekaterina Bazhenova[(✉)], Susanne Buelow, and Mathias Weske

Hasso Plattner Institute at the University of Potsdam, Potsdam, Germany
{ekaterina.bazhenova,mathias.weske}@hpi.de,
susanne.buelow@student.hpi.de

Abstract. Enterprise business process management is directly affected by how effectively it designs and coordinates decision making. To ensure optimal process executions, decision management should incorporate decision logic documentation and implementation. To achieve the separation of concerns principle, the OMG group proposes to use Decision Model and Notation (DMN) in combination with Business Process Model and Notation (BPMN). However, often in practice, decision logic is either explicitly encoded in process models through control flow structures, or it is implicitly contained in process execution logs. Our work proposes an approach of semi-automatic derivation of DMN decision models from process event logs with the help of decision tree classification. The approach is demonstrated by an example of a loan application in a bank.

Keywords: Business process · Decision management · Decision mining

1 Introduction

Business process management is widely used by many companies to run their businesses efficiently. Business process performance essentially depends on how efficiently operational decisions are managed. An interest from academia and industry in the development of decision management has lead to the recently emerged DMN standard [13] aimed to be complementary to the BPMN standard [12].

To assist companies with successful automated decision management, knowledge about "as-is" decision making needs to be retrieved. This can be done by analysing process event logs and discovering decision rules from this information. Existing approaches to decision mining concentrate on the retrieval of control flow decisions but neglect data decisions and dependencies that are contained within the logged data. To overcome this gap, we extended an existing approach to derive control flow decisions from event logs [14] with additional identification of data decisions and dependencies between them. Furthermore, we proposed an algorithm for detecting dependencies between discovered control flow and data decisions. The output of this approach is a complete DMN decision model which explains the executed decisions, which can serve as a blueprint for further decision management.

© Springer International Publishing Switzerland 2016
W. Abramowicz et al. (Eds.): BIS 2016, LNBIP 255, pp. 237–251, 2016.
DOI: 10.1007/978-3-319-39426-8_19

The remainder of the paper is structured as follows. In Sect. 2, we introduce the foundations and a running example for the paper. Section 3 presents the discovery of control flow and data decisions from the event logs. In Sect. 4 we propose an algorithm for identifying dependencies between discovered decisions. We then apply the presented concepts on the use case and introduce a proto-typical implementation of the DMN model extraction in Sect. 5. Related work is then discussed, followed by the conclusion in Sect. 7.

2   Preliminaries

2.1   Definitions

For our work, we rely on notions of process model and execution as follows.

Definition 1 (Process Model). A *process model* is a tuple $m = (N, C, \alpha)$, where $N = T \cup G$ is a finite non-empty set of control flow nodes, which comprises sets of activities T, and gateways G. $C \subseteq N \times N$ is the control flow relation, and function $\alpha : G \to \{xor, and\}$ assigns to each gateway a type in terms of a control flow construct.   ◇

Definition 2 (Process   Execution,   Process   Instance,   Activity Instance). Let m be a process model. A process execution is a sequence of activity instances $t_1 \ldots t_n$, with $n \in \mathbb{N}$ and each t_i is an instance of an activity in the set of activities T of m.   ◇

Definition 3 (Event Instance, Event Attributes, Trace, Event Log). Let E be the set of event instances and A a finite set of attributes. Each attribute $a \in A$ is associated with the corresponding domain $V(a)$, which represents a set of either numeric or nominal values. Each event instance $e \in E$ has tuples (a, v), $a \in A$, $v \in V(a)$ assigned to it. A trace is a finite sequence of event instances $e \in E$ such that each event instance appears in the trace only once. An event log L is a multi-set of traces over E.   ◇

We assume that an activity name in the process model corresponds to related event instance name in an event log. For simplification purpose, we also assume that the business processes do not contain loops, which we plan to consider in future work.

To represent the knowledge about decisions taken in business processes, we use the DMN standard, which distinguishes between two semantic levels: the *decision requirements* and the *decision logic*. The first one represents how *decisions* depend on each other and what *input data* is available for the decisions; these nodes are connected with each other through *information requirement edges*.

Definition 4 (Decision Requirement Diagram). A decision requirement diagram DRD is a tuple (D_{dm}, ID, IR) consisting of a finite non-empty set of decision nodes D_{dm}, a finite non-empty set of input data nodes ID, and a finite non-empty set of directed edges IR representing the information requirements such that $IR \subseteq D_{dm} \cup ID \times D_{dm}$, and $(D_{dm} \cup ID, IR)$ is a directed acyclic graph (DAG). ◇

A decision may additionally reference the decision logic level where its output is determined through an undirected association. One of the most widely used representation for decision logic is a decision table, which we utilize for the rest of the paper.

Definition 5 (Decision Table). Decision table $DT = (I; O; R)$ consists of a finite non-empty set I of inputs, a finite non-empty set O of inputs, and a list of rules R, where each rule is composed of the specific input and output entries of the table row. ◇

An example of the decision model is presented in Fig. 5b: *decisions* are rectangles, *input data* are ellipsis, *information requirement edges* are directed arrows. The decision logic is not presented visually in the decision requirements diagram.

2.2 Running Example

Our example process represents a loan application in a bank, as shown in Fig. 1. Although we used a Petri net for the model representation, our approach can be applied to a wider class of notations, e.g., BPMN. The process starts with the registration of the user's claim, depicted in the model by the transition *Register claim*. The claim details are recorded in the bank system as the attributes of a token produced by this transition: the *Amount* the person claims (in EUR), the desired payback *Rate* (in EUR) per month, the payback *Duration* (in months), and if the customer has a *Premium* status. Afterwards, an expert decides if a *Full check*, a *Standard check*, or a *No check* activity should be executed. The process proceeds by executing an *Evaluation* activity, deciding if the client's claim is accepted or rejected followed by sending corresponding letters to the client. This transition is followed by recording the *Risk* attribute of the token produced.

For analysing the example process, we created an event log with the help of the simulation system CPN Tools[1]. This tool uses coloured Petri nets for models'

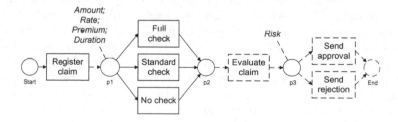

Fig. 1. Process model of the loan application in a bank

Table 1. Simulation parameters for generating the event log of the process from Fig. 1

Task/Attribute name	Simulation parameters
Trace ID	1 to 200 (incrementing)
Amount	discrete(2,99)
Premium	Random boolean
Duration	discrete(2,30)
Rate	*Amount/Duration*
Risk	if $Amount \geq 50$ and $Duration > 15$: $Risk = 4$
	if $Amount \geq 50$ and $Duration < 15$ and $Duration > 5$: $Risk = 3$
	if $Amount \geq 50$ and $Duration < 5$: $Risk = 2$
	if $Amount < 50$ and $Duration > 20$: $Risk = 3$
	if $Amount < 50$ and $Duration < 20$ and $Duration > 10$: $Risk = 2$
	if $Amount < 50$ and $Duration < 10$: $Risk = 1$
p1	if $Amount \geq 50$ and $Premium = false$: *Full check*
	if $Amount < 50$ and $Premium = false$: *Standard check*
	if $Premium = true$: *No check*
p3	if $Risk \leq 2$: *Send approval*
	if $Risk > 2$: *Send rejection*

representation which allow tokens to have data values attached to them, as in our example process. We used the simulation parameters as presented in Table 1.

Table 2 shows a fragment of the simulated event log for the process depicted in Fig. 1. For each event instance e, the event log records the event ID, the trace ID referring it to the corresponding process instance, the name of the executed task, and the set of other event attributes $a \in A$ logged when a token is produced by the corresponding transition. For example, the event instance 1 has the following attributes: *Amount* $a_1 = 84$ [EUR], *Rate* $a_2 = 2.8$ [%], *Duration* $a_3 = 30$ [Mths], *Premium* $a_4 = false$. All other information, e.g., the timestamps of event instances is discarded.

Whereas the knowledge about the process decisions can be empirically derived from the logged expert decisions depicted in Table 2 in the form of credit evaluation rules, the corresponding process model depicted in Fig. 1 does not allow for decision knowledge to be obtained. Moreover, simply applying these rules for the development of credit scoring systems can lead to the unjust treatment of an individual applicant; e.g. judging the applicant's creditability by the first letter of a person's last name [6]. Thus, tt seems reasonable to use automated credit scoring systems [5] complemented with an explanatory model, and therefore, we propose further in this paper to derive the DMN decision model from the process event log. An advantage of using such a model is that the separation of process and decision logic maximizes agility and reuse of decisions [15]. We demonstrate the DMN model derivation for the presented use case in Sect. 5.

---

[1] http://cpntools.org/.

Table 2. An excerpt of the event log for the process depicted in Fig. 1

Event ID	Trace ID	Name	Other attributes
1	1	Register claim	Amount = 84 [EUR], Rate = 2.8 [%], Duration = 30 [Mths], Premium = false
2	1	Full check	-
3	2	Register claim	Amount = 80 [EUR], Rate = 4.4 [%], Duration = 18 [Mths], Premium = true
4	2	No check	-
5	1	Evaluate	Risk = 3
6	1	Send rejection	-

3 Discovering Decisions from Event Logs

In this section we introduce different types of decisions in process models and propose ways of detecting them in process models and event logs.

3.1 Discovering Control Flow Decisions

In our previous work [3] we analysed around 1000 industry process models from multiple domains and discovered that the exclusive gateway was the most common decision pattern used in 59 % of the models. To indicate such of type of decisions, we will use the notion of *control flow decisions*, which are represented in process models by decision structures of a single split gateway with at least two outgoing control flow edges. When the control flow decision occurs, a token placed at the split gateway, fires the needed transition, to which we will refer as to a *decision outcome*. For example, the control flow decision occurs when a token, placed at the split gateway $p1$ (see Fig. 1), fires the needed transition, thus, it is decided which of the activities *Full check*, *Standard check*, or *No check* should be executed.

For deriving the control flow decisions and the decision logic behind it, we propose the following approach. Firstly, the control flow decisions are identified in the input process model. Given a process model m, we determine constructs of directly succeeding control flow nodes, that represent a gateway succeeded by a task on each of the outgoing paths. The step output is a set of M control flow decisions $P = \{p1, \ldots, pM\}$ corresponding to a process model m.

Next, we want to derive the decision logic from the event log in the form of a decision table. A decision table is a tabular representation of a set of related input and output expressions, organized into rules indicating which output entry applies to a specific set of input entries [13]. Below we introduce the notion of the decision rule.

Definition 6 (Decision Rule). Given is an event log L and a corresponding set of attributes $A = (A_1, \ldots, A_v), v \in \mathbb{N}^*$. The *decision rule* is a mapping

$$A_1 \ op \ q_1, \ldots, A_w \ op \ q_w \longrightarrow c_l, \ 1 \le w \le v \tag{1}$$

where the attributes A are decision inputs, op is a comparison predicate, $q_1, \ldots, q_w$ are constants, and $c_l \in C_{p*} = (c_1, \ldots, c_s), s, l \in \mathbb{N}^+, 1 \le l \le s$ is a decision output of the control decision $p* \in P$.   ◇

For example, the decision rule for the control flow decision $p1$ (see Fig. 1) is:

$$Premium = false, \ Amount < 50 \longrightarrow Full \ check \tag{2}$$

The control flow decision rules can be derived using the approach introduced in [14]. It is based on the idea that a control flow decision can be turned into a classification problem, where the classes are process decisions that can be made, and the training examples are the process instances recorded in the event log. All the attributes recorded in the event log *before* the considered choice construct are assumed as relevant for the case routing. In accordance with [14], we propose to use decision trees for solving the presented problem. Among other popular classification algorithms are neural networks [1] and support vector machines [16]. Pursuing the goal of deriving an *explanatory* decision model for a business process, we stick to the decision trees, as it delivers a computationally inexpensive classification based on few comprehensible business rules with a small need for customization [2].

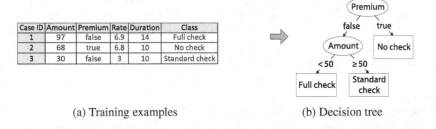

(a) Training examples   (b) Decision tree

Fig. 2. Control flow decision $p1$ represented as classification problem

The possibly influencing attributes for the control flow decision $p1$ are: *Amount, Premium, Rate, Duration*. The learning instances are created using the information from the event log as presented in the table in Fig. 2a, where lines represent process instances, and columns represent possibly influencing attributes. The *Class* column contains the decision outcome for a process instance. An example of the classification of process instances by a decision tree is presented in Fig. 2.

3.2 Discovering Data Decisions

Besides the explicit control flow decisions, process models can contain implicit data decisions. Our idea for detecting such type of decisions assumes that the values of the atoms in the decision rules can be determined for a certain process instance by knowledge of the variable assignments of other attributes. For example, in Fig. 1, it is not exhibited how the value of an attribute *Risk* is assigned, however, in practice, it depends on the values of attributes *Amount, Rate, Premium* and *Duration* recorded in the system while registering the clients claim. We distinguish the data decisions into functional and rule-based data decisions.

Definition 7 (Rule-Based Data Decision). Given is a set of attributes $A = (A_1, \ldots, A_v)$ An attribute A_j is a *rule-based decision* if there exists a non-empty finite set of rules R which relate a subset of the given set of attributes to the aforementioned attribute A_j:

$$R = \bigcup_i A_1 \ op \ q_1, \ldots, A_l \ op \ q_l \longrightarrow V^i(A_j) \tag{3}$$

where the attributes $A_1, \ldots, A_l$ are decision inputs, op is a comparison predicate, $q_1, \ldots, q_l$ are constants; hereby $v, l, i \in \mathbb{N}^+$, $1 \le l \le v$.   ◇

A rule from the rule-based data decision for *Risk* of the process model in Fig. 1 is:

$$Premium = false, \ Amount < 50, \ Duration < 10 \longrightarrow Risk = 4 \tag{4}$$

Definition 8 (Functional Data Decision). Given is a set of attributes $A = (A_1, \ldots, A_v)$ An attribute A_j is a *functional data decision* if there exists a function $f : (A_1, \ldots, A_k) \longrightarrow A_j$ which relates a subset of the given set of attributes to the aforementioned attribute A_j; hereby $v, k \in \mathbb{N}^+$, $1 \le k \le v$. ◇

A functional data decision for our running example for attribute *duration* is:

$$Duration = Amount/Rate \tag{5}$$

In Algorithm 1, we propose a way to retrieve data decisions using a process model m, an event log L, and a corresponding set of attributes of the event instances $A = (A_1, \ldots, A_v), v \in \mathbb{N}^*$ as inputs. As any of the attributes from the set A can potentially be the output of a data decision, the procedure runs for each attribute $a \in A$ (line 2).

Firstly, in line 3, a set A_{inf} of attributes possibly influencing a is determined using the assumption that all the attributes are recorded in the event log before or equal to the transition to which the attribute a is referred. Afterwards, the procedure detects whether an attribute a represents a *rule-based data decision* (lines 4–6). For this, we build a decision tree dt classifying the values of the attribute a using the set of possibly influencing attributes A_{inf} as features. If the built decision tree classifies all training instances correctly, a rule-based data decision for the attribute is yielded (here and further in the paper we assume

---

Algorithm 1. Retrieving Data Decisions

```
1: procedure FINDDATADECISIONS(processModel m, eventLog L, attributes A)
2:     for all a ∈ A do
3:         A_inf ← possibly influencing attributes for a
4:         dt ← decision tree for a using A_inf as features
5:         if dt correctly classifies all instances then
6:             return rule-based data decision for a
7:         else
8:             if a has a numeric domain then
9:                 A_infnum ← attributes from A_inf with numeric domain
10:                operators ← {+, -, *, /}
11:                funcs ← {function "a o b" | a, b ∈ A_infnum ∧ o ∈ operators}
12:                for all func ∈ funcs do
13:                    if func correctly determines value of a for all instances then
14:                        return functional data decision for a
15:                        break
```

---

that the classification correctness can be adapted for business needs by using a user-defined correctness threshold in percent). If no set of rules was found, the algorithm searches for a *functional data decision* for attribute a (lines 8–15). For this, we consider only such attributes which have a numeric domain. We determine the function form by a template representing combinations of two different attributes from A_{inf} connected by an arithmetic operator (10). The functions representing all possible combinations of such kind, are tested for producing the correct output for all known instances (lines 11–12). If that is the case, a functional data decision for a is returned (line 14). If it is determined that an attribute a is neither a rule-based, nor a functional data decision, a is discarded as a possible data decision and is treated as a normal attribute.

For our running example, the algorithm finds a rule-based data decision for attribute *Risk* depending on the attributes *Amount, Premium,* and *Duration* as presented in Fig. 3a. An instance of a functional decision is Eq. 5.

3.3   Mapping of the Discovered Decisions with DMN Model

The detected decisions represent decision nodes in the DMN decision requirement diagram which conforms to the original process model. The algorithm for constructing this diagram is straightforward. Firstly, for each discovered control flow decision $p \in P = \{p1, \ldots, pM\}$, we create a new decision node which is added to the set of decision nodes D_{dm} of the decision requirements diagram. Further, for each attribute of the event log $A_j \in A = (A_1, \ldots, A_v), v \in \mathbb{N}^*$ it is checked whether it is a rule-based or functional decision over other attributes from the set of attributes A. If this is the case, then a new decision node corresponding to this attribute A_j is added to the output decision requirements diagram DRD. Otherwise, we create a data node corresponding to this attribute A_j which is added to to the set of input data nodes ID of the decision requirements diagram. The decision nodes reference the corresponding decision tables containing the extracted rules. An example mapping is presented in Fig. 5a.

4   Discovering Decision Dependencies from Event Logs

The decisions discovered in process models as presented above, are used for creating a set of the decision nodes D_{dm} in the output decision requirement diagram DRD. For now, these decisions represent isolated nodes in the decision requirement diagram, and to "connect" them by the information requirements IR, in this section we propose an algorithm of discovering the decision dependencies from the event log.

4.1   Discovering Decision Dependencies

We propose to distinguish between two following types of decision dependencies.

Definition 9 (Trivial decision dependency). If a decision d in the set of decision nodes detected in the event log ($d \in D_{dm}$) depends on the attribute $a_k \in A$ from the event log and this attribute is detected to be a data decision ($dd \in D_{dm}$), then there is a *trivial dependency* between this decision d and the decision dd; here $k \in \mathbb{N}^+$.   ◇

Definition 10 (Non-trivial decision dependency). *Non-trivial dependencies* are dependencies between control flow decisions and other decisions.   ◇

To find the dependencies between either control flow, or data decisions detected in the event log of a process model, we propose Algorithm 2. The inputs to the algorithm are process model m, event log L, and a corresponding set D_{dm} of the detected decisions. Each decision d is tested for being influenced by other decisions (line 2). Firstly, the algorithm searches for *trivial decision dependencies* (lines 3–6). In line 3, we identify all attributes A_{inf} influencing the decision d (those are the attributes appearing the set of rules or in the function of the decision). For each attribute a_{inf} in the set of influencing attributes A_{inf}, we check if there is a data decision deciding a_{inf}. If this is the case, the algorithm yields a trivial decision dependency between this data decision on a_{inf} and the decision d (line 6).

Algorithm 2. Retrieving Decision Dependencies

1: **procedure** FINDDEPENDENCIES(processModel m, eventLog L, decisions D_{dm})
2:     **for all** $d \in D_{dm}$ **do**
3:         $A_{inf} \leftarrow$ attributes influencing d
4:         **for all** $a_{inf} \in A_{inf}$ **do**
5:             **if** D_{dm} contains a data decision dd for a_{inf} **then**
6:                 return decision dependency $dd \rightarrow d$ ▷ trivial dependency
7:         $D_{inf} \leftarrow$ possibly influencing control flow decisions for d
8:         **for all** $d_{inf} \in D_{inf}$ **do**
9:             $Feat \leftarrow A_{inf}$ - attributes influencing $d_{inf} \cup \{d_{inf}\}$
10:             $dt \leftarrow$ decision tree for d using $Feat$ as features
11:             **if** dt correctly classifies all instances **then**
12:                 return decision dependency $d_{inf} \rightarrow d$ ▷ non-trivial dependency

---

Afterwards, the algorithm searches for *non-trivial decision dependencies* (lines 7–12). We firstly identify a set of control flow decisions D_{inf} possibly influencing the decision d in line 7. Data decisions are not considered, because a data decision influencing another decision always results in a trivial decision dependency. Any decision $d_{inf} \in D_{inf}$ should satisfy two conditions: (1) d_{inf} is bound to a transition in the model that lies before or is equal to the transition of d; and (2) the set of influencing attributes of d_{inf} is a subset of the set of influencing attributes of d. Next, the algorithm detects whether there is a non-trivial decision dependency between each found possibly influencing decision d_{inf} and d. For this sake, we determine a new set of features *Feat* for the decision d (line 9). This set consists of the attributes influencing d without the attributes influencing d_{inf}, but with the output of decision d_{inf}. Using the features from *Feat*, we build a new decision tree deciding on d. If the tree is able to correctly classify all training instances, we have found a non-trivial decision dependency between d_{inf} and d.

In our example, the decision on attribute *Risk* depends on the attribute *Duration* (Eq. 4). As we identified *Duration* as being a data decision (Eq. 5), there is a trivial decision dependency between the decisions *Duration* and *Risk*. For an example of a non-trivial dependency, have again a look at the *Risk* decision in Fig. 3b. The found decision tree for *Risk* depends on the attributes *Amount, Premium* and *Duration*. We identify the control flow decision *p1* as a possibly influencing decision, as (1) the decision *p1* happens before the decision *Risk*; and (2) its influencing attributes (*Amount, Premium*) are a subset of the influencing attributes of the decision *Risk*. Further, we can build a decision tree that correctly classifies all instances by using the output of decision *p1* instead of the attributes *Amount* and *Premium*. Note that the attribute *Duration* is in the output decision tree in Fig. 3b, as it is not part of the decision *p1*.

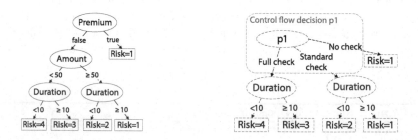

Fig. 3. Decision trees for attribute *Risk*

It might be the case that circular dependencies between attributes in the same transition in process model are discovered. For example, in Eq. 5, the discovered functional data decision *Duration* depends on *Amount* and *Rate*. However, as *Duration, Amount* and *Rate* appear in the same transition, Algorithm 2 finds the data decisions for *Amount* depending on *Duration* and *Rate*, as well as for *Rate* depending on *Duration* and *Amount*. However, in the output decision model two of these three data decisions should be discarded to avoid cyclic dependencies. We leave it to the process expert to determine which decision out of a set of cyclic data decisions is the most relevant for the output decision model.

4.2   An Improved Approach to Finding Non-trivial Decision Dependencies

The decision dependencies in Algorithm 2 are retrieved under the assumption that if an arbitrary decision d_k depends on another decision d_i $(k, i \in \mathbb{N}^+)$, then this decision d_k can *not* additionally depend on attributes that d_i depends on. This problem is illustrated in Fig. 4: here, d_i depends on attribute A' and d_k depends on decision d_i and additionally on the attributes A' and A''. The Algorithm 2 tries to rebuild the decision tree for d_k without considering the attributes of d_i, and it only utilizes attribute A''. Thus, this algorithm finds no dependency between d_i and d_k, which is not correct.

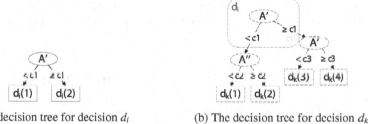

(a) The decision tree for decision d_l (b) The decision tree for decision d_k

Fig. 4. Decision d_k depends on another decision d_i, *and* on the attributes influencing d_i

To overcome this problem, we propose an alternative Algorithm 3 for finding non-trivial decision dependencies in the process event log (finding of trivial decision dependencies is equivalent to lines 3 to 6 in Algorithm 2). Firstly, the Algorithm 3 identifies a set of possibly influencing control flow decisions D_{inf} and for each $d_{inf} \in D_{inf}$ it builds a decision tree dt_{inf} containing (1) one root node, that splits according to d_{inf}; (2) as many leaf nodes as d_{inf} has decision outcomes, thereby, each of them containing a set of learning instances. Then, for each leaf node a subtree is built that decides on d for all learning instances of this leaf, thereby the features are all attributes that were used in the original found decision tree for d. In lines 7, all subtrees are attached to dt_{inf} resulting in dt_{new} which is a decision tree for d. Next, it is checked that the complexity of the newly constructed tree dt_{new} has not increased in comparison to the original decision tree for d by measuring and comparing the corresponding maximum number of nodes from the root to the leafs (lines 11–14). If this is the case, then the algorithm outputs a decision dependency between d_{inf} and d.

Algorithm 3. Alternative Retrieving of Non-trivial Decision Dependencies

1: **procedure** FINDDEPENDENCIESNEW($processModel$ m, $eventLog$ L, $decisions$ D_{dm})
2:      **for all** $d \in D_{dm}$ **do**
3:          $D_{inf} \leftarrow$ possibly influencing control flow decisions for d
4:          **for all** $d_{inf} \in D_{inf}$ **do**
5:              $dt_{inf} \leftarrow$ decision tree where the root node splits according to d_{inf}
6:              $dt_{new} \leftarrow dt_{inf}$
7:              **for all** $leaf \in dt_{inf}$ **do**
8:                  $dt_{leaf} \leftarrow$ decision tree for d classifying instances from $leaf$
9:                  $dt_{new} \leftarrow$ add dt_{leaf} to the $leaf$ of dt_{new}
10:             $levels_{inf} \leftarrow$ number of levels of dt_{new}
11:             $levels_{orig} \leftarrow$ number of levels of original decision tree for d
12:             **if** ($levels_{inf} <= levels_{orig}$ **then**
13:                 return decision dependency $d_{inf} \rightarrow d$

4.3 Mapping of Discovered Decision Dependencies with DMN Model

The trivial and non-trivial decision dependencies detected in the process event log are directly mapped to the set of information requirements IR represented by directed arrows between the discovered decision nodes D_{dm} and input data nodes ID in the output decision requirements diagram DRD. An example mapping is presented in Fig. 5b.

5   Application of the Approach on an Example Log

For evaluating our approach of the decision model discovery from the event log of a process model, we implemented it as a plug-in for the ProM framework 5.2[2] by extending the existing plug-in "Decision Point Analysis" for the discovery of control flow decision points [14] with our concepts presented in Sects. 3 and 4. The ability of the tool to derive decision models from event logs is shown in a screencast[3] using the running example from Sect. 2.2. The input for the approach is an event log of a process model simulated as discussed in Sect. 2.2, from which we mine the process model using one of the ProM process mining algorithms. As we have now both process model in the form of Petri net (Fig. 1), and corresponding event log, we can start the discovery of decisions. The screencast reflects our step-by-step approach proposed for the discovery of decisions from event logs, which is described below.

1. Discovery of control flow decisions. According to approach from Sect. 3.1, the program identifies two control flow decisions: (1) $p1$ with decision alternatives *Full check*, *Standard check*, and *No check*; and (2) $p3$ with decision alternatives *Send approval* and *Send rejection*. A decision tree constructed for $p1$ is depicted in Fig. 2.

---

[2] http://www.promtools.org/.
[3] https://bpt.hpi.uni-potsdam.de/foswiki/pub/Public/WebHome/DMNanalysis. mp4.

2. Discovery of data decisions. Executing Algorithm 1, the program finds: (1) A rule-based decision *Risk* (Fig. 3a); (2) A functional decision *Duration* (Eq. 5).

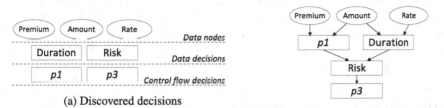

(a) Discovered decisions

(b) Discovered decision requirements diagram

Fig. 5. The discovered decisions and the DMN model for the example process

The aggregate of the decisions discovered by the program is presented schematically in Fig. 5a. Those are the elements which are used further for the construction of the decision requirements diagram: (1) Data nodes (*Premium, Amount, Rate*); (2) Data decisions (*Duration, Risk*); and (3) Control flow decisions (*p1, p3*). Additionally, the plug-in creates the decision table for each decision found (see screencast for details).

3. Discovery of decision dependencies. Further, the program executes Algorithm 2 to mine the dependencies between the discovered decisions from Fig. 5a, and it outputs the fully specified DMN decision model as depicted in Fig. 5b. Thus, the program finds the trivial dependencies between the decisions *Duration* and *Risk*, as well as between *Risk* and *p3*, also, the non-trivial dependency between the decisions *p1* and *Risk*. In case of circular dependencies, a random decision is kept. The decision dependencies discovery, whereby the influenced decision can reuse attributes from the influencing decision as described by Algorithm 3, was not implemented yet, but it is planned for future work.

The extracted decision model (Fig. 5b) shows explicitly the decisions corresponding to the process from Fig. 1, and thus, could serve for compliance checks by explaining the taken decisions. Also, the derived decision model can be executed complementary to the process model, thereby supporting the principle of separation of concerns [15].

6   Related Work

An interest from academia and industry towards exploring the advantages of separating of process and decision logic is demonstrated a number of works [8, 10,15,17]. Improving business process decision making based on past experience is described in [9], but the specifics of the DMN standard is not considered. [11] explore the possibilities of improving decisions within a DMN model, but it is not concerned with the integration of process and decision models problem. This paper extends our work in [4] and is closely related to [14], which describes the extract of decision rules for control flow decisions from event logs. However, we additionally identify data decisions, and decision dependencies. [7] extends [14], but in contrast to our paper, the authors seek to improve the performance of the rule extraction algorithm for control flow decisions, while we seek to complete the decision knowledge derived from event logs beyond control flow decisions. [3] also deals with the semi-automatic extraction of process decision logic but only on the modeling level.

7   Conclusion

In this paper we provided a formal framework enabling the extraction of complete decision models from event logs on the examples of Petri nets and DMN decision models. In particular, we extended an existing approach to deriving control flow decisions from event logs with additional identification of data decisions and dependencies between them. Furthermore, we proposed a modified approach to rebuilding decision trees to identify the dependencies between discovered decisions and overcame the problem of reusing attributes in a dependent decision. An assumption of our approach was that the decisions do not appear within loops, which we plan to investigate in future work.

The extracted DMN decision model reflects the decisions detected in the event log of a process model, which could be served as an explanatory model used for compliance checks. Additionally, executing this model complementary to the process model supports the principle of separation of concerns by providing increased flexibility, as changes in the decision model can be executed without changing the process model.

References

1. Baesens, B., Setiono, R., Mues, C., Vanthienen, J.: Using neural network rule extraction and decision tables for credit-risk evaluation. Manage. Sci. **49**(3), 312–329 (2003)
2. Baesens, B., Van Gestel, T., Viaene, S., Stepanova, M., Suykens, J., Vanthienen, J.: Benchmarking state-of-the-art classification algorithms for credit scoring. J. Oper. Res. Soc. **54**(6), 627–635 (2003)
3. Batoulis, K., Meyer, A., Bazhenova, E., Decker, G., Weske, M.: Extracting decision logic from process models. In: Zdravkovic, J., Kirikova, M., Johannesson, P. (eds.) CAiSE 2015. LNCS, vol. 9097, pp. 349–366. Springer, Heidelberg (2015)
4. Bazhenova, E., Weske, M.: Deriving decision models from process models through enhanced decision mining. In: Proceedings of 3th International Workshop on Decision Mining and Modeling for Business Processes. Springer (2015, accepted for publication)
5. Board: Report to the Congress on Credit Scoring and Its Effects on the Availability and Affordability of Credit. Technical report (2007)
6. Capon, N.: Credit scoring systems: a critical analysis. J. Mark. **46**(2), 82–91 (1982)

7. de Leoni, M., Dumas, M., García-Bañuelos, L.: Discovering branching conditions from business process execution logs. In: Cortellessa, V., Varró, D. (eds.) FASE 2013 (ETAPS 2013). LNCS, vol. 7793, pp. 114–129. Springer, Heidelberg (2013)
8. Debevoise, T., Taylor, J.: The Microguide to Process Modeling and Decision in BPMN/DMN. CreateSpace Independent Publishing Platform, Seattle (2014)
9. Ghattas, J., Soffer, P., Peleg, M.: Improving business process decision making based on past experience. Decis. Support Syst. **59**, 93–107 (2014)
10. Kornyshova, E., Deneckère, R.: Decision-making ontology for information system engineering. In: Parsons, J., Saeki, M., Shoval, P., Woo, C., Wand, Y. (eds.) ER 2010. LNCS, vol. 6412, pp. 104–117. Springer, Heidelberg (2010)
11. Mertens, S., Gailly, F., Poels, G.: Enhancing declarative process models with DMN decision logic. In: Gaaloul, K., Schmidt, R., Nurcan, S., Guerreiro, S., Ma, Q. (eds.) BPMDS 2015 and EMMSAD 2015. LNBIP, vol. 214, pp. 151–165. Springer, Heidelberg (2015)
12. OMG: Business Process Model and Notation (BPMN), v. 2.0.2 (2013)
13. OMG: Decision Model And Notation (DMN), v. 1.0 - Beta 2 (2015)
14. Rozinat, A., van der Aalst, W.M.P.: Decision mining in ProM. In: Dustdar, S., Fiadeiro, J.L., Sheth, A.P. (eds.) BPM 2006. LNCS, vol. 4102, pp. 420–425. Springer, Heidelberg (2006)
15. Von Halle, B., Goldberg, L.: The Decision Model: A Business Logic Framework Linking Business and Technology. Taylor and Francis Group, Boca Raton (2010)
16. Walder, C.J: Support vector machines for business applications. In: Business Applications and Computational Intelligence, p. 267 (2006)
17 Zarghami, A , Sapkota, B., Eslami, M.Z., van Sinderen, M.: Decision as a service: separating decision-making from application process logic. In: EDOC. IEEE (2012)

Governing IT Activities in Business Workgroups—Design Principles for a Method to Control Identified Shadow IT

Stephan Zimmermann[1(✉)], Christopher Rentrop[1], and Carsten Felden[2]

[1] HTWG Konstanz - University of Applied Sciences, Konstanz, Germany
{stephan.zimmermann,
christopher.rentrop}@htwg-konstanz.de
[2] Technische Universität Bergakademie Freiberg (Sachsen), Freiberg, Germany
carsten.felden@bwl.tu-freiberg.de

Abstract. The IT unit is not the only provider of information technology (IT) used in business processes. Aiming for increased work performance, many business workgroups autonomously implement IT resources not covered by their organizational IT service management. This is called shadow IT. Associated risks and inefficiencies challenge organizations. This study proposes design principles for a method to control identified shadow IT following action design research in four organizational settings. The procedure results in an allocation of task responsibilities between the business and the IT units following risk considerations and transaction cost economics. This contributes to governance research regarding business-located IT activities.

Keywords: Shadow IT · Workaround · Governance · Action design research

1 Introduction

Two-thirds of managers acknowledge the existence of shadow IT in their company [1]. This occurs as business units follow their own implementation of information technology (IT) outside of formal IT processes [2]. Shadow IT, as a kind of workaround [3], promises flexibility and performance gains for a business and may encourage innovation [2]. However, the occurrence of risks and inefficiencies [2] form a managerial burden and define the need for conceptual support. Studies show that 55 % of shadow IT is mission-critical, which concerns managers [4]. To address these concerns, this study aims at design principles to handle any identified shadow IT.

Processes in research and practice point to IT governance to approach this [5, 6]. Studies describe a broad spectrum of organizational behavior to handle shadow IT, involving further accountability in the business, a handover of partial tasks (e.g., provision of data or application programming) to the IT unit, or a complete transfer [7, 8]. While researchers are aware of this task sharing, practitioners struggle with a procedure on how to approach it in the optimum way. We address this based on action design research (ADR) [9] and provide design principles to deal with identified shadow IT by

© Springer International Publishing Switzerland 2016
W. Abramowicz et al. (Eds.): BIS 2016, LNBIP 255, pp. 252–264, 2016.
DOI: 10.1007/978-3-319-39426-8_20

allocating task responsibilities between a business and the IT unit. Task analysis and synthesis [10] were the bases for the proposed method. To achieve the allocation, we considered the risks of shadow IT and workarounds [3] and applied principles from transaction cost economics (TCE) [11]. It leads to an IT service governance contributing to the IT governance discussion on IT task responsibilities [12].

The following chapter formulates the problem and derives the research question. Based on the ADR approach, we present the chosen organizational settings and the theoretical basis. Next, we describe the results from the study and discuss the design principles. The study closes with conclusions and directions for further research.

2 Formulating the Problem of Managing Shadow IT

Organizations started to transfer the usage of IT and minor development tasks to end users in the 1980s. While management initiated this concept and was fully aware of end-user computing (EUC), over recent decades, employees have started to deploy IT resources without this awareness resulting in large-scale non-transparent IT applications [13, 14]. The term shadow IT defines this phenomenon[1].

Research defines shadow IT as IT solutions autonomously deployed in business departments to support their processes [15, 16]. In contrast to formal IT, the resulting instances are not embedded in the organizational IT service management [15]. Typical occurrences are autonomously sourced software [17], applications based on EUC platforms such as spreadsheets [14], self-programmed applications, cloud services [18, 19], mobile devices [8], self-sourced hardware, and combinations thereof [15].

Shadow IT relates to workarounds, which present deviations from existing work systems [3], and to un-enacted IT projects [20]. The implementation can be prompted by workgroups or individuals [2, 3, 16, 19] and mainly results from emerging IT needs in combination with a perceived lower expenditure compared with a formal implementation [3, 5, 15]. The underlying need can result from an inadequate formal IT [21]. Users perceive a relative advantage in looking for alternative solutions [18]. Finally, they implement shadow IT if a formal solution seems too expensive and time-consuming compared with their own, hidden implementation [22]. Perceived transaction and production costs in the business and IT units dictate these considerations [12, 15]. Available resources and expertise in the business reflect preconditions [7, 22].

Shadow IT challenges include inefficient and nonprofessional implementation [14, 21], security risks [17, 18], and compliance issues [23]. At the same time, it enables adaptability and user-driven innovation [2, 19]. While previous studies introduced steps to evaluate these effects [6, 24], a comprehensive approach for handling shadow IT is not apparent, thereby challenging IT management. Researchers consider this challenge to be within IT governance [5, 8, 15, 19], which specifies accountability to increase IT

---

[1] We reviewed prior literature to build a basis for our research. We queried EBSCOhost, ScienceDirect, IEEE Xplore, AISeL, Jstor based on abstract, title, and keywords. Employing the four-eye principle, we removed duplicates and irrelevant papers. The search terms *shadow IT, shadow systems, feral systems, gray IT, rogue IT,* and *hidden IT* combined with *IT, information services, information systems,* and *information security* resulted in 29 papers.

control [25]. Several criteria seem to require a specific control of shadow IT [26]. Shadow IT can be valuable for an organization [2, 3, 5]. Thus, the necessity of forbidding business-
located IT is doubtful. Moreover, shadow IT is evidence of available IT resources and knowledge present in the business. In contrast, it is understandable that, e.g., IT managers, are prejudiced against business-located IT [5]. As aforementioned, risks and inefficiencies occur. In terms of risk reduction and efficient and adaptive governance [27], organizations need to solve this. Research provides, e.g., first approaches to deter and prevent hidden IT solutions [8]. Furthermore, it is necessary to control identified shadow IT and its effects [6].

A possible solution is to reallocate responsibilities and share the tasks of identified shadow IT between the business and the IT units [7]. Regarding this allocation, researchers recently addressed the different involvement of business and IT units in IT task responsibilities [12]. The researched spectrum of organizational behavior to handle shadow IT occurrences supports this approach of task sharing [6, 7]. In summary, research has identified specific governance choices for identified shadow IT [6, 7]. Nevertheless, no studies exist that cover a guided application of these task-sharing choices. However, it does seem necessary that organizations should be able to decide which task allocation is reasonable. This leads to the study's research question:

How should organizations allocate IT task responsibilities for identified shadow IT instances between a business workgroup and the IT unit?

3   Action Design Research

To address the research question, we use ADR that combines design and action research and aims to generate prescriptive design knowledge through building and evaluating IT artifacts in organizations. It deals with two seemingly disparate challenges: (1) addressing a problem situation encountered in a specific organizational setting by intervening and evaluating and (2) constructing and evaluating an IT artifact [9].

Our research goal links to these challenges. Using an action research project, we address the control of identified shadow IT in four organizational settings and, following design science, aim at constructing and evaluating guidance for a method design [28]. Choosing ADR is particularly appropriate as shadow IT is a complex phenomenon and managing it means considering several perspectives. ADR provides a way to address this, as it assumes that valid knowledge comes from various sources and the insights of researchers and practitioners are relevant to successful research [29].

ADR follows several stages [9]. The first stage is problem formulation. We formulated our research problem based on literature as seen in Sect. 2. It focuses on the governance challenge for organizations of controlling identified shadow IT tasks when allocating IT task responsibilities. Coupled with initial empirical investigations, we were also able to determine the scope, roles, and practitioner participation. Therefore, it became more and more evident due to several influencing, complex criteria of shadow IT that a selection of more than one organizational setting would be necessary to increase the reliability of the results. This resulted in an ADR procedure concentrating on different processes in four organizations. Related theories as justificatory knowledge provided a further basis for creating the new artifact [9]. Based on this, the second stage of ADR requires an iterative process of *building* the artifact, *intervention* in the organization, and *evaluation* (BIE) [9]. We describe this in the four settings, focusing on different theory-based aspects of an IT service governance for identified shadow IT. Next, in the reflection and learning stage, we discuss the resulting design principles. Finally, we generalize the learning and the gained knowledge [9].

3.1   Four Organizational Settings for Practice-Inspired Research

To constitute the selection of the multiple organizational settings with regard to the ADR principle of practice-inspired research, we used the insights from literature on case study research [30]. We selected different business processes (Table 1) in four companies from various industries and started an ADR project one after another to achieve an iterative refinement of our artifact. We aimed for variation in the selection to provide more valid results, i.e., the chosen processes and underlying IT requirements were quite standardized in Company A but very compliance and security critical; in Company B, frequently highly business-specific, and in Company C they were affected by high uncertainty owing to rapidly changing conditions. Finally, our choice of Company D allowed a more robust evaluation, as the company had previous experience in governing shadow IT. Participants from this company served as experts [31].

Table 1. Organizational settings (Company profiles all anonymized)

Company	A	B	C	D
Industry	Insurance	Engineering	Electronics	Finance
Country	Switzerland	Germany	Germany	Germany
Staff	1.300	11.500	5.500	500
Selected processes	Benefits statements	Order management	Corporate marketing	Risk management and reporting
Shadow IT	6 instances	52 instances	41 instances	102 instances
Participants	2 Department heads; 1 IT manager	2 Department heads; 3 Employees; 1 Site director; 5 IT managers; 1 CIO	3 Department heads; 3 IT managers	10 Business team heads; 5 employees; 1 Division head; 3 IT managers

All the ADR projects were built on a prior identification of shadow IT using interviews with business members [6]. Authorized by the management, design principles should be derived to control the occurrences. We identified three stakeholders within the settings: Management, IT units, and the business workgroups with shadow IT. In each company, the management appointed one project manager from the IT unit who, together with the researchers, formed the ADR team. Participants from the different units served as end users, contributed to the build process, and evaluated the design.

As sources of evidence within the BIE procedure, we used data from prior semi-structured interviews that were conducted during the shadow IT identification process, documents, technical artifacts, and contextual observations. Further interviews, group discussions, and assessment and feedback loops with the participants supported the BIE iterations (Table 1). In addition, ADR team members discussed the artifact with the management. We started the ADR projects based on initial concept designs at the different developmental stages. The theoretical grounding for these follows.

3.2   Theory-Ingrained Artifact: Allocating Task Responsibilities for Shadow IT Based on Risk and Transaction Cost Economics

Research on application governance proves the benefit of the involvement of both business and IT units in task execution for IT applications [12]. A shared involvement to control a prior shadow IT service defines an IT service governance. To define the degree of involvement and allocate task responsibilities, we focused on several steps.

To validate the allocation, an initial structuring of tasks was necessary. Following task analysis and synthesis [10], the structuring of an overall task to sub-tasks progressed according to characteristics such as the type of execution or involved objects. The succeeding synthesis joins tasks to structured units and allocates these to individuals or workgroups. We applied this to the allocation of IT task responsibilities.

The risk considerations of shadow IT form one basis for the execution of this allocation [6, 7]. These considerations refer to the underlying theory of workarounds and define the creation of hazards or errors, impacts on subsequent activities, and regulation issues [3]. Internal risk of not achieving strategic goals, security problems, and compliance issues may result. With respect to risk analysis, the high impact of a shadow IT on organizational goals and relatively low quality of such an instance leads to high internal risk, which will need to be addressed by an organization [6].

When addressing inefficiencies [3], the relation between the business and the IT units when exchanging IT services justifies the application of TCE [15]. As stated in [12] "Business units either enter into a contract with (…) IT units (…) or coordinate (…) operations hierarchically". In the business/IT relationship, transaction costs exist for processing and organizing the exchange of an IT service, in addition to its production cost. Experience from practice shows that business workgroups will obtain shadow IT if they assume the total cost of applying a formal service to be higher than the initial production costs for own-implementation [15]. However, bounded rationality and opportunism [11] influence this assumption and the business neglects own governance costs. Solely allocating tasks to the business may be inappropriate [15, 25].

TCE can be used to overcome this issue [6, 12]. Its logic provides alternative governance modes to decide whether services should be produced internally or externally [27]. This means that with regard to the business/IT exchange relationship and identified shadow IT, business workgroups can either keep IT tasks and components or obtain them from the IT unit. Thus, the procedure enables a sharing of tasks for the required IT service forming a hybrid governance mode of co-operation [32]. TCE support the decision regarding such arrangements and determine whether an internal service task allocation or one from the IT unit (external from a business view) is more beneficial [27]. From an organizational meta-level, the theory aims for optimizing the total internal and external costs in the relationship. Several cost drivers are involved.

Asset specificity influences the relationship between internal and external costs [11]. An increasing specificity implies a more specialized investment. Production costs for an exchange partner rise as reusability for other customers declines. From a business perspective, external transaction costs with the IT unit rise compared with internal governance costs owing to more complex selection and negotiation processes. In total, internal coordination becomes more efficient. These considerations are also applicable to the involvement of business and IT units in IT task executions [12].

Environmental uncertainty is another dimension. It implies that "environment is characterized by uncertainty with respect to technology, demand, local factor supply conditions, inflation, and the like" [11] and is important in the case of shadow IT, which often starts as a prototype with uncertain requirements due to changing external conditions [5, 15]. Uncertainty moderates the influence of asset specificity on internal and external costs [11]. In case of non-specific assets, uncertainty has no effect. If specificity rises, external costs increase, e.g., negotiating or contracting becomes more complex, and internal organization becomes more efficient. This is in particular relevant for a mixed specificity "incorporating standardized and customized elements" [33] and a high uncertainty. It provides a further basis to allocate shadow IT tasks.

Finally, frequency describes the recurrence of transaction activities and moderates the influence of specificity on governance arrangements [11]. In the IT domain, an appropriate determination of this dimension is the scope of use of an IT service [12]. A higher scope increases the IT unit involvement in task responsibilities based on costs and risk considerations [12]. We take the theoretical concepts discussed in this section into account during the further development of the artifact.

4   Building, Intervention and Evaluation of a Method to Control Identified Shadow IT Instances by Allocating IT Tasks

The framed problem together with theoretical principles provided us with a preliminary artifact design. We further shaped the principles behind this by applying them successively in the four companies and in subsequent design cycles [9]. We present this process and show sequent developmental stages focusing on different constructs.

4.1   Risk in Company A

Swiss insurance Company A started the ADR project in their Benefit Statements department to control six identified shadow IT instances. The ADR team applied the initial method design and analyzed consequences for the task allocation.

Application results defined a complete task transfer to the IT unit for two of the instances due to security and compliance risks. Participants tended, e.g., to re-engineer an end-user computing (EUC) tool-based workflow system by the IT unit. This provided evidence on how organizations can deal with the method's risk construct. Two further instances remained in the business due to low relevance and no perceived cost benefit. However, IT participants claimed the necessity of documentation and maintenance. For the two remaining instances, participants pursued a balance between *complete* and *no* transfer. This assumed the involvement of both the business and IT units.

The participants evaluated the general method approach in the ADR project as positive. An interviewed IT manager stated, *"It is helpful to regard the relevance of a shadow IT instance in considering internal IT system risk and cost aspects to decide if the IT unit needs to adopt the service. Not all solutions need to be necessarily delivered by the IT unit."* One of the business department heads agreed with this, *"The idea of the approach provides a better basis to allocate responsibilities for former shadow IT."* Based on this feedback, we introduced a task analysis and synthesis in our method design to build a basis for a hybrid involvement of business and IT units and for explaining the arrangement in between the two alternatives of transferring the whole instance to the IT unit or retaining it in the business. We advanced this for a larger number of shadow IT instances by starting an ADR project in Company B.

4.2   Specificity and Scope of Use in Company B

Company *B* executed the ADR project on the order management process of a German manufacturing plant belonging to a corporate group. Of the 52 identified shadow IT instances, 32 required a task reallocation. All remaining instances stayed in the business unit due to low risks and no expected cost benefits. However, these were registered with the IT service management for transparency reasons. For the reallocation, participants were able to cluster and integrate solutions with similar functionalities.

As participants often regarded a total transfer of shadow IT tasks to the IT unit as inappropriate, we used our approach to find reasonable sharing of tasks. Thereby, the investigated shadow IT enabled us to focus (besides risk considerations as used in Company *A*) on the TCE dimensions of IT task specificity and scope of use, while uncertainty was principally low and less relevant. Table 2 reflects a typical example.

The Table 2 contains the structuring of tasks and components as well as the responsibility allocation between the business and IT unit. In general, besides tasks with high risks, non-specific tasks as well as tasks with a mixed specificity were transferred to the IT unit due to standardization and reusability reasons. Furthermore, if IT tasks remained in a business workgroup, a large scope of use of the solution required a consultation with the IT unit to ensure efficiency and a further risk reduction. Participants valued this procedure highly for reducing risks and inefficiencies due to shadow IT. With regard to the described instance in Table 2, the business department head appreciated that *"it is more reasonable to transfer server and database administrations, as well as access control tasks and interface management to the IT unit to reduce risks and to use synergies."* The IT unit supported this allocation for specificity, quality, and flexibility reasons. To improve this stage of the method with regard to the uncertainty construct, we started an ADR project in Company *C*.

Table 2. Shadow IT instance in Company B with reallocated responsibilities and task sharing

Shadow IT	IT service governance explanation	Specified task responsibilities
Order processing program consisting of a self-developed web-based application, with a database, hosted on an externally provided server. The engineering unit developed the solution to process orders with drawings, calculations, & scheduling.	Due to non-acceptable security risks and standardization, database- & server-related tasks as well as access control procedures were transferred to the IT unit. In addition, participants transferred interface programming to the IT unit based on the business requirements due to a mixed specificity (standardized and customized elements) but a low uncertainty. Other tasks stayed in the business because of a high human asset specificity: specific skills for programming drawings, calculations, etc. However, so far underrepresented tasks like documentation and testing were also defined. Furthermore, it is necessary to consult with the IT unit due to a broad scope of use to reduce risks if, e.g., business users resign.	*(nested table)* Service component / Execution type / Task responsibility. **Application**: User support, Requirements engineering, Documenting, Testing, Implementing changes, Software engineering → Business workgroup (consults IT unit); Access control, Interface programming → IT unit. **Database**, **Server** → IT unit.

4.3 Uncertainty in Company C

The German electronics Company *C* has experienced intense growth in recent years. Business units have developed shadow IT on a large scale. Management was searching for an approach to deal with this. We started the ADR project in the Marketing Department. Due to flexibility, a fast changing environment, and time to market, as well as assumed high cost when using the IT unit, marketing workgroups implemented shadow IT to test new IT ideas or solutions. In the beginning, the implementation was often uncertain regarding requirements and future usage following on the changing external conditions. This influenced the allocation of IT service tasks, especially those with mixed specificity, and enabled us to provide evidence for the uncertainty construct in our method design. Table 3 provides a typically example of this.

Table 3. Shadow IT instance in Company C with remaining task allocation to the business

Shadow IT	IT service governance explanation	Specified task responsibilities
Event management tool: Marketing sourced this cloud service. Data transfer from the enterprise system into the solution based on spreadsheets. The usage happened in a prototype stage with unclear requirements.	Management claimed for transparency & professionalism. Company *C* achieved this by embedding the solution in the IT service management and by adding testing and documentation. Thereby, the execution of all tasks stayed in the business due to acceptable risks, a mixed specificity, but a high uncertainty regarding requirements and future usage, and a small scope. Transparency facilitates possible future task transfers if uncertainty decreases.	*(nested table)* Service component / Execution type / Task responsibility. **Cloud service**: User support, Requirements engineering, Sourcing, Testing, Documenting, Organizing data transfer from enterprise system → Business workgroup.

Both business and IT staff agreed that the arrangements, as described in Table 3, were reasonable. A member of the IT unit stated, *"There is no necessity to spend resources from the IT unit's side until the business is certain about future usage. If it is more certain it will become, e.g., the task of the IT unit to provide an appropriate interface to core systems."* Besides this initially complete task allocation to the business, other instances exceeded the limits of uncertainty. For these, a reallocation of service tasks took place according to risk, specificity, and scope of use. Similar to Company *B*, e.g., infrastructure and database components were centralized.

Of the 41 identified shadow IT instances, 25 resulted in a reconsideration of task allocation. The participants of Company *C* evaluated the method as high quality. A management representative stated that, *"the method enables us to find a balance between flexibility, needed within the fast growing company environment, and efficiency."* To achieve a more robust evaluation we concluded with ADR in Company *D*.

4.4 Setting of Expertise in Company D

The ADR project in Company *D* enabled the final development and evaluation of the design principles in an environment of expertise on governing shadow IT. In this German finance company, we analyzed the Risk Management processes with 102 business-located IT solutions. These had previously been requested during audits. The project encompassed all constructs of task allocation (risk, specificity, scope of use, uncertainty) discussed so far. In total, 84 instances required task reallocation.

After discussion with participants, minor interventions were made to the method. Revision of the design principles was necessary when using the combined constructs for the task allocation. Overall, the experts emphasized the high quality of the procedure. They validated the method as an improvement to their prior governance regulations. An expert from the IT unit stated, *"Business-located IT can constitute an asset; however, it also causes problems. Therefore, it is important to maintain transparency and define task responsibilities. This method increases our current regulations."* A statement by a departmental head supports this: *"The approach provides a good way to increase the quality of IT solutions implemented by business workgroups without reducing the advantages."* A representative from a business workgroup added: *"The method is a good way to show necessary tasks and those currently missing. Besides creating this awareness, we welcome the allocation of tasks to the IT unit. However, we also affirm the importance of retaining particular service tasks in our group to ensure flexibility and usage of our specific knowledge."* Based on this positive evaluation, the company started to apply the method to other departments.

5   Reflection and Learning

This ADR stage moves conceptually from building an artifact for particular settings to applying that learning to a broader class of problems [9]. While the initial design principles constituted less developed and detached constructs, the BIE process and learning from the participants led to refinement and combination of these. We now present the final design and the learning outcomes structured by the study's research question: *How should organizations allocate IT task responsibilities for identified shadow IT instances between a business workgroup and the IT unit?*

A reasonable allocation follows risk- and TCE-related principles as shown in Fig. 1. The decision of who is in charge leads to IT service governance forming a co-operation of business and IT units with shared tasks [32]. As a precondition, the structuring of components and subtasks [10] is necessary. Furthermore, sufficient IT knowledge in the business and by the users is essential if IT task responsibilities are to be allocated to them, which needs to be respected in the risk analysis regarding the quality of the shadow IT [6]. A further procedure determines the allocation.

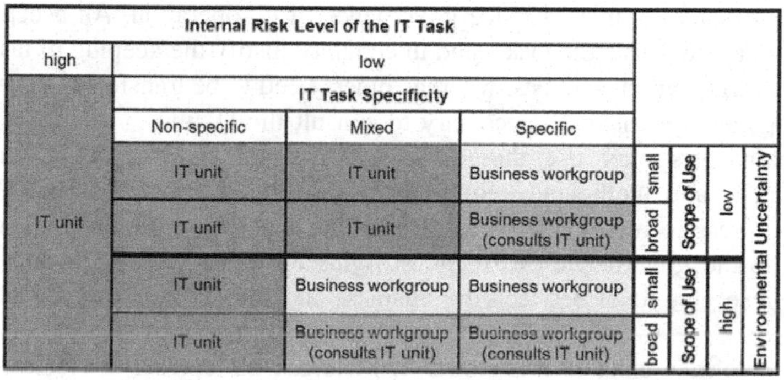

Internal Risk Level of the IT Task						
high	low					
	IT Task Specificity					
	Non-specific	Mixed	Specific			
IT unit	IT unit	IT unit	Business workgroup	small / broad — Scope of Use	low — Environmental Uncertainty	
	IT unit	IT unit	Business workgroup (consults IT unit)			
	IT unit	Business workgroup	Business workgroup	small / broad — Scope of Use	high — Environmental Uncertainty	
	IT unit	Business workgroup (consults IT unit)	Business workgroup (consults IT unit)			

Fig. 1. Design principles to allocate task responsibilities for identified shadow IT.

Initially, in cases of high risk, tasks are allocated to the IT unit to ensure a high level of control. This relates to the internal risks that shadow IT present to goals, security, and compliance [3, 6]. Transfer to the IT unit and even re-engineering based on the business requirements is reasonable – which may also lead to an abandonment of a shadow IT – if highly relevant and critical shadow IT is of low quality. TCE provide another criteria and address inefficiencies [15, 21]. In cases with acceptable risk, the specificity of IT tasks is at first decisive. An assignment of non-specific tasks to the IT unit will happen, because they are suitable for several solutions or reusable. Transaction costs for such assignments are low. Specific tasks tend to stay in the business owing to the related idiosyncratic knowledge, the transfer of which would be too costly. In cases with low uncertainty, the transfer of tasks with a mixed specificity to the IT unit follows the same reasoning as that for non-specific tasks. This inverts for high uncertainty. Business users may want to experiment with an idea that has uncertain requirements resulting from environmental influences. Thereby, a transfer of knowledge to the IT unit would cause too high transaction costs. Finally, the scope of use is an influencing factor. If a service has a broad scope, the business requires IT consultation to ensure efficient task execution and further risk reduction.

This IT service governance method addresses shadow IT challenges. First, it uncovers missing or unprofessionally provided tasks and inadequate components. Second, risk mitigation occurs. Third, by building allocations on the specificity of tasks moderated by the uncertainty, organizations can raise efficiency, while they keep the adaptability. Including the scope of use in this process supports a proper task execution due to the IT unit's consultation function. The detailed guidelines for allocating IT task responsibilities extend prior research regarding managing shadow IT [6–8, 24] and contribute to the research on related IT governance questions [5, 12, 15].

6   Conclusion

To address the problems of business-located IT activities, this study proposes design principles for a method of sharing IT task responsibilities for identified shadow IT between a business and the IT unit. The allocation procedure is based on risk and TCE constructs and leads to an IT service governance of co-operation. An acceptable risk level, highly business-specific tasks, and uncertainty justify the keeping of tasks within the business, if IT expertise exists, whereas others need to be transferred to the IT unit. A large scope always makes it necessary to consult the IT unit.

Decision-makers may use these results to handle shadow IT in practice by decreasing risks and inefficiency. Simultaneously, organizations maintain the adaptability of business-located IT activities. Regarding theoretical implications, this study contributes to the governance of execution rights for IT services. Allocating responsibilities at a service level between the business and the IT units may be adapted for formal IT.

Some limitations exist to this study. We address the control of shadow IT at an initial level. More research is necessary to specify the constructs, e.g., how risks can be categorized and what dimensions exist of the scope of use. It is also necessary to identify other possible factors, such as supporting technical solutions and competence of the IT unit. This is connected to the inclusion of external suppliers, e.g., for Software as a Service. Finally, considering IT architecture, questions appear regarding the integration and monitoring of business-located IT in larger networks. In carrying out further research, scientists may be able to advance a long-term control of shadow IT.

References

1. Chejfec, T.: Shadow IT survey v3. http://chejfec.com/2012/11/03/shadow-it-infographic/shadow-it-survey-v3/. Accessed 12 Mar 2016
2. Behrens, S.: Shadow systems: the good, the bad and the ugly. Commun. ACM **52**(2), 124–129 (2009)
3. Alter, S.: Theory of workarounds. Commun. Assoc. Inf. Syst. **34**(1), 1042–1066 (2014)
4. Rentrop, C., Zimmermann, S.: Schatten-IT. Informatik Spektrum **38**(6), 564–567 (2015)
5. Györy, A., Cleven, A., Uebernickel, F., Brenner, W.: Exploring the shadows: IT governance approaches to user-driven innovation. In: Proceedings of the 20th European Conference on Information Systems, Barcelona, Spain (2012)
6. Zimmermann, S., Rentrop, C., Felden, C.: Managing shadow IT instances - a method to control autonomous IT solutions in the business departments. In: Proceedings of the 20th Americas Conference on Information Systems, Savannah, Georgia, USA (2014)

7. Chua, C., Storey, V., Chen, L.: Central IT or Shadow IT? Factors shaping users' decision to go rogue with IT. In: Proceedings of the 35th International Conference on Information Systems, Auckland, New Zealand (2014)
8. Beimborn, D., Palitza, M.: Enterprise app stores for mobile applications - development of a benefits framework. In: Proceedings of the 19th Americas Conference on Information Systems, Chicago, IL (2013)
9. Sein, M., Henfridsson, O., Purao, S., Rossi, M., Lindgren, R.: Action design research. Manag. Inf. Syst. Q. **35**(1), 37–56 (2011)
10. Picot, A.: Division of labour and responsibilities. In: Grochla, E., Gaugler, E., et al. (eds.) Handbook of German Business Management, pp. 745–752. Poeschel, Stuttgart (1990)
11. Williamson, O.E.: The Economic Institutions of Capitalism. Firms, Markets, Relational Contracting. Free Press, New York (1985)
12. Winkler, T.J., Brown, C.V.: Horizontal allocation of decision rights for on-premise applications and software-as-a-service. J. Manag. Inf. Syst. **30**(3), 13–48 (2014)
13. Ferneley, E.H.: Covert end user development: a study of success. J. Organ. End User Comput. **19**(1), 62–71 (2007)
14. Panko, R.R., Port, D.N.: End user computing: the dark matter (and dark energy) of corporate IT. J. Organ. End User Comput. **25**(3), 1–19 (2013)
15. Zimmermann, S., Rentrop, C.: On the emergence of shadow IT – a transaction cost-based approach. In: Proceedings of the 22nd European Conference on Information Systems, Tel Aviv, ISR (2014)
16. Haag, S., Eckhardt, A., Bozoyan, C.: Are shadow system users the better IS users? – insights of a lab experiment. In: Proceedings of the 36th International Conference on Information Systems, Fort Worth, TX (2015)
17. Silic, M., Back, A.: Shadow IT – a view from behind the curtain. Comput. Secur. **45**, 274–283 (2014)
18. Haag, S.: Appearance of dark clouds? - an empirical analysis of users' shadow sourcing of cloud services. In: Wirtschaftsinformatik Proceedings, Osnabrück, Germany (2015)
19. Zainuddin, E.: Secretly saas-ing: stealth adoption of software-as-a-service from the embeddedness perspective. In: Proceedings of the 33rd International Conference on Information Systems, Orlando, FL (2012)
20. Buchwald, A., Urbach, N.: Exploring the role of un-enacted projects in IT project portfolio management. In: Proceedings of the 33rd International Conference on Information Systems, Orlando, FL (2012)
21. Jones, D., Behrens, S., Jamieson, K., Tansley, E.: The rise and fall of a shadow system: lessons for enterprise system implementation. In: Proceedings of the 15th Australasian Conference on Information Systems, Hobart, Australia (2004)
22. Behrens, S., Sedera, W.: Why do shadow systems exist after an ERP implementation? Lessons from a case study. In: Proceedings of the 8th Pacific Asia Conference on Information Systems, pp. 1712–1726, Shanghai, China (2004)
23. Gozman, D., Willcocks, L.: Crocodiles in the regulatory swamp: navigating the dangers of outsourcing, SaaS and shadow IT. In: Proceedings of the 36th International Conference on Information Systems, Fort Worth, TX (2015)
24. Fürstenau, D., Rothe, H.: Shadow IT systems: discerning the good and the evil. In: Proceedings of the 22nd European Conference on Information Systems, Tel Aviv, ISR (2014)
25. Weill, P., Ross, J.W.: IT Governance. How Top Performers Manage IT Decision Rights for Superior Results. Harvard Business School Press, Boston (2004)
26. Tiwana, A.: Governance-knowledge fit in systems development projects. Inf. Syst. Res. **20**(2), 180–197 (2009)

27. Williamson, O.E.: The economics of governance. Am. Econ. Rev. **95**(2), 1–18 (2005)
28. Braun, C., Wortmann, F., Hafner, M., Winter, R.: Method construction - a core approach to organizational engineering. In: ACM Symposium 2005, pp. 1295–1299 (2005)
29. Moloney, M., Church, L.: Engaged scholarship: action design research for new software product development. In: Proceedings of the 33rd International Conference on Information Systems, Orlando, FL (2012)
30. Yin, R.K.: Case Study Research. Design and Methods. Sage Publ., Los Angeles (2014)
31. Flick, U.: An Introduction to Qualitative Research. Sage Publ., Los Angeles (2009)
32. Powell, W.W.: Neither market nor hierarchy: Network forms of organization. Res. Organ. Behav. **12**, 295–336 (1990)
33. McIvor, R.: How the transaction cost and resource-based theories of the firm inform outsourcing evaluation. J. Oper. Manag. **27**(1), 45–63 (2009)

A Formalization of Multiagent Organizations in Business Information Systems

Tobias Widmer$^{(\boxtimes)}$, Marc Premm, and Stefan Kirn

Department of Information Systems 2, University of Hohenheim,
Stuttgart, Germany
{tobias.widmer,marc.premm,stefan.kirn}@uni-hohenheim.de

Abstract. Multiagent (MA) organizations can be regarded as a functional part in business information systems, in which software agents negotiate conditions for participation in the organization. How the strategic behavior of self-interested agents and MA-Organizations affects the formation process, however, is still not known. This research is concerned with the specification of MA-Organizations in business information systems and the design of negotiation protocols for determining the agents participation conditions. We draw on mechanism design to model the participation decision of the agent and the organization as a bilateral trading game. In a simulation experiment we find that a rather simple manipulation scheme provides a suitable approximation for the equilibrium strategies employed by the agents.

Keywords: Software agents · Multiagent organization · Auctions

1 Introduction

A key concern in multiagent systems (MAS) research is the cooperation and coordination among autonomous agents [1]. The design of MAS focuses on maximum flexibility; they emerge for single problems and dissolve without leaving a trace. However, this kind of flexibility is not applicable to many business structures because human organizations generally aim at reliability facing legal issues such as accounting policies. Organization theory provides means to integrate reliability and stability into the cooperation of both human and software agents [2].

The formation of MA-Organizations and the negotiation of terms and conditions for participating in MA-Organizations play a central role in the structuring process. However, since software agents need strictly formalized concepts of the MA-Organization to perform participation actions, models from management science are not sufficient to meet these requirements [3]. In particular, it is not clear how the decision process for participation must be designed such that the outcome is socially optimal.

The formation of MA-Organizations requires potential members to make decisions about their participation. Formation problems as such have been subject of inquiry in prior research. Current approaches in MA coalition formation [4], however, either assume complete information among the agents [5] or the

© Springer International Publishing Switzerland 2016
W. Abramowicz et al. (Eds.): BIS 2016, LNBIP 255, pp. 265–276, 2016.
DOI: 10.1007/978-3-319-39426-8_21

agents are not concerned with their individual payoffs [6]. In MA-Organization literature, existing approaches mainly focus on the definition of constructs and models but do not address the decision problem in detail [7]. The design of distinct negotiation protocols for participating in MA-Organizations, however, is crucial for modeling the decision-making behavior of software agents.

We address the problem of designing an interaction protocol that enables the negotiation for participation between a MA-Organization and a software agent. In our framework, a preexisting MA-Organization offers an open position, while an outside software agent applies for this position. We draw on mechanism design, a branch of game theory, to model the participation decision of the agent and the organization. Game theory provides an established basis for modeling and analyzing the interactions and decision making of self-interested agents within a MAS [8]. While applying the bilateral bargaining framework introduced by [9] to our setting, we study the strategic behavior of the software agent and the MA-Organization when negotiating the conditions for participation. We show that a rather simple manipulation scheme on both sides provides a suitable approximation of the equilibrium strategies identified by [9].

Negotiating participation conditions in MAS have been addressed in prior research [10]. Widmer et al. [11] study the formation problem of agent-based virtual organizations using auctions. Their approach applies the concept of multi-attribute auctions to determine the optimal sourcing strategies in MA-Organizations. However, the proposed auction settings is limited to single-sided competition only. Double-sided competition for bilateral models was studied in economic theory [12] and operations research [9]. Our research applies the results of [9] to the decision-making problem in MA-Organizations. The objectives are to (1) provide a UML specification of MA-Organizations in business information systems, (2) apply a bilateral bargaining mechanism to the MA-Organization formation process, and (3) demonstrate the efficacy of the proposed mechanism by a simulation experiment.

The remainder of this paper is structured as follows. Section 2 discusses the theoretical background to our research. Section 3 presents the organizational model. The auction-based approach is presented in Sect. 4. Section 5 reports on the evaluation. Section 6 concludes.

2   Theoretical Background

2.1   Multiagent Systems and Multiagent Organizations

An *autonomous software agent* is an encapsulated software system that is situated in an environment and that is capable of autonomous action in that environment in order to meet its delegated objectives [1]. From a technical perspective, the environment consists of anything an agent can perceive through its sensors and act on through its effectors [13]. Apart from objects, agents have control over their internal state and their own behavior; that is, they possess autonomy over their choices. Multiagent systems (MAS) are shared by multiple agents and multiagent environments provide communication means to enable agent inter-action. Further, MAS are typically open without a central designer (i.e., there are *multiple loci of control*) [1]. While MAS typically emerge when agent need to handle a single task and dissolve after the task has been successfully solved, Distributed Problem Solvers in contrast have an a-priori known problem that even influences the development of the agents itself.

Research Directions in Multiagent Organizations: The emerging organi-zational context between agents defines the agents' relationship with each other [14]. Agent organizations provide a framework of constraints and expectations about the agents' behavior with focus on decision-making and action of spe-cific agents [15]. Horling and Lesser present an overview over various multiagent organizational paradigms and organization formation methods [2].

The notion of *coalitions* constitutes a prominent example of organizations within MAS [4]. The concepts used in designing coalition formation are mainly drawn from cooperative game theory and employ automated negotiation among self-interested agents to form coalitions that are stable with regards to some appropriate metric. Coalition formation approaches are closely related to the formation of MA-Organizations. Generally, coalition formation comprises the formation of agent coalitions in situations of partial conflict of interest, though complete information is traditionally assumed [5]. Other approaches consider coalition formation where agents cooperate that are not concerned with their personal payoffs, that is, without any conflict of interest [6]. In contrast, in MA-Organizations, members are bound by contracts, have incomplete information, and do not necessarily pursue compatible goals.

Modeling Perspectives: In MAS research, various perspectives on MA-Organizations exist. Depending on the type and structure, a MA-Organization can be regarded as a system with a given aim that the participating agents try to fulfill or it may result from an emergent process of multiple independently acting agents. Two of the most important perspectives are the *task fulfillment* view with a focus on the overall goal of the MA-Organization as well as the *resource-based view* focusing on the usage of the available resources of the MA-Organization.

Task fulfillment view. In contrast to MAS, Cooperative Distributed Problem Solvers (CDPS) have an a-priori known task to solve that is usually divided into subtasks distributed among the agents [15]. As these subtasks are usu-ally interdependent, the agents solving the subtasks have to cooperate to cope with these interdependencies. The need for cooperation among the agents distin-guishes CDPS from their non-cooperative counterpart. From an organizational perspective, the fulfillment of the overall task is the goal of the MA-Organization. The establishment of an organizational structure may help the MA-Organization to cope with stability issues of a system populated by autonomous agents.

Resource-based view. While the task fulfillment view assumes that an a-priori known task or overall goal is given, the resource-based view focuses on the available resources of a MA-Organization. This viewpoint may be compared with the MAS paradigm, where several agents collaborate to solve a single problem that is in general unknown at design time. Every member of the MA-Organization provides resources that can be used by the MA-Organization to handle single tasks [2]. However, in contrast to MAS that dissolve when the task is fulfilled – like in the task fulfillment view – MA-Organizations need to provide stability for economic applications. The key challenge in the resource-based view is the establishment of stable organizational structures that still provide the necessary flexibility to handle new types of problems or tasks.

2.2 Organizational Theory

We refer to organizational theory to define relevant terms in the context of MA-Organizations and the negotiation process with a potential member. These definitions are independent of the modeling perspective and may be used for a task fulfillment or resource-based view. A *position* in an organization is the smallest autonomously acting organizational unit [16]. As a matter of principle, a position is always created independently of the agent occupying the position [3]. The cardinality of the connection between a position and the occupying agent is usually defined to be at maximum 1. Hence, a position can only be occupied by a single agent. A position is inevitably linked to one or more tasks, competences and responsibilities:

1. **Task.** A *task* is a target performance that is linked to the position and has to be reached by the agent occupying the position. To handle a task the agent in charge has to provide a set of capabilities, e.g., communication, calculation, or learning capabilities. In addition to capabilities, the agent requires access to available resources allowing the agent to handle the task and use its capabilities.
2. **Competence.** The right to act in a certain way is denoted as *competence* in organizational literature [17]. Competences are the formal basis for position-specific influence on the work of agents. Types of competences may differ in various way. For instance, these types include the right to handle an assigned task with a certain degree of freedom or the right to solely be involved in decisions without a direct decision competence.
3. **Responsibility.** The obligation to act in certain way (especially concerning an assigned task) is denoted as *responsibility*. The responsibility has a executive and a legal dimension: The agent obliged to perform a specific action is on the one hand responsible to execute the task. On the other hand, the agent must bear the legal consequences for failure or negligence [17].

Once a task is assigned to a position, the position automatically requires the corresponding competences to handle the task. It bears the responsibility of handling the task in a proper way. In organizational theory, the corresponding match between these terms is known as the *congruency* between task, competence, and responsibility. We assume that each position with an assigned task also receives the appropriate competences to solve the task and bears the corresponding responsibility. The assignment of a task to a position and thus to an agent implies that other members of the MA-Organization expect the agent to handle this task. Expectations in MA-Organizations are called *roles* and may be divided in three different types [17]: (i) The expectations concerning a certain task are denoted as *task-specific roles*. (ii) Expectations may be related directly to the position in form of a *position-specific role*; for instance, the expectations of the top management may be to follow a long-term strategy. (iii) The agent as a member of a MA-Organization can be confronted with *individual-specific roles* resulting from former behavior that is expected for future actions.

2.3 Auctions

Auctions can be used to negotiate terms and conditions for participation in MA-Organizations. Widmer et al. [11] propose a combinatorial auction that is used for the formation process of agent-based virtual organizations. Agents submit bids with multiple attributes specifying the resource quality they are to bring into the organization. These attributes are then aggregated using a scoring function, which is then used to determine the winners of the auction. General scoring auctions with multiple attributes have been proposed by the seminal work of Che [18]. In these auction settings, however, private information exists on the bidder side only. Since in our model, private information exists on both sides (i.e., agent side and MA-Organization side), mechanisms for bilateral trade environments are of particular interest. One seminal piece of work in economy theory considers one buyer and one seller that negotiate for the allocation of a single object [12]. Related work in operations research identifies the optimal strategies of the agents when they are engaged in the so-called "split-the-difference" game [9]. Satterthwaite and Williams [19] generalize their approach by studying the efficiency properties of the bilateral trade model using the k-pricing scheme. In the k-pricing scheme, the buyer's and the seller's payments are based on a share (determined by the value of k) of the total gains of trade. [19] find that none of these mechanisms are ex ante efficient whenever $k \in (0, 1)$. However, if $k \in \{0, 1\}$, mechanisms with k-pricing payments exist that are ex ante efficient. In the context of BIS, research in automated negotiations among multiple agents has been applied to solve coordination problems in task allocation, resource sharing, or surplus division [20]. In this research, we apply the results of the split-the-difference game [9] to model the decision-making process of an agent and a MA organization regarding the participation in the organization.

3 Model

As MA-Organizations usually involve software agents as well as human agents, models of MA-Organizations have to link information systems modeling perspectives with those of economics. Hence, to model this socio-technical system we first provide a model using UML and transfer this to an economic model that is used later for decision making on participation.

3.1   UML Specification

The Unified Modeling Language (UML) provides means to model business information systems as a conceptualization for later implementation. Therefore, UML is used to model MA-Organizations from a software agents perspective. The terms introduced in Sect. 2.2 are transferred to a UML class diagram. Figure 1 gives an overview on the main classes of the UML specification.

A member of a MA-Organization is linked to a specific position that consists of several tasks, competences and responsibilities. The member provides a set of capabilities and resources that are useful for the MA-Organization. This "usefulness" is manifested in the requirements that are necessary for handling a specific task. Hence, this task is also linked to some capabilities and resources required for execution. Three types of subclasses for class *role* represent expectations: (i) Individual-specific role linked directly to the member, (ii) position-specific role linked to a position and (iii) task-specific role linked to a task.

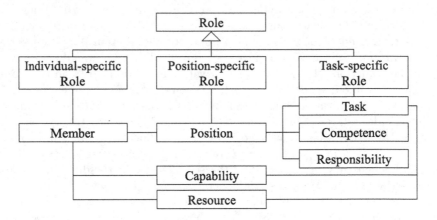

Fig. 1. Overview UML-specification

3.2   Economic Model

We consider a participation mechanism where agent a_i applies for a position in the existing MA organization $\mathcal{M}$, which offers an open position. Position $\wp$ is defined by a task t, which requires capabilities c and resources r for execution. Formally, a position can be described by a 3-tuple defined by the task, the required capabilities and the required resources to complete the task; that is, $\wp = (t, c, r)$. Thus, the proposed model captures only the relevant parts of the UML description provided above.

Agent a_i and MA organization $\mathcal{M}$ privately observe their preferences regarding capabilities and resources for the position $\wp$. Assume that task t is publicly known to both parties. Let vector $\wp_\theta = (\theta_c, \theta_r)$ denote the private information that agent a_i has regarding its capabilities and resources for task t. The capabilities and resources of a_i are private to a_i itself; that is, $\mathcal{M}$ is unaware of a_i's information. Further, let vector $\wp_\sigma = (\sigma_c, \sigma_r)$ denote the private information of the MA organization $\mathcal{M}$ regarding the open position. Hence, $\mathcal{M}$ privately assigns the required capabilities and resources to its open position. Agent a_i does not observe this information. In this framework, capabilities and resources are modeled as real numbers. This model facilitates comparable preference representations of the private information. By assumption, all private information follows a given probability distribution with realizations in the unit interval. In the following, all definitions use vector-valued notation and component-wise vector operations.

For describing the participation decision of agent a_i, we use the binary variable $x(\wp_\theta, \wp_\sigma) \in [0, 1]$. This variable gives the probability that agent a_i is actually joining the MA organization $\mathcal{M}$. We assume that agent a_i derives a utility from participation in $\mathcal{M}$, which is given by

$$u_a(\wp_\theta, \wp_\sigma) = \wp_\theta x(\wp_\theta, \wp_\sigma) - p(\wp_\theta, \wp_\sigma), \tag{1}$$

where $p(\wp_\theta, \wp_\sigma)$ is the certainty equivalent of the payment that a_i makes to $\mathcal{M}$ for joining the organization. Notice that p is a 2-vector. Likewise, $\mathcal{M}$ derives a utility from occupying its open position by agent a_i given by

$$u_M(\wp_\sigma, \wp_\theta) = p(\wp_\theta, \wp_\sigma) - \wp_\sigma x(\wp_\theta, \wp_\sigma). \tag{2}$$

We assume that all parties are risk neutral such that the utilities take a quasi-linear form. The objective is the maximization of the social welfare, which is defined as the sum of all utilities. Clearly, the ex post social welfare is maximized if and only if $\wp_\theta \geq \wp_\sigma$.

4   Decision Making on Participation

We describe the process of decision making on participation as a bilateral negotiation auction. Agent a_i submits its sealed bid $\hat{\wp}_\theta = (\hat{\theta}_c, \hat{\theta}_r)$ specifying its preferences for the open position. Similarly, MA organization $\mathcal{M}$ submits its sealed bid $\hat{\wp}_\sigma = (\hat{\sigma}_c, \hat{\sigma}_r)$ describing its preferences for the required candidate. The bids are submitted simultaneously. Based on the reported bids, the social planner determines whether agent a_i participates in $\mathcal{M}$, and how high the certainty equivalent is that a_i has to pay to $\mathcal{M}$. Consider the following participation mechanism. The decision rule is given by

$$x(\hat{\wp}_\theta, \hat{\wp}_\sigma) = \begin{cases} 1 & \text{if } \hat{\wp}_\theta \geq \hat{\wp}_\sigma \\ 0 & \text{otherwise} \end{cases} \tag{3}$$

with certainty equivalents transferred between $\mathcal{M}$ and a_i of

$$p(\hat{\wp}_\theta, \hat{\wp}_\sigma) = \begin{cases} \frac{1}{2}(\hat{\wp}_\theta + \hat{\wp}_\sigma) & \text{if } \hat{\wp}_\theta \geq \hat{\wp}_\sigma, \\ (0, 0) & \text{otherwise.} \end{cases} \tag{4}$$

Notice that the relational operator "$\geq$" used in (3) and (4) is a component-wise operator. The decision rule of the participation mechanism allocates agent a_i to MAS $\mathcal{M}$ if and only if both the reported capabilities and resources of agent a_i is greater or equal the corresponding reports of MA organization $\mathcal{M}$. The associated pricing scheme used in this mechanism is based on the k-double auction; in particular, the certainty equivalents defined in (4) are calculated with $k = 0.5$, which constitutes the average of the two bids [19, p. 108].

By construction, the mechanism never runs a deficit; that is, it balances the budget. In addition, the mechanism definition ensures that each party is never worse-off after the mechanism is run; that is, the mechanism is individually rational for all parties. Since participation takes place whenever a_i's valuation is greater or equal $\mathcal{M}$'s valuation, the mechanism maximizes the social welfare ex post. However, standard impossibility results from mechanism design assert that such mechanisms cannot be incentive compatible [12].

Example 1. Suppose agent a_i's private information regarding the open position for task t is given by $\wp_\theta = (0.7, 0.8)$; that is, a_i's capabilities are expressed by 0.7 and its resources by 0.8. Similarly, $\mathcal{M}$'s private information is $\wp_\sigma = (0.5, 0.6)$. Assume that both a_i and $\mathcal{M}$ report their private information truthfully. Since $(0.7, 0.8) \geq (0.5, 0.6)$ component-wise, a_i is hired by $\mathcal{M}$. The certainty equivalent is calculated as $p(\hat{\wp}_\theta, \hat{\wp}_\sigma) = \frac{1}{2}[(0.7, 0.8) + (0.5, 0.6)] = (0.6, 0.7)$. Notice that the certainty equivalent is defined component-wise; that is, the transfer from a_i to $\mathcal{M}$ is realized as two separate payments. The utility of a_i is then $u_a(\wp_\theta, \wp_\sigma) = (0.7, 0.8) - (0.6, 0.7) = (0.1, 0.1)$ and the utility of $\mathcal{M}$ is $u_M(\wp_\sigma, \wp_\theta) = (0.6, 0.7) - (0.5, 0.6) = (0.1, 0.1)$.

Example 1 assumes truthful bidding of both parties. However, the following example shows that the strategic behavior of each party can influence the certainty equivalent.

Example 2. Assume with Example 1 that $\wp_\theta = (0.7, 0.8)$ and $\wp_\sigma = (0.5, 0.6)$. Let $\mathcal{M}$ report its private information truthfully; that is, $\hat{\wp}_\sigma = \wp_\sigma$. Suppose, a_i manipulates its bid by understating its true preferences by 0.1 in each component such that $\hat{\wp}_\theta = (0.6, 0.7)$. Trade takes place because $\hat{\wp}_\theta \geq \hat{\wp}_\sigma$, and the certainty equivalent is calculated based on the reports as $p(\hat{\wp}_\theta, \hat{\wp}_\sigma) = (0.55, 0.65)$. The utility of a_i is now given by $u_a(\hat{\wp}_\theta, \hat{\wp}_\sigma) = (0.7, 0.8) - (0.55, 0.65) = (0.15, 0.15) > (0.1, 0.1)$. Hence, agent a_i is able to increase its utility by 0.05 in each component when understating its true valuation by 0.1. The manipulation of agent a_i caused a decrease in the certainty equivalent, which in turn raised its utility. A similar argument shows that MA organization $\mathcal{M}$ can also increase its utility by overstating its true valuation, assuming that a_i reports truthfully. This example confirms that the mechanism is not incentive compatible.

The preceding discussion shows that the proposed participation mechanism maximizes the social welfare ex post, guarantees individual rationality and satisfies budget balance. However, the mechanism is not incentive compatible because it cannot prevent the parties from manipulating their bids. Research in auction theory finds that k-pricing mechanisms for bilateral trading environments are always inefficient whenever $k \in (0,1)$ [19]. When $k \in \{0,1\}$, however, these mechanisms achieve ex ante efficiency. Ex ante efficiency means that agent a_i on average cannot be made better off without making $\mathcal{M}$ on average worse off, and vice versa [19, p. 115], while incentive compatibility, individual rationality, and budget balance are satisfied. Theorem 2.2 in [19] provides techniques necessary for making the proposed mechanism ex ante efficient.

5   Evaluation

This section reports an experimental evaluation of the participation mechanism developed in this proposal in the form of a proof-of-concept. The evaluation setup is based on the concept of the labor market auction model, in which businesses bid for employees (e.g., [21]). In our research, each business is represented by a MA-Organization and each potential employee is represented by a single software agent. The agent applies for an open position in the MA-Organization by reporting the amount of resources it is willing to invest into the MA-Organization. In turn, the MA-Organization reports the amount of resources it requires for the open position. In the following, we describe the setup, report the results, and discuss the findings.

5.1   Experimental Setup

The experimental evaluation considered uniformly distributed private information of agent a_i and MA-Organization $\mathcal{M}$. For ease of exposition, we modeled the offered/requested *resources* as the single parameter; that is, $\theta_r \sim U(0,1)$ and $\sigma_r \sim U(0,1)$. The MA-Organization offers an open position $\wp$ by submitting a sealed bid $\hat{\sigma}_r$, which reflects $\mathcal{M}$'s required resources for the position. Simultaneously, agent a_i submits a bid $\hat{\theta}_r$, reflecting the maximum amount of resources a_i is willing to bring into the MA-Organization. Based on the reported bids, the mechanism calculates the certainty equivalent that is transferred from a_i to $\mathcal{M}$ if a contract is established. In particular, the utilities of each party were calculated in two settings. First, we assumed truthful bidding; that is, $\hat{\theta}_r = \theta_r$ and $\hat{\sigma}_r = \sigma_r$. The utilities of truthful bidders served as a benchmark to measure the utility gain of manipulating agents. Second, we assumed manipulating bidders. By assumption, both parties manipulate their bids by a linear manipulation function in the following manner. Agent a_i understates its true resource offer by manipulation factor $\mu_a \in (0,1)$; that is, a_i's manipulation function is $\hat{\theta}_r = \mu_a \theta_r$. In contrast, $\mathcal{M}$ overstates its true resource requirement by manipulation factor $\mu_m \in (1,2)$ such that $\hat{\sigma}_r = \mu_m \sigma_r$. The utility gain ratio is then calculated as the utility with manipulating bidders divided by the utility of truthful bidders. Similarly, the total efficiency loss is given by the ratio between the sum of manipulating utilities and the sum of truthful utilities.

5.2   Results

Figure 2 illustrates the impact of manipulation on the performance of the participation mechanism. The left side shows the utility gain and the overall efficiency loss of the mechanism when agent a_i misrepresented its resource offer, while $\mathcal{M}$ reported truthfully. A utility gain ratio (red graph) of greater 1 indicates that a_i was able to increase its utility by manipulation. For $\mu_a = 0.7$, a_i attained its maximum gain in utility on average (i.e., understatement by 30 %). The overall efficiency loss (blue graph) decreased monotonically once a_i's manipulation activity increased. Obviously, full efficiency is achieved for truthful reporting (i.e., $\mu_a = 1$). The right side of Fig. 2 shows the reversed case. Here, when $\mathcal{M}$ manipulated its required resource, while a_i reported truthfully, the utility gain ratio was always greater 1. The maximum utility gain is reached at $\mu_m = 1.3$; hence, overstating by 30 % yielded the highest utility gain for $\mathcal{M}$.

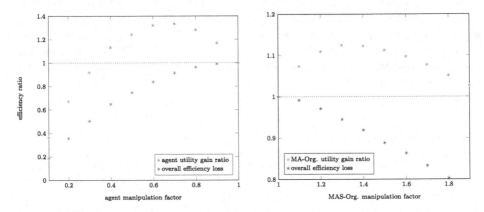

Fig. 2. Efficiency loss and utility gain ratio as a function of manipulation factor. (Color figure online)

5.3   Discussion

Our experiment demonstrates the impact of manipulation on the agents' utilities and provides an estimation of the mechanism's overall efficiency loss. These findings provide evidence for the efficacy of the proposed mechanism.

Our findings can be compared to the results obtained by Chatterjee and Samuelson [9], who studied the bilateral bargaining game with incomplete information. In particular, they derived the equilibrium strategies of both buyer and seller. Using the same pricing scheme as the one in our work ($k = 1/2$), the seller overstates its true resource requirement by $2/3\sigma_r + 1/4$. In our work, $\mathcal{M}$ obtains the highest utility gain when he chooses the strategy $\mu_m = 4/3$ (cf. right side of Fig. 2). These two strategies intersect at $\hat{\sigma}_r = 1/2$. Although our manipulation scheme does not implement the equilibrium strategies (see [9]), our results imply that the proposed mechanism provides a good approximation for the optimal strategic behavior of $\mathcal{M}$.

For the buyer, [9] showed that the equilibrium strategies are given by $2/3\theta_r + 1/12$. In our proposal, agent a_i attains the highest utility gain when he reports $\hat{\theta}_r = 2/3\theta_r$. Notice that this manipulation factor just differs from the optimal strategy by $1/12$. In fact, both strategies are parallel and shifted by $1/12$. Again, this result implies that the agent's manipulation scheme proposed in this work provides a suitable approximation for the equilibrium strategies found by [9].

At the optimal manipulation factor of MA-Organization $\mathcal{M}$, the total efficiency loss was approximately 5%, when a_i reported truthfully. However, once a_i misrepresented its value by its optimal manipulation factor $\mu_a = 2/3$, the total efficiency loss was close to 10%. The reason for the higher efficiency loss is that a_i's utility gain ratio is higher than $\mathcal{M}$'s utility gain ratio. In fact, a_i was able to increase its utility by approximately 33% by manipulating its true value, while $\mathcal{M}$'s maximal utility increase accounted for 12%. The higher utility gain of a_i as compared to $\mathcal{M}$ is compensated by a potential utility loss for $\mu_a < 1/3$. In contrast, $\mathcal{M}$ never experienced any utility loss for all manipulation factors. This permanent utility gain implies that it is always advantageous for $\mathcal{M}$ to misrepresent its true reservation values.

6   Conclusion

The contribution of this research is twofold. First, we provide a UML specification of a MA-Organization in a business information system. Second, we propose a bilateral negotiation mechanism to model the decision-making process of an agent and the MA-Organization for participation. In our experiment, we found that a rather simple strategic manipulation scheme employed by both sides provides a suitable approximation for the equilibrium strategies identified by [9]. This finding is significant because it indicates that agents and MA-Organizations can focus on surprisingly simple strategy calculations without deviating too much from the actual equilibria. Future research might be pursued into the direction of multi-attribute decision-making. In this work, the agent's capabilities and resources for the desired position in the MA-Organization are interpreted as two separate components in the negotiation mechanism. This limitation could be removed by using a multi-attribute auction approach with scoring functions similar to other approaches in auction theory [11,18].

Acknowledgments. This work has been supported by the project InnOPlan, funded by the German Federal Ministry for Economic Affairs and Energy (BMWi, FKZ 01MD15002).

References

1. Jennings, N.: On agent-based software engineering. Artif. Intell. **117**, 277–296 (2000)
2. Horling, B., Lesser, V.: A survey of multiagent organizational paradigms. Knowl. Eng. Rev. **19**(4), 281–316 (2004)

3. Picot, A.: Organisation. Schäffer-Poeschel Verlag, Stuttgart (2005)
4. Sandholm, T., Lesser, V.: Coalitions among computationally bounded agents. Artif. Intell. **94**, 99–137 (1997)
5. Kahan, J.P., Rapoport, A.: Theories of Coalition Formation. L. Erlbaum Associates, London (1984)
6. Shehory, O., Kraus, S.: Methods for task allocation via agent coalition formation. Artif. Intell. **101**(1–2), 165–200 (1998)
7. Wooldridge, M., Jennings, N.R., Kinny, D.: The Gaia methodology for agent-oriented analysis and design. Auton. Agent. Multi-Agent Syst. **3**(3), 285–312 (2000)
8. Bulling, N.: A survey of multi-agent decision making. Künstliche Intelligenz **28**, 147–158 (2014)
9. Chatterjee, K., Samuelson, W.: Bargaining under incomplete information. Oper. Res. **31**(5), 835–851 (1983)
10. Premm, M., Widmer, T., Karänke, P.: Bid-price control for the formation of multiagent organisations. In: Klusch, M., Thimm, M., Paprzycki, M. (eds.) MATES 2013. LNCS, vol. 8076, pp. 138–151. Springer, Heidelberg (2013)
11. Widmer, T., Premm, M., Karaenke, P.: Sourcing strategies for energy-efficient virtual organisations in cloud computing. In: Proceedings of the 15th IEEE Conference on Business Informatics (CBI 2013) (2013)
12. Myerson, R.B., Satterthwaite, M.A.: Efficient mechanisms for bilateral trading. J. Econ. Theor. **29**, 256–281 (1983)
13. Russell, S.J., Norvig, P.: Artificial Intelligence: A Modern Approach, 2nd edn. Prentice Hall, Upper Saddle River (2003)
14. Ferber, J., Gutknecht, O.: A meta-model for the analysis and design of organizations in multiagent systems. In: Proceedings of the Third International Conference on Multi-agent Sytstems (ICMAS 1998) (1998)
15. Bond, A.H., Gasser, L. (eds.): Readings in Distributed Artificial Intelligence. Morgan Kaufmann Publishers, San Mateo (1988)
16. Thom, N.: Stelle, Stellenbildung und -besetzung. In: Frese, E. (ed.) Handwörterbuch der Organisation, 3rd edn, pp. 2321–2333. C.E. Poeschel Verlag, Stuttgart (1992)
17. Hill, W., Fehlbaum, R., Ulrich, P.: Organisationslehre, 5th edn. Paul Haupt, Berlin (1994)
18. Che, Y.K.: Design competition through multidimensional auctions. RAND J. Econ. **24**, 668–680 (1993)
19. Satterthwaite, M.A., Williams, S.R.: Bilateral trade with the sealed bid k-double auction: existence and efficiency. Econ. Theor. **48**(1), 107–133 (1989)
20. de la Hoz, E., López-Carmona, M.A., Marsá-Maestre, I.: Trends in multiagent negotiation: from bilateral bargaining to consensus policies. In: Ossowski, S. (ed.) Agreement Technologies. Springer, Dordrecht (2012)
21. Kelso, A., Crawford, V.P.: Job matching, coalition formation, and gross substitutes. Econometrica **50**(6), 1483–1504 (1982)

Bridging the Gap Between Independent Enterprise Architecture Domain Models

Thomas Stuht[1(✉)] and Andreas Speck[2]

[1] PPI AG Informationstechnologie, 22301 Hamburg, Germany
Thomas.Stuht@ppi.de
[2] Institute of Computer Science, Christian-Albrechts-University of Kiel,
24098 Kiel, Germany
aspe@informatik.uni-kiel.de

Abstract. An Enterprise Architecture (EA) provides a holistic view about the domains to support planning and management tasks. The creation can be made more efficient if present domain models and measures are integrated. But these sources often lack coordination and thus are rather isolated. The paper proposes an approach based on Semantic Web technologies to combine these sources. A key aspect is the indirect connection through bridging elements to reduce the effort to establish an EA. These elements and a small EA vocabulary are the basis for an integrated data pool being a "mash up" instead of a new data silo.

Keywords: Enterprise architecture · Domain models · Bridging elements · Semantic web · Integration · Alignment

1 Introduction

Enterprises ace significant challenges to operate in complex business ecosystems. There is a strong need to holistically coordinate their domains, e.g. information technology (IT), business and risk management [1]. The domains document their information in separate models and often with no central ownership [2]. Specialized tools are used to document the models [3]. Moreover, these domain often have different terminologies and understandings making it complicated to decide from a holistic view [4]. An Enterprise Architecture (EA) promises to offer such a holistic view. It captures the elements of an enterprise and their relations [2]. In most approaches the EA contains only aggregated artifacts, e.g. [5]. Detailed elements as well as measures are thus not present when accessing the EA later.

The study in [6] demonstrated that filling the EA is predominantly a manual task. However, more than 90 % of the participants are faced with one or more challenges, e.g. laborious data collection, low data quality and integration problems. Following the study the EA implementation would benefit from more automated approaches. Heterogeneity and missing connections of domain models [1] lead to the challenge how to integrate these models to form a consolidated EA. Literature shows that the problem of data integration in an EA context is still a relevant topic for research and improvement, e.g. [1,7].

© Springer International Publishing Switzerland 2016
W. Abramowicz et al. (Eds.): BIS 2016, LNBIP 255, pp. 277–288, 2016.
DOI: 10.1007/978-3-319-39426-8_22

This is consistent with the authors' observations of industrial projects they participated. Present approaches often assume a fixed meta model or homogeneous domain models with extensively connected elements. Furthermore, detailed information or measures are rarely integrated. Starting from these considerations, the authors' interest is guided by two research questions: *RQ1: How can detailed domain models without nearly any interconnection but with heterogeneous terminologies be integrated in a joined EA model? Can this be achieving with limited effort by a generic approach? RQ2: How can the EA be efficiently accessed using an arbitrary terminological level on top of the diverse domain languages?*

We chose a design science research method following Hevner et al. [8]. The remainder of this article is outlined as follows. Section 2 presents the foundations and the approach objectives. Section 3 describes the proposed approach. Followed by its demonstration in Sect. 4. Section 5 presents the related work and Sect. 6 discusses the approach. Section 7 concludes the paper.

2   Theoretical Background

The first subsection focus on the relationship between the domain models and the holistic EA. The integration of domain models into the EA result in different challenges. For the technical foundation Semantic Web technologies are recommendable. Integration is a main ability within the Semantic Web as described in the second subsection. Another very important feature in contrast to database management systems is the explicit representation of meaning. Queries can utilize these semantics for smart answering.

2.1   Fragmentation of an EA and Its Domain Models

The EA benefits arise from the **documentation** of an enterprise and the **analysis** of this information [9]. A distinction is made between the **domain models** as the primary source of detailed information and the EA as a holistic view. The content of an EA is not generally determined but is often documented on an aggregated level. It highly depends on the used framework and the individual focus of an enterprise which information is required (see e.g. [5,9]).

The interaction between enterprise domains may be problematic in industrial practice. Different terms, understandings and methods makes them rather **isolated** to each other. Furthermore, the domain models may be stored within different application systems, databases, spreadsheets or other sources. This hinders the aggregation and relating the various artifacts becomes a challenge [1,2,5].

A common EA concept in literature and industry is the notion of **capabilities**. They represent an ability an enterprise can perform and for which several components (processes, applications, people and so on) are combined. This decouples the involved components on a logical level. The sets of capabilities are comparable within the same industry sector. So templates could be used to create the very own capability set (see e.g. [10]).

Based on these aspects and authors' experiences from projects at financial institutions there are main **objectives** of an integration approach. It has to accept the logical isolation of the domain models. And it has nevertheless to overcome this issue and create a holistic view based on the given information. It should efficiently address industry needs and support an arbitrary EA terminology.

2.2 Foundations of Semantic Web with Focus on Integration

The Semantic Web represents the vision that content can be processed by machines or applications. **Meaning** is made explicit for example by formulating inference rules to automatically draw **conclusions** on present information. This enables a reasoner to deduce new knowledge [11,12]. Several basic **technologies** form the basis of the Semantic Web. This section describes the relevant technologies and main aspects that are required for the presented approach.

The **data model** of the Semantic Web is RDF (Resource Description Framework, see [13]). RDF represents a piece of information as a **triple**, also called statement, similar to the sentence structure using subject, verb and object. The components are normally identified through a globally **unique IRI** (Internationalized Resource Identifier), similar to an URL [11–13].

The triple structure is very simple but it allows a flexible documentation of information. It can be considered as a **graph** structure with the subject and object resources as nodes and the verbs (or predicates) as edges between them [13].

On top of RDF another vocabulary has been built, called RDF Schema or RDFS (see [14]). It allows to define a **custom data model** and the terminology to be used for a domain [12]. RDFS differentiates between classes and their instances [14]. RDFS provides a good balance between expressiveness and reasoning performance [15]. **Querying** is supported by the SPARQL standard which is in a way comparable to SQL [12,15].

Further specialized languages exist, for example SKOS (Simple Knowledge Organization System, see [16]) for describing **controlled vocabularies**. SKOS is based on RDF so the information can be linked with other RDF data [16].

A frequently used technique is mapping different graph structures to perform integration. This enables the integration of information sources (concepts and individuals), represented in RDF [12]. **Merging** is also easy to achieve because the RDF data sources form graphs and the IRIs are globally unique. Therefore the graphs can be combined [15].

3 Alignment of Separate EA Domain Models

This section presents the approach to integrate independent EA domain models. It consists of six steps as visualized in Fig. 1. However, the steps are not performed only once. Using an EA requires up-to-date information so that the steps are to be performed frequently - preferably automatically.

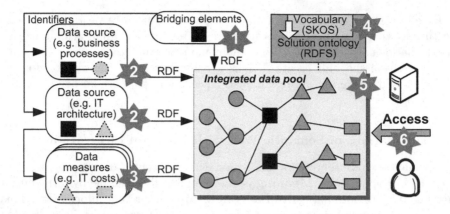

Fig. 1. Integration approach for independent domain models consisting of six steps

Step 1: Provide Bridging Elements for Indirect Relationships. Initially a set of bridging elements have to be defined: concepts on the ontological level and instances for the connection. These are used for the indirect relationship between domain models. Therefore, domain models do not have to be extensively interconnected and can be independent except for sharing the bridging elements.

This kind of indirection is not unique to Semantic Web. For example [17] use an alignment architecture to the align business processes and IT applications. Functional services are used as the connecting elements, but they do not exist within the detailed models. However, the approach at hand does not focus on decoupling connected models but on bridging independent models. Moreover, the set of bridging elements is not restricted. Business capabilities, business objects, etc. may be candidates. The bridging instances should be stable and have to be uniquely identifiable. It is useful to coordinate the set with domain stakeholders because the elements will be referenced within the domain models later.

The step's output is twofold. First a shared set of identifiable bridging elements is provided. Second, an RDF representation is created that can be imported.

Step 2: Prepare the Domain Models. The domain models are the information base. It is hence required to make the data available to the EA. This is often done based on aggregated information only. Details for more comprehensive analyses are thus not present in the EA (e.g. see [1,5,9]).

The domain models are the main input of this step. They may be stored in various formats or source types, for example relational databases, applications or spreadsheets. But the sources may use different terminologies. Moreover, it is assumed that they lack comprehensive interconnections. Therefore, the second input is the set of bridging element identifiers from step 1. Their aim is to indirectly connect the models. Embedding them depends on the data source.

The instances within the domain model are linked to the bridging instances they belong to. Possible sources are tools storing the IT architecture or risk aspects. Another example is a process management tool. The bridging instances could be embedded within BPMN process models. They may be used as another level of nested lanes so the tasks and other elements are implicitly related to their corresponding bridging instance.

The step's goal is the transformation of each domain model into RDF using common or individual ontologies. Solutions are available to support this [12,15]. Hence, the RDF representations including the bridging instances are the output.

Step 3: Prepare the Sources of Data Measures. EA approaches often focus on the concepts stored in domain models and seem to exclude measures. Examples are process times or costs. Including measures would augment the holistic view and allow for more detailed analyses. Source types can be e.g. databases or spreadsheets. The transformation into RDF is analogous to domain models. The output of this step therefore is also a RDF representation of each source.

In contrast to step 2, this step does not use the bridging instances. The data values directly refer to the elements of the domain models. For example, the IT personnel costs are directly assigned to the application for contract management. Besides, the units should be explicitly added in the transformation process. Different units may thus be automatically handled at the integration in step 5. A recommendation is to use common agreed IRIs to reference a specific unit, e.g. based on the known dbpedia[1] as a pragmatic solution.

Step 4: Use a Vocabulary to Produce a Solution Ontology. An EA vocabulary helps to ease the communication between people from different domains with individual terminologies and understandings. It contains main concepts and relations. The vocabulary acts as an overlaying individual terminological layer and it forms the basis for the EA. The domain models retain their leading role for documentation and each model may use its own terminology. The integrated data pool is thus not a new data silo but just a "mash up". A draft of the vocabulary can be created within a workshop and may be refined later at any time. It requires less effort as creating an extensive ontology or EA meta model and it is allowed to be more dynamic than a formalized EA meta model.

Using SKOS to represent the EA vocabulary is preferable because it is a popular Semantic Web technology and several management tools exist.[2] No broad technical knowledge is required. A term hierarchy for each source model may be defined. SKOS contains predefined means to represent the parent-child relationships. Furthermore, concepts can refer to other concepts with a neutral "related"-relationship instead of diverse relationships like "used by" or "calls".

In addition, domain concepts link to corresponding data types, for example using IRIs from dbpedia to make this relationship explicit.

The SKOS vocabulary should be augmented to a RDFS based solution ontology as shown in Fig. 1. The terms and relationships can be supplemented by

---

[1] For example http://dbpedia.org/resource/Euro references the European currency. (Accessed Dec 17, 2015).

[2] See e.g. TemaTres http://www.vocabularyserver.com/. (Accessed Dec 28, 2015).

classes and relationships for convenience, e.g. relationships with meaningful names and standard RDFS relationships to mirror the parent-child relationships. The resulting solution ontology is the output of this step and at the same time the terminological basis for the integrated data pool established in the next step.

Step 5: Align and Combine the Sources with the Ontology. All previous outputs (bridging elements, transformed data sources, solution ontology) are joined together in this step as shown in Fig. 1. The solution ontology acts as the shared terminological layer. It is sufficient to establish mappings from each source just to the solution ontology. There are no mappings between sources among one another. The task of matching ontologies is a research topic on its own.

A key aspect is the indirect connection based on the bridging elements. The bridging concepts link to the corresponding concepts within the solution ontology and domain models. The bridging instances play the role of a hub when joining together the data sources. They indirectly establish the identity connections. All other relevant classes from the sources are mapped to their correspondences in the solution ontology. If a class from a domain model has related data types and values then are mapped to the solution ontology, too. As a result, the solution ontology is the central access point to the joined data pool. An example scenario with two domain models and one source of data measures all mapped to the solution ontology is shown in Fig. 2.

The mapping may be supported by using design patterns for alignments, e.g. [18]. They provide patterns of typical alignment constellations. Moreover, mappings between properties make relationships explicit and accessible using the EA vocabulary later. In the case of measure mappings the linked units facilitate automatic conversions.

After the specification of mappings they have to be processed to actually integrate the information. One option is to transform them to a Semantic Web rule language. Together with the basic RDFS inference rules a reasoner can use that to infer new RDF statements that realize the mappings. Mapping modifications are only necessary if the structures of the sources or the EA vocabulary change. But they can be processed frequently to keep the data pool updated.

At the end of this step all information have been integrated and mapped. This results in the integrated data pool as the main output of this step. It is the starting point of all further use cases based on this data pool.

Step 6: Access the Integrated Data Pool. The integrated data pool contains all information of each domain model and source of data measures. This allows to use domain model internal relationships not directly mapped to the solution ontology. Because all information is stored as a graph these relationships can be traversed transparently. Moreover, more detailed assessments are possible.

The bridging instances indirectly link the models. And the solution ontology facilitates to access the data pool using a neutral terminology. Despite the abstraction through the bridging elements comprehensive queries are possible with SPARQL. The graph can be large so the processing may be optimized. One

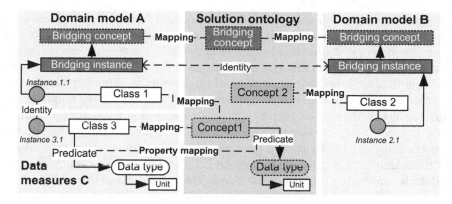

Fig. 2. Mapping scenario with two domain models and one data measure source

example is to cut off search paths if certain criteria are met. This depends on the specific use case based on a heuristic.

4 Showcase for an Exemplary Application of the Approach

The application of the approach should be demonstrated by a showcase taken from a project one of the authors worked for. The client was a medium-sized German insurance company. The business organization department aims to build up an EA for a holistic view. The departments (IT department and different segments) have established methods for documenting their domains. But the contents have not been coordinated and thus the models were not aligned. The implementation of a central EA tool is not possible without resistance. At this point the present approach could be used for a first alignment of the models offering a basic holistic view with limited effort.

The first input is a collection of business capabilities for the insurance sector stored within a spreadsheet. It acts as the set of bridging instances. This structure can be transformed into RDF, e.g. with a tool like "RDF123"[3]. The processes are documented using BPMN and thus the proposed extension with lanes for the bridging instances is applied. BPMN is based on XML so the transformation to RDF can be performed with a converter. The IT artifacts are documented in an IT management tool storing the information in a relational database. The tool "D2RQ"[4] creates a RDF representation of that database. For simplicity the data measures are stored in spreadsheets. Each row refer to an artifact (e.g. concrete process name or IT system) and specify the measure values. A short workshop prepares an EA vocabulary and documents it in a SKOS tool.

---

[3] RDF123: http://ebiquity.umbc.edu/project/html/id/82/RDF123. (Accessed Dec 13, 2015).

[4] D2RQ Platform: http://d2rq.org/. (Accessed Dec 22, 2015).

A prototype based on a Java web framework and Apache Jena[5] (framework for applications based on Semantic Web technologies) has been developed. It provides a broad range of functions, e.g. handling RDF, inferencing and querying. The prototype can import the transferred data sources and the SKOS vocabulary. Figure 3 shows the alignment page of the prototype. The user graphically aligns the data sources' classes with the vocabulary concepts. Each mapping can be further detailed through restrictions or property mappings by opening an edit dialog box. The processing of the alignments happens in the background to create the integrated data pool. The user can access it on another view to ask for e.g. assignments of artifacts using the EA vocabulary. We cannot go more into details concerning the prototype due to the limited space available.

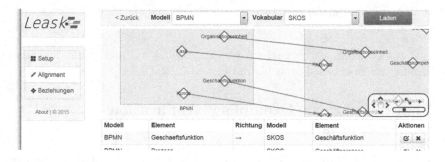

Fig. 3. Screenshot of the prototype: The alignment view enables the user to define the mappings between the data source's classes (left) and vocabulary concepts (right)

5   Related Work

To provide a holistic view about an enterprise using an EA is a common technique. Integration is thus a key problem in particular if the goals are to achieve efficiency and practicability as well as the requirement of up-to-date information.

The authors in [19] also use Semantic Web technologies. Whereas they start with a present EA model. They describe how to transform the model with manual effort into a semantic representation providing access to the information base. In contrast to that, the paper at hand assumes no previously defined consistent EA model and aims at reducing the manual effort.

The aim of [20] is to integrate main artifacts from data sources into the EA through creating a semantic representation. The central artifacts of the data sources are formalized using an ontology and extensively linked with the imported artifacts from the other data sources. The sources are therefore linked one another which determine the available terms. A difference to the present approach is the higher effort for extensive linkage compared to the use of the bridging elements as well as more technical expertise for using logical languages.

---

[5] Apache Jena: https://jena.apache.org. (Accessed Dec 23, 2015).

The authors in [21] utilize ArchiMate as a consistent meta model. Domain-specific models are directly mapped to this meta model. The alignment of concepts and instances is done using direct links leading to a high effort. [22] also starts from a harmonized EA model, again ArchiMate, and transforms it into an ontology. Concrete instances are not in the focus of the paper. [23] propose a methodology to completely annotate models using a highly abstracted meta model. Model information and relationships should be documented in a machine-processable way. This could act as a foundation for further integration, but the work does not describe an EA integration process. [24] offers an EA toolset with a best practice meta model. The administration tools are technically oriented whereas the separated viewing components are focused on business users. The integration of existing sources is possible but requires manual effort, technical expertise and knowledge about the meta model.

The review shows that Semantic Web technologies are commonly used for integration aspects. But most approaches assume a consistent EA model as a starting point or that the elements have to be mapped directly. This entails a certain effort in the initial phase as well as in later use. Moreover, measures or detailed information are usually not taken into account. The present approach contributes to the state of the art by proposing indirect connections to reduce the effort. Besides, it is independent from a certain meta model and uses only a vocabulary. Furthermore, the consideration of measures is an explicit part.

6 Discussion and Limitation

The research questions in Sect. 1 have guided the research work. RQ1 focuses on a generic approach to efficiently integrate detailed domain models into an EA. RQ2 deals with the access to this EA by means of a terminological layer.

Addressing RQ1, the approach offers an alignment of previously isolated domain models. Due to the stable bridging elements there is no need of extensive direct connections between the data sources as present approaches often assume (see Sect. 5). The augmented EA vocabulary is the basis for an improved inter-communication between the stakeholders. This aspect targets RQ1 and RQ2. The vocabulary is independent of an EA framework in opposite to other approaches where often a fixed meta model is used. The industrial practice motivates this flexibility as the variety of custom or reference meta models show [25].

The approach aims to ease the EA implementation and to deliver benefits quickly. [26] states that the establishment takes a long time in industrial practice. This includes the creation of a complete meta model. Using a simple vocabulary could reduce this effort and ease the maintenance when the terminology develops. The integration also eliminates the laborious manual creation of the EA, e.g. [26]. Semantic Web technologies facilitates this by means of explicit semantics and machine-processable information. Reducing the effort is important because [7] states that the additional effort for stakeholders could have a negative impact on EA acceptance. In addition, [7] reports on failed initiatives because the documentation take too much effort or the maintenance was not organized.

The mappings between the domain model concepts and the vocabulary terms must obviously be created. But this happens ideally once because they are relatively stable. It also may be supported by matching algorithms. The bridging elements make direct connections between elements of different domain models unnecessary. It must be ensured that the elements are related to the bridging instance within the domain model. This might be seen as a limitation because it must be possible to link a domain element with a bridging element. Although a mapping table may be used as a substitute to document the relation.

Indirect linkage offers the advantage to decouple domain models as proven by e.g. [17]. It helps to overcome the different rates of the changes occurring in the domains [4] as well as the problem to quickly adjust the EA content according to those changes [6]. The domain models can develop independently but at the same time the EA consists of up-to-date information about the current state.

Reflecting the relationship between domain models and the EA in general there are resistances conceivable. Stakeholders may assume a reduction of autonomy if the EA is the new planning basis. According to [27] this may lead to resistance against an enterprise-wide coordination. The approach could potentially reconcile the interests because the domain models are the integral part of the EA data pool. At the same time a holistic view and coordination is possible.

On first view the dependence from the domain models and their heterogeneous terminologies might be problematic. This may hinder effective collaboration if understandings vary [28]. Although the approach could positively influence enterprise collaboration. The consolidation by using the EA vocabulary seems less serious than to introduce a comprehensive meta model from the perspective of effort and acceptance. Besides, it is not reasonable to imagine that all domain tools, used for specialized tasks [3], can either be completely replaced by one consolidated EA tool or entirely customized to adhere to one unique meta model.

An assumption of the solution is that indirect connections and thereby an abstraction is suitable for the EA use case. Another assumption is that the domain models are either directly accessible or can be exported in order to transfer them to RDF. Whereas a domain tool that cannot extract its data should be called into question from a general point of view. In summary the approach aims to lower the effort creating a first usable EA and to offer EA benefits early.

7   Conclusion

It seems reasonable to use existing sources whenever possible instead of creating a new data silo for the EA repository with a laborious data acquisition. But in the industrial practice there are often obstacles that hinder easy realization. Domain models are often neither coordinated nor compatible to each other. That has been the motivation for the proposed approach to align domain models forming an integrated data pool. An important aspect is the notion of bridging elements to indirectly connect the domain models. Semantic Web technologies are the basis of this approach. In the integration step the domain models are aligned with the solution ontology building the integrated data pool.

The related work show the increasing attention being paid to EA supported by Semantic Web technologies. But this has not reached a high level of maturity thus the research have to be focused on practicability. The presented approach contributes by making a proposal to facilitate the EA establishment and reduce the required effort and time. It provides the basis for further use cases. Therefore the further work will be twofold. On the one hand it should be focused on the application and evaluation of the approach in more scenarios. On the other hand usage scenarios have to be investigated that use the integrated data pool.

References

1. Lankhorst, M.: Enterprise Architecture at Work: Modelling, Communication and Analysis, 3rd edn. Springer, Heidelberg (2013)
2. Jonkers, H., Lankhorst, M.M., ter Doest, H.W.L., Arbab, F., Bosma, H., Wieringa, R.J.: Enterprise architecture: management tool and blueprint for the organisation. Inform. Syst. Front. **8**(2), 63–66 (2006)
3. Hanschke, I.: Enterprise Architecture Management - einfach und effektiv: Ein praktischer Leitfaden für die Einführung von EAM. Hanser, München (2012)
4. Sandkuhl, K., Wißotzki, M., Stirna, J.: Unternehmensmodellierung: Grundlagen, Methode und Praktiken. Springer, Heidelberg (2013)
5. Winter, R., Fischer, R.: Essential layers, artifacts, and dependencies of enterprise architecture. J. Enterp. Archit. **3**(2), 7–18 (2007)
6. Roth, S., Hauder, M., Farwick, M., Breu, R., Matthes, F.: Enterprise architecture documentation: current practices and future directions. In: Wirtschaftsinformatik Proceedings 2013, Paper 58, pp. 911–925, Leipzig (2013)
7. Löhe, J., Legner, C.: Overcoming implementation challenges in enterprise architecture management: a design theory for architecture-driven IT Management (ADRIMA). Inform. Syst. e-Bus. Manage. **12**(1), 101–137 (2013)
8. Hevner, A.R., March, S.T., Park, J., Ram, S.: Design science in information systems research. MIS Q. **28**(1), 75–105 (2004)
9. Niemann, K.D.: Von der Unternehmensarchitektur zur IT-Governance: Bausteine für ein wirksames IT-Management. Vieweg, Wiesbaden (2005)
10. Beimborn, D., Martin, S.F., Homann, U.: Capability-oriented modeling of the firm. In: Proceedings of the IPSI 2005 Conference, Amalfi/Italy (2005)
11. Berners-Lee, T., Hendler, J., Lassila, O.: The semantic web: a new form of web content that is meaningful to computers will unleash a revolution of new possibilities. Sci. Am. **284**(5), 28–37 (2001)
12. Antoniou, G., van Harmelen, F.: A Semantic Web Primer, 2nd edn. MIT Press, Cambridge (2008)
13. Cyganiak, R., Wood, D., Lanthaler, M.: RDF 1.1 Concepts and Abstract Syntax: W3C Recommendation, 25 February 2014. http://www.w3.org/TR/2014/REC-rdf11-concepts-20140225/. Accessed 12 Nov 2015
14. Brickley, D., Guha, R.V.: RDF Schema 1.1: W3C Recommendation, 25 February 2014. http://www.w3.org/TR/2014/REC-rdf-schema-20140225/. Accessed 15 Nov 2015
15. Hebeler, J., Fisher, M., Blace, R., Perez-Lopez, A.: Semantic Web Programming. Wiley, Indianapolis (2009)

16. Miles, A., Bechhofer, S.: SKOS Simple Knowledge Organization System Reference: W3C Recommendation, 18 August 2009. http://www.w3.org/TR/2009/REC-skos-reference-20090818/. Accessed 02 Dec 2015
17. Aier, S., Winter, R.: Virtuelle Entkopplung von fachlichen und IT-Strukturen für das IT/Business Alignment – Grundlagen, Architekturgestaltung und Umsetzung am Beispiel der Domänenbildung. WIRTSCHAFTSINFORMATIK 51(2), 175–191 (2008)
18. Scharffe, F., Zamazal, O., Fensel, D.: Ontology alignment design patterns. Knowl. Inf. Syst. 40(1), 1–28 (2014)
19. Osenberg, M., Langermeier, M., Bauer, B.: Using semantic web technologies for enterprise architecture analysis. In: Gandon, F., Sabou, M., Sack, H., d'Amato, C., Cudré-Mauroux, P., Zimmermann, A. (eds.) ESWC 2015. LNCS, vol. 9088, pp. 668–682. Springer, Heidelberg (2015)
20. Ortmann, J., Diefenthaler, P., Lautenbacher, F., Hess, C., Chen, W.: Unternehmensarchitekturen mit semantischen technologien. HMD Praxis der Wirtschaftsinformatik 51(5), 616–626 (2014)
21. Antunes, G., Barateiro, J., Caetano, A., Borbinha, J.: Analysis of Federated Enterprise Architecture Models. In: ECIS 2015 Completed Research Papers, paper 10 (2015)
22. Bakhshadeh, M., Morais, A., Caetano, A., Borbinha, J.: Ontology transformation of enterprise architecture models. In: Camarinha-Matos, L.M., Barrento, N.S., Mendonça, R. (eds.) DoCEIS 2014. IFIP AICT, vol. 423, pp. 55–62. Springer, Heidelberg (2014)
23. El Haoum, S., Hahn, A.: Using metamodels and ontologies for enterprise model reconciliation. In: van Sinderen, M., Oude Luttighuis, P., Folmer, E., Bosems, S. (eds.) IWEI 2013. LNBIP, vol. 144, pp. 212–224. Springer, Heidelberg (2013)
24. Enterprise Architecture Solutions Ltd: The Essential Project (2015). http://www.enterprise-architecture.org/. Accessed 22 Dec 2015
25. Keller, W.: IT-Unternehmensarchitektur: Von der Geschäftsstrategie zur optimalen IT-Unterstützung. 2, überarb. und erw. Auflage. dpunkt.verlag, Heidelberg (2012)
26. Barkow, R.: EAM-Strategie. In: Keuntje, J.H., Barkow, R. (eds.) Enterprise Architecture Management in der Praxis: Wandel, Komplexität und IT-Kosten im Unternehmen beherrschen, pp. 49–79. Symposion, Düsseldorf (2010)
27. Abraham, R., Aier, S., Labusch, N.: Enterprise architecture as a means for coordination–an empirical study on actual and potential practice. In: MCIS Proceedings, paper 33 (2012)
28. Nakakawa, A., van Bommel, P., Proper, H.A.: Challenges of Involving stakeholders when creating enterprise architecture. In: van Dongen, B.F., Reijers, H.A. (eds.) 5th SIKS/BENAIS Conference on Enterprise Information Systems, pp. 43–55, Eindhoven, November 2010

A Usage Control Model Extension
for the Verification of Security Policies
in Artifact-Centric Business Process Models

Ángel Jesús Varela-Vaca$^{(\boxtimes)}$, Diana Borrego, María Teresa Gómez-López,
and Rafael M. Gasca

University of Seville, Seville, Spain
{ajvarela,dianabn,maytegomez,gasca}@us.es
http://www.idea.us.es/

Abstract. Artifact-centric initiatives have been used in business
processes whose data management is complex, being the simple activity-
centric workflow description inadequate. Several artifact-centric initiatives pursue the verification of the structural and data perspectives of
the models, but unfortunately uncovering security aspects. Security has
become a crucial priority from the business and customer perspectives,
and a complete verification procedure should also fulfill it. We propose an
extension of artifact-centric process models based on the Usage Control
Model which introduces mechanisms to specify security policies. An automatic transformation is provided to enable the verification of enriched
artifact-centric models using existing verification correctness algorithms.

Keywords: Artifact-centric business process model · Verification ·
Security · Declarative security policy · Usage control model

1 Introduction

Nowadays, organizations model their operations with business processes. To
ensure the proper operation of the companies, it becomes necessary the verification of those processes to avoid unexpected errors at runtime, which may deal
to inconsistent situations that cannot reach a business goal successfully. Therefore, it is more suitable to detect possible anomalies in the model at design-time
before the processes are enacted, so preventing errors at runtime.

Traditionally, business processes are modeled as activity-centric business
process models [1], where data are used as inputs and outputs of the activities.
The activity-centric proposals describe at design-time the imperative workflow
that an instance can follow. However, for some types of scenarios, it is very difficult to include the data state transitions into the activity point of view, specially
for complex data models. For this reason, the artifact-centric methodology (data-
centric approach) has emerged as a new paradigm to support business process
management, where business artifacts appeared for the necessity of enriching
the business process model with information about data [2], providing a way for

© Springer International Publishing Switzerland 2016
W. Abramowicz et al. (Eds.): BIS 2016, LNBIP 255, pp. 289–301, 2016.
DOI: 10.1007/978-3-319-39426-8_23

understanding the interplay between data and process. Artifacts are business-relevant objects that are created, evolved, and (typically) archived as they pass through a business, combining both data aspects and process aspects into a holistic unit [3].

Artifact-centric modeling establishes data objects (called artifacts) and their lifecycles as focus of the business process modeling. This type of modeling is inherently declarative: the control flow of the business process is not explicitly modeled, but follows from the lifecycles of the artifacts. The lifecycle represents how the state of an artifact may evolve over the time. The different activities change the state of the artifact and the values of the data associated to each artifact; these may be manual (i.e. carried out by a human participant of the process) or automatic (i.e. by a web service). The evolution of the artifacts implies a change of the state and the values of the data, until a goal state of an artifact is reached. One of the reasons why the artifact-centric paradigm facilitates the process description is the capacity to model the relations between objects with different cardinalities, not only 1-to-1 relations. This modeling capabilities are not entirely supported in activity-centric scenarios. For instance, BPMN 2.0 [4] (currently wide accepted activity-centric notation) allows to easily represent multi-instance activities and pools (processes), but with some limitations, such as the relations between different processes can only be expressed as hierarchies.

On the other hand, artifact-centric models allow 1-to-N and N-to-M relations between artifacts. Then, when more than one artifact is involved in the process, it is possible that a combination of services and data values violate the policies of the business. In order to avoid this situation at runtime, it is necessary to detect some of these possible errors at design time.

To our best knowledge, there are no pure works that consider security issues at artifact-centric business process models. However, the artifact-centric methodology needs for a way to express the security aspects that are not natively considered in the artifacts, such as *Subjects*, *Rights* and specific *Predicates*, and these security aspects need also to be included in the artifact verification since they can change the state of the artifacts.

The goal of this paper is, on the one hand, to provide an artifact-centric business process model specification of security features and constraints. On the other hand, this paper aims to address the automatic transformation of these enriched models into standard artifact-centric model where the design time verification techniques found in the literature can be applied including the security perspective.

The rest of the paper is structured as follows: Section 2 presents a brief review of the notions of artifact and artifact union as a formal model for artifact-centric business process models. Section 3 provides an extension of the existing artifact-centric model to enrich it with the security perspective. Section 4 explains the model transformation so that it is verifiable at design time by means of previous presented verification mechanisms. Section 5 presents an overview of related work found in the literature. Finally, conclusions are drawn and future work is proposed in Sect. 6.

2   Artifact-Centric Business Process Model in a Nutshell

Artifact-centric paradigm facilitates the process creation oriented to the description of the data object evolution during a process execution. The formalization of the artifact-centric business process model to be extended is presented and widely explained in [5]. Nevertheless, the main notions are listed below in order to facilitate the understanding, using an adaptation of the example in [5] to support the concepts. Although the complete example describes the handling of a conference by an organizing committee, in this paper we focus on the registration of participants, review of papers, and paper submission by means of artifacts. These three artifacts are shown in Fig. 1, where solid circles, squares and arrows represent states, services and flows within an artifact respectively, whereas dotted elements are included to represent structural dependencies between artifacts. Likewise, attributes are listed on the right of each artifact.

As reflected in [5], artifacts are represented as specified in the framework BALSA [6] as a basis. That way, the formalization of the model includes:

- **Structural perspective**, identified by the tuple $G = \langle St, Ser, E \rangle$, representing the set of states (St, circles in Fig. 1), the set of services (Ser, squares in Fig. 1) and the set of edges connecting them (E, arrows in Fig. 1), which form the lifecycle of each artifact;
- **Data perspective**, identified by the tuple $Data = \langle id, at, pre, post \rangle$, representing the identifier (id), the set of attributes (at), and the pre and postconditions (pre and $post$) of the services in Ser;

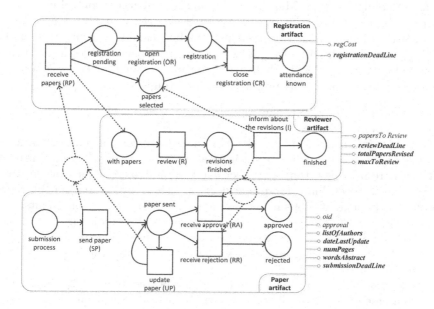

Fig. 1. Example of artifact union

- **Goal states**, identified by Ω, which is a set composed of subsets of St representing the end points in the lifecycle of the artifact.

Likewise, as also explained in [5], the global model is defined as the union of all artifacts composing it. That union is established by two types of policies, which limit the coordinated execution of artifacts lifecycles: (1) **Structural Policies**, expressing constraints on the relation between states and/or services of different artifacts (dotted elements in Fig. 1); and (2) **Data Policies**, expressing invariant conditions over the data (i.e. attributes) managed by the different artifacts in the complete model. For instance, the example in Fig. 1 counts on the data policy *All submitted papers should be reviewed*.

The constraints that should be satisfied during the execution of the services (that is, preconditions, postconditions and policies) are linear or polynomial equations or inequations over artifact instances and the attributes in At. That set of constraints is generated by the grammar presented in [5].

3   Usage Control Model (UCON$_{ABC}$) Extension for Artifacts

As previously commented, artifact-centric models do not provide a way to include security policies, a crucial aspect to ensure the artifact correctness. One way to incorporate them is following the UCON$_{ABC}$ model. UCON$_{ABC}$ model [7] has emerged as a generic formal model to represent complex, adaptable and flexible security policies in new environments such as Internet of Things (IoT). For instance, Digital Right Management (DRM) is an access control mechanism which can be modeled by UCON$_{ABC}$. Moreover, other traditional access control and trust management mechanisms can be defined by using this model. UCON$_{ABC}$ model consists of eight components: Subjects, Objects, both Subject and Object Attributes, Rights, Authorizations, Obligations and Conditions. These components can be divided into various groups:

1. **Components** are defined and represented by their attributes. There are two types of components:
   - *Subjects* is a component which holds or exercise certain rights on objects.
   - *Objects* is an entity which a subject can access or usage with certain rights.
2. **Rights** are privileges that a subject can hold and exercise on an object.
3. **Predicates** to evaluate for usage decision. These predicates can be represented without limitations using the same grammar (i.e. by constraints) proposed in [5]. UCON$_{ABC}$ model defines three types of predicates:
   - *Authorizations (A)* have to be evaluated for usage decisions and return whether the subject (requester) is allowed to perform the requested rights on the object.
   - *Obligations (B)* represent functional predicates that verify mandatory requirements a subject has to perform before or during a usage exercise.
   - *Conditions (C)* evaluate environmental or systems factors to check whether relevant requirements are satisfied or not.

All these predicates can be evaluated before or during the rights are exercised. In that case, $UCON_{ABC}$ model splits each predicate into two types of sub-predicates depending on when it must be evaluated: (1) pre-Authorization (preA) is evaluated before a requested right is exercised; and (2) on-Authorization (onA) is performed while the right is exercised. Likewise, obligations and conditions can be divided into pre- and on-predicates. Regarding attributes, $UCON_{ABC}$ model introduces the concept of mutability which indicates whether certain attributes can be modified or not during the usage decision process. This concept can be modeled using specific predicates as part of the usage decision predicates.

We have used $UCON_{ABC}$ components to extend the original artifact model in order to achieve a secure-extended artifact model which includes a news perspective called Security Perspective. The Security Perspective that extends the artifact models is formalized as follows:

– Security Perspective is identified by the tuple $Sec = \langle R, Sub, Pol \rangle$, where R represents the set of rights, Sub represents the assignments of subjects, and Pol is the set of predicates that define the security policy.

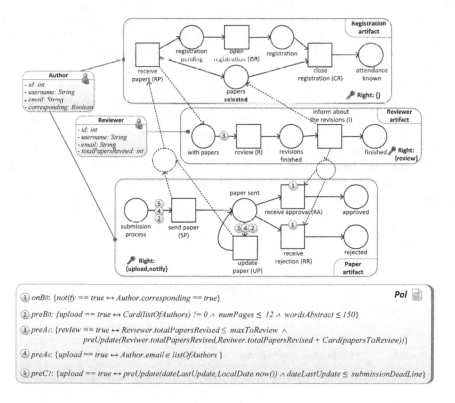

Fig. 2. Example of subjects, rights and security policy.

Objects are the artifacts or artifact unions in which security policies should be defined. Nevertheless, artifact models do not provide elements to define *Subject* concerns. The provided extension enables the specification of *Subjects* in the artifact model. Thus, artifacts can be performed by one or various subjects. When it comes to represent a subject, there are two possibilities: (a) a *Subject* is defined for the complete artifact; and (b) different *Subjects* are defined for each specific services within the artifact. In Fig. 2, the example shows two different *Subjects*: (a) 'Author' are assigned to the Paper and Registration artifact; (b) 'Reviewer' is assigned to reviewer artifact. The *Subjects* are formalized by a set of attributes in the same way than an artifact, as shown in Fig. 2. These attributes have an associate semantic that is used to the evaluation process of the predicates. An example of attribute description for the Author *Subject* is listed in Table 1.

Regarding *Rights*, a set of them has to be defined for each artifact. They represent the different *Rights* that a subject may exercise in the artifact. In Fig. 2, no *Rights* are defined in the registration artifact, the reviewer artifact has only one right called 'review', enabling a reviewer to carry out a review process. Likewise, in the paper artifact there are two *Rights*: '*upload*', enabling the files uploading; and '*notify*', which enables an author to be notified or not.

As it was aforementioned, *Pol* is a security policy represented by a set of *Predicates*. The UCON$_{ABC}$ model introduces multiple kinds of *Predicates* depending on the type of *Predicate* and where the *Predicate* has to be evaluated. Our extension enables the specification of *Predicates* throughout the different parts of the artifact model. There are several places where *Predicates* can be located: (1) transitions between states and services, where all types of *pre-Predicates* can be checked; and (2) in the services, where all type of *on-Predicates* can be checked; and (3) invariants defined for a complete artifact. Thus, these predicates do not require to be checked in a specific place but that have to be checked in every moment. In this case, all types of *Oblitations* and *Conditions* predicates can be defined and checked. For instance, an invariant could be an *Obligation* predicate which indicates *Author* subject must be the same for *Registration* and *Paper* artifact. This type of constraint may define as a invariant which indicates *Author.id* attribute in *Registration* artifact have to be equal to the *Author.id* in *Paper* artifact.

Table 1. Subject attribute description

Author	
id	The author's identification
username	The author's username
email	The author's email
corresponding	The author which submits the paper is the author to be notified

In Fig. 2, there is a unique security policy defined for the three artifacts. This policy is encompassed of five *Predicates* in order to illustrate the main *Predicates* of UCON$_{ABC}$:

1. An *Obligation* predicate (cf. onB_0) enables the notification *Right* whether the author to be notified is established as a corresponding author.
2. A *Conditional* predicate (cf. $preB_0$) enables the upload *Right* whether any author is established for the paper, its number of pages are less than twelve and the abstract is composed of less than one hundred and fifty words.
3. An *Authorization* predicate (cf. $preA_1$) enables the review *Right* whether the reviewer *Subject* has a '*totalPapersRevised*' less than or equal to a 'max-ToReview' established as attribute of the artifact. This predicate introduces a *preUpdate* predicate that establishes an update for the attribute '*totalPapersRevised*'. The *preUpdate* predicate establishes an update prior to the usage by means of an increment of '*totalPapersRevised*' with the number of papers from the list of papers within '*papersToReview*' [5].
4. An *Authorization* predicate (cf. $preA_0$) enables the upload *Right* whether the author's email belongs to one of the authors in the list of author of the paper (cf. '*listOfAuthors*').
5. A *Condition* predicate (cf. $preC_1$) enables the upload options whether the local date when the paper is being uploaded or updated, is less than the submission date established as deadline in the conference. This predicate introduces a *preUpdate* predicate that establishes an update for the attribute '*dateLastUpload*'. This predicate attempts to establish a new value for the '*dateLastUpload*' with the local time.

The relation of predicates and the elements of the artifact model are depicted by circled-numbers attached to the transitions and services (notice that the numbers do not indicates the order of the policy). In Fig. 2, one *pre-Authorization* predicate (cf. $preA_1$) is included in reviewer artifact previous to the '*review(R)*' service is carried out. Other eight *Predicates* are included in *Paper* artifact: one *pre-Conditional, pre-Authorization* and *pre-Obligation* (cf. $preA_0$, $preB_0$, $preC_1$) are established previous the '*send paper(SP)*' and '*update paper(UP)*' are carried out. Two *on-Conditional* predicate are (cf. onB_0) established during the services of '*receive approval (RA)*' or '*receive rejection(RR)*'.

4 Transformation for the Verification of Security Policies in Artifacts

The inclusion of a security perspective in artifact models aims to provide mechanisms to verify at design time the correctness of security policies, as well as structural and data policies. The verification may consider many aspects, and we propose to follow the verification ideas and algorithms introduced in [5] where two types of correctness are carried out:

- Reachability, which checks whether there is a possible trace of execution where every state can be reached, so there is an evolution of the lifecycle in which

the state is available taken into account the pre and post conditions of the possible service executions.
- Weak-termination, which is a correctness criterion that ensures that a goal state is always reachable from every reachable state.

These existing design time verification algorithms are prepared to verify an artifact model that only contains structural and data policies. Nevertheless, security features are now included, therefore they have to be taken into consideration for the verification. Then, we propose an automatic transformation of the secure-extended artifact model to a simple artifact model as shown in Fig. 3. Thus, we propose to transform automatically the three types of components provided by $UCON_{ABC}$ model (*Subjects*, *Rights* and *Security Policies* (*Predicates*)) into artifact elements such as described in literature and summarized in Sect. 2.

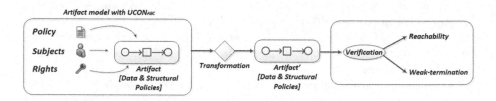

Fig. 3. Verification process of an extended $UCON_{ABC}$ artifact.

The transformations proposed for each component are as follows:

- *Subjects* are transformed into attributes of the related artifact. For instance, Author is linked to *Registration* and *Paper* artifact, hence it is transformed into an attribute inside of these artifacts.
- *Rights* are transformed into attributes of the related artifact. For instance, upload is a *Right* defined in the *Paper* artifact, hence it is transformed into an attribute within this artifact.
- *Security policy (Predicates)* are transformed into constraints to be checked along with the pre and postconditions of the artifact.

Likewise, it is necessary to consider when some *Predicates* have to be evaluated. That is, previous or after pre and postconditions:

1. *Predicates* to be evaluated before preconditions:
   $pre'(n) \rightarrow \langle preA(n) \wedge preB(n) \wedge preC(n) \wedge pre(n) \rangle$
2. *Predicates* that have to be evaluated just after preconditions and previous to postconditions:
   $post'(n) \rightarrow \langle onA(n) \wedge onB(n) \wedge onC(n) \wedge post(n) \rangle$

For instance, the *Reviewer* artifact at the '*review (R)*' service contains the next precondition (cf. Table 2 in [5]):

$pre(n) : Card(papersToReview) \geq 4$

The new precondition after the transformation looks like:

$pre'(n)$: $upload == true \rightarrow \langle Reviewer.totalPapersRevised \geq maxToReview$
$\wedge\ Reviewer.totalPapersRevised\ \ ==\ \ Reviewer.totalPapersRevised'\ +$
$Card(papersToReview)\rangle \wedge Card(papersToReview) \geq 4$

The *preUpdate* predicate has been adapted to a constraint that compares the value of *Reviewer.totalPapersRevised* with its previous value updated, *Reviewer.totalPapersRevised'*. These new pre (cf. $pre'(n)$) and postconditions (cf. $post'(n)$) replace the pre (cf. $pre(n)$) and postconditions (cf. $post(n)$) through the artifact. That is, pre and postconditions of services are replaced by enriched constraints that also consider security concerns.

In summary, the formalization of the artifact model changes as follows:

- **Data perspective**, identified by the tuple $Data = \langle id, at, pre, post \rangle$, the set of attributes (at) are extended with the objects *Subjects* and *Rights*, and the pre and postconditions (*pre* and *post*) of the services in *Ser* are extended as aforementioned.
- **Data Policies**, the inclusion of new attributes in *Data* may require a set of new invariants related to those attributes. As aforementioned, there are some *Predicates* established as invariants. These invariants are transformed into invariants of the Data Policies.

With this guidelines, after the model is transformed, the algorithms presented in [5] can be applied, getting a complete verification covering the three perspectives (data, structural and security) in the artifact-centric process models. The perspectives are formulated into a Constraint Satisfaction Problem (CSP), and the algorithms add constraints to determine both correctness for each state in the process. Both algorithms are complete: neither false positives nor false negatives are generated. Moreover, the algorithms offer precise diagnosis of the detected errors, indicating the execution causing the error where the lifecycle gets stuck.

The execution time is linked to the complexity of the resolution of the CSP, as it was discussed in [8]. In general, no affirmation about the efficiency or scalability of our proposal can be given, mainly because the scalability could be affected by a large increase in the number of constraints and/or variables wrt the number of states. However, this is not usual in real life artifact-centric business processes [9]. Furthermore, owing to the search methods used by CSP solvers, and to the constraints limited by a grammar, the increase in the number of constraints and/or variables could not affect the execution time. As a concrete example, the verification of the reachability of the motivating example takes less than 3 s[1].

5   Related Work

The compliance of security issues in business process models is studied in [10]. Although the authors only focused on activity-centric process models,

---

[1] The test case is measured using a Windows 7 machine, with an Intel Core I7 processor, 3.4 GHz and 8.0 GB RAM.

they provide a LTL formalism to define security compliance rules for business processes. The authors also indicate the difficulty of using artifact-centric models since there exists no well-defined operational semantics for directly executing the defined models. They underlying the impossibility of artifact perspective to introduce rules that enable to establish conditional activities however our contribution enable to establish conditional execution of tasks based on a security policy.

Security has been considered in other several stages of business process management, [11]. A vast number of works provide several ways to represent and verify security requirements at the modeling stage, such as [12–14]. These works enable to generate security components from the business process models. Currently, monitoring and process mining techniques are new trends in order to detect whether certain security requirements are complied by analyzing event logs [15]. Nevertheless, these works are carried out taking into consideration just the activity-centric perspective skipping the artifact-centric perspective.

Regarding UCON$_{ABC}$, the UCON$_{ABC}$ model provides an advantage with regard to traditional access control models since it covers a wide spectrum of security issues such as access controls, trust management, and DRM in a systematic manner for protecting digital resources. However, the UCON$_{ABC}$ model also presents several limitations such as how the UCON$_{ABC}$ can handle the contextual information of the scenarios or the lacks of the UCON$_{ABC}$ to support the complex usage modes that are required in modern computing scenarios. These limitations have been detailed and discussed in [16].

Related to what artifact model is more appropriate to include security aspects, we realize that most previous works in the literature do not take into account numerical data verification in the artifact-centric model. The paper [17] performs a formal analysis of artifact-centric processes by identifying certain properties and verifying their fulfillment, such as persistence and uniqueness. Although it is [18] who performs a static verification of whether all executions of an artifact system satisfy desirable correctness properties. In that work services are also specified in a declarative manner, including their pre and postconditions.

However, they fail in the presence of even very simple data dependencies or arithmetic, both crucial to include security policies. This problem is addressed and solved in [19], where data dependencies (integrity constraints on the database) and arithmetic operations performed by services are considered. To verify the behavior of an artifact system, contribution [20] transforms the GSM model into a finite-state machine and systematically examines all possible behaviors of the new model against specifications. Likewise, the approach in [21] observes two deficiencies in the GSM approach, and resolves them. They also observe that GSM programs generate infinite models, so that they isolate a large class of amenable systems, which admit finite abstractions and are therefore verifiable through model checking.

The field of compliance for artifact-centric processes has been addressed in [22]. The authors extend the artifact-centric framework by including the modeling of compliance rules, and obtain a model that complies by design. This way, the runtime verification of compliance is not required. The contribution [23] checks for conformance between process models and data objects at design time. They propose a notion of weak conformance, which is used to verify that the correct execution of a process model corresponds to a correct evolution of states of the data objects. Although is in [5] where the reachability and weak-termination is verified combined structural and data information. This verification approach integrates some requirements necessary for security perspective: pre and post-conditions defining the behavior of the services, numerical data verification when the model is formed by more than one artifact, and handling 1-to-N and N-to-M associations between artifacts.

6   Conclusions and Forthcoming Work

To ensure the correctness of artifact-centric business process models, their security perspective should be considered. Therefore, we propose an extension of a previous artifact-centric business process model related to Usage Control Model (UCON$_{ABC}$) that introduces mechanisms to specify modern security polices and constraints that help the coverage of the security perspective besides the aspects in the structural and data perspectives.

Since the existing techniques and algorithms do not provide the possibility to manage security policies, we transform automatically the enriched model into a artifact model to be verified avoiding its manual performance which is time-consuming and error-prone.

To the best of our knowledge, this paper presents the first approach for artifact-centric business process models that integrates security aspects, which define the behavior of the artifact dealing with security restrictions.

As future work, we plan to offer additional feedback in case of a violation, making easier the job of fixing the problem causing the error. Furthermore, we plan to deploy diagnosis algorithms in order to explain how and why violations are produced. This work is only focused on the design perspective of artifacts although the same ideas can be adapted to be applied at runtime. That is, we can extract runtime event logs that can be matched with the artifact policies in order to check the compliance of the three perspectives.

Acknowledgement. This work has been partially funded by the Ministry of Science and Technology of Spain (TIN2015-63502) and the European Regional Development Fund (ERDF/FEDER).

References

1. Weske, M.: Business Process Management: Concepts, Languages, Architectures. Springer, New York (2007)
2. Nigam, A., Caswell, N.S.: Business artifacts: an approach to operational specification. IBM Syst. J. **42**(3), 428–445 (2003)
3. Cohn, D., Hull, R.: Business artifacts: a data-centric approach to modeling business operations and processes. IEEE Data Eng. Bull. **32**(3), 3–9 (2009)

4. OMG: Object Management Group, Business Process Model and Notation (BPMN) Version 2.0. OMG Standard (2011)
5. Borrego, D., Gasca, R.M., Gómez-López, M.T.: Automating correctness verification of artifact-centric business process models. Inf. Softw. Technol. **62**, 187–197 (2015)
6. Hull, R.: Artifact-centric business process models: brief survey of research results and challenges. In: Meersman, R., Tari, Z. (eds.) OTM 2008, Part II. LNCS, vol. 5332, pp. 1152–1163. Springer, Heidelberg (2008)
7. Park, J., Sandhu, R.: The UCON ABC usage control model. ACM Trans. Inf. Syst. Secur. **7**(1), 128–174 (2004)
8. Gómez-López, M.T., Gasca, R.M., Pérez-Álvarez, J.M.: Compliance validation and diagnosis of business data constraints in business processes at runtime. Inf. Syst. **48**, 26–43 (2015)
9. Chinosi, M., Trombetta, A.: BPMN: an introduction to the standard. Comput. Stand. Interfaces **34**(1), 124–134 (2012)
10. Reichert, M., Weber, B.: Enabling Flexibility in Process-Aware Information Systems - Challenges, Methods, Technologies. Springer, Heidelberg (2012)
11. Leitner, M., Rinderle-Ma, S.: A systematic review on security in process-aware information systems - constitution challenges, and future directions. Inf. Softw. Technol. **56**(3), 273–293 (2014)
12. Salnitri, M., Brucker, A.D., Giorgini, P.: From secure business process models to secure artifact-centric specifications. In: Gaaloul, K., Schmidt, R., Nurcan, S., Guerreiro, S., Ma, Q. (eds.) BPMDS 2015 and EMMSAD 2015. LNBIP, vol. 214, pp. 246–262. Springer, Heidelberg (2015)
13. Wolter, C., Menzel, M., Schaad, A., Miseldine, P., Meinel, C.: Model-driven business process security requirement specification. J. Syst. Archit. **55**(4), 211–223 (2009)
14. Jürjens, J.: Developing secure systems with UMLsec — from business processes to implementation. In: Fox, D., Köhntopp, M., Pfitzmann, A. (eds.) Verlssliche IT-Systeme 2001. DuD-Fachbeiträge, pp. 151–161. Springer, Verlag (2001)
15. Accorsi, R., Wonnemann, C., Stocker, T.: Towards forensic data flow analysis of business process logs. In: 2011 Sixth International Conference on IT Security Incident Management and IT Forensics, Institute of Electrical & Electronics Engineers (IEEE), May 2011
16. Grompanopoulos, C., Mavridis, I.: Challenging issues of UCON in modern computing environments. In: Proceedings of the Fifth Balkan Conference in Informatics. BCI 2012, pp. 156–161. ACM, New York (2012)
17. Gerede, C.E., Bhattacharya, K., Su, J.: Static analysis of business artifact-centric operational models. In: SOCA, pp. 133–140. IEEE Computer Society (2007)
18. Deutsch, A., Hull, R., Patrizi, F., Vianu, V.: Automatic verification of data-centric business processes. In: ICDT, pp. 252–267 (2009)
19. Damaggio, E., Deutsch, A., Vianu, V.: Artifact systems with data dependencies and arithmetic. ACM Trans. Database Syst. **37**(3), 22 (2012)
20. Gonzalez, P., Griesmayer, A., Lomuscio, A.: Verifying GSM-based business artifacts. In: Goble, C.A., Chen, P.P., Zhang, J. (eds.) ICWS, pp. 25–32. IEEE Computer Society (2012)
21. Belardinelli, F., Lomuscio, A., Patrizi, F.: Verification of GSM-based artifact-centric systems through finite abstraction. In: Liu, C., Ludwig, H., Toumani, F., Yu, Q. (eds.) Service Oriented Computing. LNCS, vol. 7636, pp. 17–31. Springer, Heidelberg (2012)

22. Lohmann, N.: Compliance by design for artifact-centric business processes. In: Rinderle-Ma, S., Toumani, F., Wolf, K. (eds.) BPM 2011. LNCS, vol. 6896, pp. 99–115. Springer, Heidelberg (2011)
23. Meyer, A., Polyvyanyy, A., Weske, M.: Weak conformance of process models with respect to data objects. In: Proceedings of the 4th Central-European Workshop on Services and their Composition, ZEUS-2012, Bamberg, pp. 74–80, 23–24 February 2012

Overcoming the Barriers of Sustainable Business Model Innovations by Integrating Open Innovation

Jad Asswad$^{(\boxtimes)}$, Georg Hake, and Jorge Marx Gómez

Business Information Systems, Carl von Ossietzky University Oldenburg,
26129 Oldenburg, Germany
{jad.asswad,georg.hake,jorge.marx.gomez}@uni-oldenburg.de

Abstract. A profitable and renowned business model no longer guarantees a strong position against competitors in the future, as disruptive technologies and new businesses gain momentum. The society, politics and industry consortia shape the market further and their claims are way beyond the economic interests of current industries, but inhere environmental and societal desires as well. This work takes the new balance of forces into account by illustrating the obstacles of a sustainable business model (SBM) innovation and demonstrating how to overcome them by making use of the toolset of an open innovation approach.

Keywords: Business model innovation · Sustainable business model innovation · Open innovation · Triple bottom line

1 Background

No enterprise of today can decline the growing importance and dependence on their business partners and stakeholders. Businesses rely on their suppliers as well as on their customers, while staying in a continuous competition with rivals that can inhere both of these roles. Maintaining tight bonds with business partners represents a critical element in order to remain competitive in the market. However, those alliances between companies not only exchange goods along their supply chain, but also share knowledge, information, licenses and services [1].

Nevertheless, even in coordination with all business partners of the network, a long-term, and especially a SBM change requires a constant collaboration with all partners that is both continual and results in a benefit for all participating parties in the network. Hence, not only the profit of the individual, but the overall utility of the network of coopetitors - the business ecosystem as a whole - needs to result in a sustainable state that will endure in the long view. Therefore, a SBM transformation has to incorporate not only economic aspects, but needs to be in line according to the triple bottom line of sustainable development that includes an ecological and a social dimension to the economic perspective [2].

As a result, this work shines a light why traditional business model innovation processes founder on the transformation process. In order to illustrate

© Springer International Publishing Switzerland 2016
W. Abramowicz et al. (Eds.): BIS 2016, LNBIP 255, pp. 302–314, 2016.
DOI: 10.1007/978-3-319-39426-8_24

the obstacles that companies encounter during a typical innovation process, in chapter 4, it is shown which areas companies misdetermine stakeholder interests and how each of these decisions jeopardizes a sustainable state with regard to the three perspectives of the triple bottom line: economical, ecological and social sustainability. Ultimately, in chapter 5, countermeasures are presented for each area based on the principles of open innovation that factors in all stakeholder interests and takes a sustainable development into account. Based on this problem depiction the following research question arises: *What are the barriers of a SBM innovation and what countermeasures of open innovation are required to overcome these obstacles?*

2   Research Methodology

Goal of the long term research process within the project eCoInnovateIT is the development and analysis of a software system that supports the requirements of open innovation in all phases of the life cycle model, in particular for information and communication technology (ICT) devices. This work aims at building the requirements specification for one element, the business model, of the overall system model depicted in Fig. 1. In the first step, a literature review was conducted to derive the overall requirements for a SBM innovation. In the second step, an argumentative-deductive analysis followed, which identified barriers hindering the fulfillment of each category. Finally, based on a second literature review and using argumentative reasoning, possible solution concepts from the open innovation toolset have been identified and assigned to their corresponding SBM archetype in order to overcome the presented barriers. The open innovation approach was chosen to follow the research stream of IT-enabled business model development, which "elucidates how an organization is linked to external stakeholders, and how it engages in economic exchanges with them to create value for all exchange partners" [3, p. 181]. Further scholars emphasize the IT-enabled interfirm exchange even further by stating its measures go beyond firm-specific factors and traditional industry characteristic [4, 5]. Therefore, this work adapts to the current research stream of IT-enabled business models and extends its notion by integrating the concept of SBM innovation.

3   Sustainability-Oriented Business Models

3.1   Business Models and Business Model Innovation

History and literature have proven that ideas and technologies are not enough to make a company successful, no matter how effective and creative they are. We live in an era of technological advances and globalization, and a good technology or idea needs an innovative business model to see the light and differentiate its position in the market. Chesbrough emphasized the importance of a good business model over a good technology by stating that a better business model will often beat a better idea or technology [6] and that an average technology or idea wrapped in a great business model might be more valuable than a great idea or technology that operates within an average business model [7] or, in other words: "Products and services can be copied; the business model is the differentiator" [8, p. 27].

In general, scholars do not agree on what a business model is and that the concept has been interpreted differently through the lenses of different researchers [9]. However, it can be agreed that a business model is about creating and capturing value for the organization. Mitchell and Coles defined a business model as "the combination of *who, what, when, where, why, how,* and *how much* an organization uses to provide its goods and services and develop resources to continue its efforts" [10, p. 17]. Other scholars viewed business models from an activity system perspective [11] or described them by their characteristic elements [12].

Finally, Chesbrough and Rosenbloom provided a detailed and operational definition of a business model by defining its functions [13]. This description includes the (i) articulation of the value proposition, (ii) the identification of a market segment (iii) and the structure of the value chain. Furthermore, (iv) a specification of the revenue mechanism is required and (v) an estimation of the cost structure and the profit potential has to be given. Finally, (vi) the position of the firm within the value network (ecosystem) has to be defined and (vii) the competitive structure needs to be formulated.

Nevertheless, no matter how successful a company is, innovation is the key for the continuous success of any company. Traditionally, a companys innovation meant innovating within the body of the company by performing operational innovations or innovations in products and services, especially by investing in technology and R&D. However, an American Management Association study back in 2008 determined that less than 10 % of innovation investment at global companies is focused on developing new business models [12]. At the same time in the last decade, companies came to realize the importance of innovating their business models and that the value gained from such innovations is as much valuable as products and services innovations. A 2005 survey by the Economist Intelligence Unit reported that more than 50 % of executives believe that business model innovation will become even more important for the success of a company than products and services innovations [12]. In 2006, IBMs CEO study interviewed 765 corporate and public sector leaders around the world on the subject of innovation [14]. According to the report, upon the new opportunities and threats presented by globalization and technological advances, CEOs are focusing on business model innovation to create sustainable competitive advantage and differentiate themselves in the marketplace. In fact, business model innovation is now given a high priority and is considered as important as innovations around products and services [15].

3.2 SBM Innovation

While the elements of a business model innovation process outline the important aspects that have to be considered to integrate and maintain the iterative process of change, it is indeed a different thing to sustain the change in the long term.

Famous examples such as Polaroid, Kodak, Blockbuster, Nokia, Sun Microsystems, Yahoo or Hewlett-Packard, although having successful core businesses, all stopped dead in their tracks, while new innovations changed the market and new and innovative businesses took over.

In order to integrate the problem of a long-term development with ecological and social considerations, the concept of SBMs was developed. Bocken et al. underline in [16] that a "SBM can serve as a vehicle to coordinate technological and social innovations with system-level sustainability" [16, p. 44]. Such a SBM can be defined as "a business model that creates competitive advantage through superior customer value and contributes to a sustainable development of the company and society" [17, p. 23]. Hence, a SBM targets towards a sound position in the market while maximizing the utility of the customer and taking the demands of society into account, which defines the objective a sustainable business innovation aims towards. In order for a business model innovation to be sustainable, it has to incorporate the dimensions of the triple bottom line. Traditionally, the concept of the triple bottom line distinguishes three perspectives of sustainable development: economic aspects, environmental issues and social factors. The driving force towards such a sustainable business state is the innovation process [18]. Bocken et al. outline in [16] that business model innovations for sustainability are "innovations that create significant positive and/or significantly reduced negative impacts for the environment and/or society, through changes in the way the organisation and its value-network create, deliver value and capture value (i.e. create economic value) or change their value propositions" [16, p. 44]. Therefore, a SBM innovation towards sustainability changes the way a business creates value and integrates a new perspective that includes the value-network and its environment.

4 Business Model Innovation Barriers

Every SBM innovation needs to be primarily guided towards the three dimensions of the triple bottom line, which distinguishes an economic, environmental and social perspective. These three dimensions encompass each decision taken during the design and implementation of a SBM innovation. Taking the dimensions of the triple bottom line into consideration in each decision keeps a balance on all aspects necessary to fulfill the broad spectrum of stakeholder demands associated with the product, while avoiding an exclusive profit oriented planning process [18].

Hence, a SBM innovation process has to be guided by the dimensions of the triple bottom line. In Fig. 1, this influence is depicted on the lifecycle of a generic sustainable product or service. The lifecycle itself is directly influenced by the distinct elements of sustainable oriented innovation presented by [20]. While the first two of the three elements, technological innovations and product service systems (PSS), have a direct impact on the lifecycle, the business model as the third one, only indirectly influences the lifecycle, as a change in the business model first has to be initiated through the channels of open innovation and only then have an effect on the product lifecycle. Open Innovation, in this context, was chosen as enabler to realize business model change.

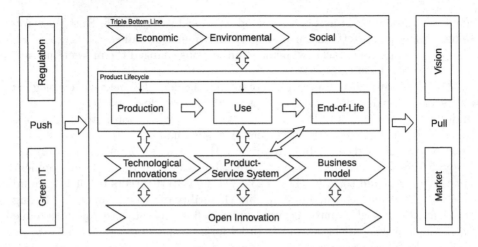

Fig. 1. Integration of Sustainability-Oriented Innovation in the Life-Cycle of ICT [19]

Finally, the overall process is driven by push- and pull-drivers that reflect external forces triggering an overall change in the model. From a horizontal viewpoint, the two mutual reinforcing sides push and pull the innovation process of the model by forcing a business to react to regulations coming from outside the ecosystem a business interacts in (push) or by changing the variables of the ecosystem a business is situated and forcing it to adapt to the new environment (pull). Pushing drivers of a sustainable innovation process can be new or optimized technological elements influencing the product, like material or energy efficiency, as well as regulations that force a company to comply to rules passed to it in a top-down manner. From the other side, pulling drivers include the demands from the market that a business has to fulfill, such as the customer demand, its image, labor costs, new markets or their market share [21].

The pushing and pulling drivers force a business to act or to follow the given external variables. Hence, a business model has to be designed such that it incorporates the environmental factors of the ecosystem it interacts in, and by that, consequently the desires and demands of all stakeholders affected by its business activities. Hence, a SBM innovation process has to factor in multiple dimensions to avoid a vulnerability in its design and put the long term development in danger. In a comprehensive literature review [16], Bocken et al. evaluated the dimensions presented by scholars until now and condensed the various perspectives into eight distinct business model archetypes that are required to derive a SBM (see Table 1). The archetypes again are categorized as technological, social and organizational archetypes. The technological category encounters an archetype for (i) material and energy efficiency, (ii) the reevaluation of waste, (iii) and the substitution of elements in the process with renewable or natural resources. The social dimension covers archetypes that focus on the (iv) change from ownership towards functionality oriented use of products. Moreover, the business is encouraged to (v) occupy a stewardship role and (vi) encourage sufficiency from

Table 1. The Barriers of SBM Innovation

SBM archetypes [16]	Barriers preventing SBMs	Effects of the barrier
Maximize material and energy efficiency	Adds no visible value in the eye of the customer	Substitute efficiency for cheap materials/energy
Create value from waste	See waste as worthless/ problem	Dismiss waste and loose valuable resources
Substitute sources with renewables	Changes are expensive and not visible	See environmental damages as cost factor
Functionality instead ownership	Path dependency	Ignore changing consumption patterns
Adopt a stewardship role	Missing foresight of what will be a future standard	Comply only to current standards
Encourage sufficiency	Lack of framework to communicate changes	Positioning only by directly visible attributes
Repurpose for society/ environment	Problem in determining stakeholder demands	Focus solely on economic profit
Develop scale up solutions	Collaboration takes effort and requires openness	Find solutions isolated and miss synergy effects

a marketing and educational perspective. Finally, the organizational category emboldens to (vii) repurpose the business model for the society and the environment and in order to do that (viii) to develop scale up solutions for a long term development.

These archetypes condense the various perspectives on sustainability endeavors for business models in a distinguished way and allow to evaluate for each archetype individually what obstacles could evolve when trying to accomplish each individual dimension as a goal in a SBM innovation process. As today's business models have been mainly evolved or designed from an economic perspective, they are not aimed to integrate social and environmental aspects. Hence, the transition from the traditional as-is state to a sustainable to-be state requires to integrate the missing dimensions from the triple bottom line. This transformation is impaired by overwhelming short term investments and the lack of a long term commitment towards the transition process.

Therefore, based on previous research in [22], for each archetype a barrier that interferes with the development of each individual SBM archetype has been identified, the possible consequences have been evaluated and solutions that enable the overcoming of each barrier have been developed. The identification of the obstacles and their conquests was done by a comprehensive literature review and the reevaluation of business cases that encountered comparable barriers [7,9,16,17,20,21]. The identified obstacles and their solutions raise no claim to completeness and are not intended to be exhaustive, but to tackle each SBM archetype individually.

The first category of archetypes, the technical archetypes, mainly focuses on the optimization of technical aspects of the product and the elements it is composed out of. Therefore, in the first category, internal optimization processes are required to reach for a more sustainable product. Hence, a change in these areas would lack predominately a visible value in the eye of the customer, while at the same time the potential of unused resources have to be communicated on a vertical scale along the company. The second dimension, the social archetype category, demands a change in the way the company sells and the consumer purchases a physical good. However, both sides are stuck in their entrenched habit on how a transaction between the two takes place. Both parties have path dependencies such that a change in their production or consumption pattern cannot be implemented with justifiable expenditure. Ultimately, the third category, the organizational archetypes, dictates a fundamental change in the purpose a company is expected to do business. New business models are supposed to reduce their profit oriented focus and to re-orientate their processes towards the support of society and environment and developing scale up solutions in collaboration with multiple partners. Such a demand requires a business to exactly determine its stakeholder's needs and wants within larger collaborations, which requires a long term effort and more transparency from all partners. Hence, we argue that all affected stakeholders are required to open up to their business partners in order to overcome these barriers. In order to derive a solution concept, in the following, the barriers are discussed for each SBM archetype individually and the toolset of the open innovation approach is used to present possible ways to allow business partners more openness toward each other.

5   Overcoming the Obstacles of Business Model Innovation Through Open Innovation

The identified barriers show, innovating new business models can be risky and expensive, as it requires companies to invest in internal R&D, explore new markets, and optimize operational processes. In order to externalize and reduce these issues, Zott and Amit presented in [11] the possibility of so-called 'open business models' that enable the company to expand its boundaries and open up its business model to rely on resources and capabilities from outside and benefit from external ideas and technologies. Similar to this concept, Miles, Miles, and Snow introduced the concept of collaborative entrepreneurship, which is "the creation of something of economic value based on new jointly generated ideas that emerge from the sharing of information and knowledge" [23, p. 2].

In the new model, companies open up their boundaries, rather than relying only on internal ideas and resources, in order to leverage internal and external sources of ideas and technologies and at the same time to expand the markets for external use of innovation [24]. The new open innovation paradigm secured its importance as the new driver for business model innovation. Two important factors were presented by Chesbrough in [25] supporting this view. The first factor is the rising cost of technology development, and the second is the shortening life cycles of new products. These two factors are increasing the difficulty to justify investments in innovation. Open innovation fills the gap by reducing the costs of innovation through the greater use of external technologies in the company's own R&D process, and it expands the markets that the company serves through licensing fees, joint ventures and spinoffs.

Open innovation is summarized in three types: outside-in (inbound), inside-out (outbound) [24, 26–28], and coupled open innovation [28–30]. The outside-in type involves opening up the company's innovation process to many kinds of external inputs and contributions, where the inside-out type allows the company to exploit its internal knowledge and its unused or underutilized ideas through external markets and other business models. The coupled open innovation links outside-in and inside-out processes. It combines knowledge inflows and outflows between the stakeholders of the innovation process. In the following, we will spot the light on how open innovation can contribute not only to facilitate innovating business models, but also to overcome the barriers on establishing a SBM.

Barrier 1 - Material/Energy efficiency adds no visible value - Problem: There are no reliable measurements available to distance a more sustainable product from unsustainable one.

Open innovation plays an important role in filling the gap between consumers and companies. Through the mechanisms of inbound open innovation, a company can let its stakeholders be part of the product lifecycle while ensuring their value proposition. A great example of how open innovation can overcome this barrier, is the Fairphone. This fair social enterprise has encouraged stakeholders to be part of its open-design platform, where it raised awareness to the connection between conflict minerals and consumer electronics by engaging designers, creatives, and experts in the process of making a phone based on fair principles. The concept of the Fairphone helped the costumers not only to be part of the change but also to draw a relation with their product and the concept itself as the production of the phone is based on crowdfunding [31].

Barrier 2 - See waste as worthless/problem - Problem: The end-of-life phase in a product life-cycle marks a hassle and cost factor for most companies.

Investing in developing waste management and recycling solutions can be expensive and time consuming. Besides, more and more technologies, initiatives and university programs are focusing on this issue. Open innovation provides companies with the possibility to license such technologies and enable them to collaborate with intermediaries and even competitors that share the same concern. In doing so, companies will be able to make use of ideas and technologies not only to manage the waste and recycling problem, but also to develop sustainable products by designing for sustainability as in the previous example of the Fairphone.

Barrier 3 - Changes take long, are expensive and include risks - Problem: Long term investments such as the use of renewables or adjusting processes to reduce environmental footprint are still assessed solely on their economic value rather than their impact on climate or society.

As if innovating a business model is not hard enough, innovating a business model to be not only economical but also ecological and social sustainable can be risky, expensive and time consuming. One idea that has been discussed in research, is the approach to conquer barriers of business model innovation through the commitment to experimentation [7,25]. Open experimentations allow companies to have insight on their own business models and discover their latent opportunities. Such experimentations are risky themselves and might fail or generate spillovers. Such spillovers were considered as the cost of innovation in the closed innovation paradigm. On the other hand, open innovation with its outbound type treats spillovers as a consequence of the company's business model and considers them as an opportunity to expand the business model or spin off a technology outside the firm to a different business model [26]. A good look on this subject is presented in [13] by highlighting Xerox's experience with selected spin offs and ventures that commercialized technology emanating from Xerox's research laboratories.

Barrier 4 - Path dependency - Problem: It is difficult to pull off the new growth that business model innovation can bring [12]. In addition, innovative business models may often conflict with existing industry structures and threaten the ongoing value of the company [32].

The path dependency barrier can be also managed through open experimentation. Experimenting on providing services is not expensive and leads to better understanding of how to develop the new business model or adapt it with the existing one. An example from the music sector is the release of Radiohead's CD 'In Rainbows' in 2007 on the band's website instead of releasing it with the band's record company. The band invited its fans to pay whatever they want for the tracks. The Experiment went successfully as 3 million visited the website in the first 60 days after the release where 1/3 paid nothing and 2/3 paid an average of 4 pounds [7].

Barrier 5 - Missing foresight of what will be standard in the future - Problem: Disruptive technology and disruptive innovation can force a change on the exciting market without fundamentally changing the company's own business model [33].

Developing a stewardship role can't be easily assured, especially with the continuous occurrence of new and disruptive technologies in the market. In addition to the mechanisms of inbound and outbound innovation, coupled open innovation plays an important role overcoming such a barrier. Coupled open innovation involves two or more partners giving each other the opportunity to share knowledge flows across their boundaries through different mechanisms, such as strategic alliances, joint ventures, consortia, networks, platforms, and ecosystems [27]. Intel for example, and despite the great success that they have, didn't limit themselves to their internal R&D or the external ideas that they leverage through different channels, rather they constructed an ecosystem around their technologies by building platforms for others to build upon and take advantage of Intel technology [34].

Barrier 6 - Lack of framework to seize and communicate changes - Problem: There is, as of today, no agreement on a common framework or standard in industry or in research on how sustainability measures can be classified and communicated between all stakeholders [35].

Regulations are pushing economy and society towards sustainable standards. Expanding the boundaries of ecosystems to include a wider range of stakeholder can play a crucial role in developing common sustainable frameworks and standards. Academic institutions as well as research projects are increasingly addressing the subject of sustainable development. Companies can collaborate with such institutions as well as with policy makers to take a step towards defining sustainable standards. At the same time, companies can contribute by adopting more transparency and sharing their sustainability reports.

Barrier 7 - Problem in determining stakeholder demands - Problem: Companies focus on economic profit instead of determining how to position themselves in the market in order to suit social and environmental requirements.

Most of the companies that address sustainability within their business model consider it as an economical advantage or as a marketing campaign to commercialize themselves as green companies. Opening the business model to be transparent to stakeholders creates positive impact and connects both producers and consumers to the product and its lifecycle, like in the example of Fairphone. They did not only provide an open design platform but also made the entire value chain transparent to the stakeholders and together developed a connection that extends along the product lifecycle [31].

Barrier 8 - Collaboration takes effort and requires to open up - Problem: Avoiding collaboration for more sustainability has no measurable effect in short term.

Coupled open innovation is becoming a must for companies to sustain their business economically as well as socially and ecologically. The IBM's global CEO report [15] shows that major strategic partnerships and collaborations are on the top of the list of most significant business model innovations. It also indicates that companies innovating through such ecosystems and cooperation models enjoy the highest operating margin growth. Finally, businesses can also advantage from their ecosystems and alliances to develop a common sustainable base.

6   Conclusion

In summary, the measures of open innovation allow businesses to reduce the risks and investments involved in a business model innovation, while at the same time, let them benefit from the opportunities of the collaboration. Nevertheless, the barriers that stand against innovating SBMs, are an undeniable problem that requires all stakeholders within a business ecosystem to cooperate. In this work we outlined the barriers that need be addressed in a SBM innovation process and described ways to vanquish each obstacle respectively. As a toolset, the concept of open innovation has been applied. By opening the innovation process and guiding collaborations by an open innovation platform, an initiative could be get pushed forward. The barriers of business model innovations and in some occasions, SBM innovations, have been discussed beforehand in literature, and different scholars suggested countermeasures or strategies to overcome these barriers. The roadmap and the solutions presented in this work are the results of an extensive literature review of the previous solutions and the researches in the fields of SBMs, business model innovation and open innovation in general.

This work is part of a requirement analysis for a platform that implements the concepts of sustainability-oriented innovations in the life cycle as well as the business models of ICT through open innovation. The concept of the platform is to enable both stakeholders and organizations to be on the same level in establishing sustainable ICT products and adapting their life cycle in all phases, from production to use to the end-of-life phase. The findings of the current work will be considered as the basic to include the mechanisms of open innovation in the platform's functionality. The platform should facilitate the innovation process and enable organizations to benefit from the ecosystem that surround them by cooperating with stakeholders through the mechanisms of open innovation provided by the platform. In a previous work we developed a model that integrates the concept of open innovation along with sustainability-oriented innovation and product's life cycle in general and ICT as a special case. In summary, we presented a roadmap to overcome the barriers of SBM innovations, which will serve as an artifact towards the development of a platform that facilitates the integration of sustainability in organizations' business models.

Acknowledgement. This work is part of the project *"Sustainable Consumption of Information and Communication Technology in the Digital Society - Dialogue and Transformation through Open Innovation"*. The project is funded by *the Ministry for Science and Culture of Lower Saxony and the Volkswagen Foundation (Volkswagen Stiftung)* through the *"Niedersächsisches Vorab"* grant programme (grant number *VWZN3037*).

References

1. Basole, R.C., Park, H., Barnett, B.C.: Coopetition and convergence in the ICT ecosystem. Telecommun. Policy **39**(7), 537–552 (2015)
2. Schumacher, E.F.: Small is Beautiful: A Study of Economics as if People Mattered. Random House, New York (2011)
3. Zott, C., Amit, R.: Business model design and the performance of entrepreneurial firms. Organ. Sci. **18**(2), 181–199 (2007)
4. Rai, A., Tang, X.: Research commentary—information technology-enabled business models: a conceptual framework and a coevolution perspective for future research. Inf. Syst. Res. **25**(1), 1–14 (2013)
5. Veit, P.D.D., Clemons, P.E., Benlian, P.D.A., Buxmann, P.D.P., Hess, P.D.T., Kundisch, P.D.D., Leimeister, P.D.J.M., Loos, P.D.P., Spann, P.D.M.: Business models. Bus. Inf. Syst. Eng. **6**(1), 45–53 (2014)
6. Chesbrough, H.: Business model innovation: it's not just about technology anymore. Strategy Leadersh. **35**(6), 12–17 (2007)
7. Chesbrough, H.: Business model innovation: opportunities and barriers. Long Range Plan. **43**(2–3), 354–363 (2010)
8. Giesen, E., Berman, S.J., Bell, R., Blitz, A.: Three ways to successfully innovate your business model. Strategy Leadersh. **35**(6), 27–33 (2007)

9. Zott, C., Amit, R., Massa, L.: The business model: recent developments and future research. J. Manage. **37**(4), 1019–1042 (2011)
10. Mitchell, D.W., Coles, C.B.: Business model innovation breakthrough moves. J. Bus. Strategy **25**(1), 16–26 (2004)
11. Zott, C., Amit, R.: Business model design: an activity system perspective. Long Range Plan. **43**(2–3), 216–226 (2010)
12. Johnson, M.W., Christensen, C.M., Kagermann, H.: Reinventing your business model. Harv. Bus. Rev. **86**(12), 57–68 (2008)
13. Chesbrough, H., Rosenbloom, R.S.: The role of the business model in capturing value from innovation: evidence from xerox corporation's technology spin-off companies. Ind. Corp. change **11**(3), 529–555 (2002)
14. Palmisano, S.: Expanding the innovation horizon: the global CEO study 2006. Report, IBM Global Business Services, Somers, NY (2006)
15. Pohle, G., Chapman, M.: IBM's global CEO report 2006: business model innovation matters. Strategy Leadersh. **34**(5), 34–40 (2006)
16. Bocken, N.M.P., Short, S.W., Rana, P., Evans, S.: A literature and practice review to develop sustainable business model archetypes. J. Cleaner Prod. **65**, 42–56 (2014)
17. Lüdeke-Freund, F.: Towards a Conceptual Framework of 'Business Models for Sustainability'. SSRN Scholarly Paper ID 2189922, Social Science Research Network, Rochester, NY, September 2010
18. Elkington, J.: Cannibals with Forks: The Triple Bottom Line of 21st Century Business. New Society Publishers, Oxford (1998)
19. Asswad, J., Hake, G., Marx Gómez, J.: Integration von open innovation in der entwicklung nachhaltiger IKT. In: Tagungsband der MKWI 2016, Ilmenau (2016)
20. Hansen, E.G., Grosse-Dunker, F.: Sustainability-Oriented Innovation. SSRN Scholarly Paper ID 2191679, Social Science Research Network, Rochester, NY, December 2012
21. Rennings, K.: Redefining innovation eco-innovation research and the contribution from ecological economics. Ecol. Econ. **32**(2), 319–332 (2000)
22. Asswad, J., Hake, G., Marx Gómez, J.: The obstacles of sustainable business innovations. In: Conf-IRM 2016 Proceedings, Cape Town (2016)
23. Miles, R.E., Miles, G., Snow, C.C.: Collaborative entrepreneurship: a business model for continuous innovation. Organ. Dynam. **35**(1), 1–11 (2006)
24. Chesbrough, H.: Open innovation: The New Imperative for Creating and Profiting from Technology. Harvard Business Press, Boston (2003)
25. Chesbrough, H.: Why companies should have open business models. MIT Sloan Manage. Rev. **48**(2), 22–28 (2007)
26. Chesbrough, H.: Open innovation. Res. Technol. Manage. **55**(4), 20–27 (2012)
27. Chesbrough, H., Bogers, M.: Explicating Open Innovation: Clarifying an Emerging Paradigm for Understanding Innovation. SSRN Scholarly Paper ID 2427233, Social Science Research Network, Rochester, NY, April 2014
28. West, J., Salter, A., Vanhaverbeke, W., Chesbrough, H.: Open innovation: the next decade. Res. Policy **43**(5), 805–811 (2014)
29. Gassmann, O., Enkel, E.: Towards a theory of open innovation: three core process archetypes. In: R&D Management Conference, vol. 6 (2004)
30. Bogers, M.: Knowledge Sharing in Open Innovation: An Overview of Theoretical Perspectives on Collaborative Innovation. SSRN Scholarly Paper ID 1862536, Social Science Research Network, Rochester, NY, January 2012

31. Wernink, T., Strahl, C.: Fairphone: sustainability from the inside-out and outside-in. In: D'heur, M. (ed.) Sustainable Value Chain Management: Delivering Sustainability Through the Core Business. CSR, Sustainability, Ethics & Governance, pp. 123–139. Springer, Switzerland (2015)
32. Amit, R., Zott, C.: Value creation in e-business. INSEAD (2000)
33. Christensen, C.: The Innovator's Dilemma: The Revolutionary Book that Will Change the Way You Do Business (Collins Business Essentials). Harper Paperbacks, Scarborough (1997)
34. Chesbrough, H.: Open platform innovation: creating value from internal and external innovation. Technol. J. **7**(3), 5–9 (2003)
35. Schaltegger, S., Lüdeke-Freund, F., Hansen, E.G.: Business cases for sustainability: the role of business model innovation for corporate sustainability. Int. J. Innovation Sustain. Dev. **6**(2), 95–119 (2012)

An Ontological Matching Approach for Enterprise Architecture Model Analysis

Marzieh Bakhshandeh[1,3]($\boxtimes$), Catia Pesquita[2,3], and José Borbinha[2,3]

[1] INESC-ID - Information Systems Group, Lisbon, Portugal
marzieh.bakhshandeh@ist.utl.pt
[2] LaSIGE, Faculdade de Ciências, Universidade de Lisboa, Lisbon, Portugal
cpesquita@di.fc.ul.pt, joao.borbinha@ist.utl.pt
[3] Instituto Superior Técnico, Universidade de Lisboa, Lisbon, Portugal

Abstract. Enterprise architecture aligns business and information technology through the management of different elements and domains. Performing an integrated analysis of EA models using automated techniques is necessary when EA model representations grow in complexity, in order to support, for example, benchmarking of business processes or assessing compliance with requirements. Moreover, heterogeneity challenges arise from the frequent usage of multiple modelling languages, each based on a specific meta-model that cross-cuts distinct architectural domains. The motivation of this paper is, therefore, to investigate to what extent ontology matching techniques can be used as a means to improve the execution of automated analysis of EA model representations, based on the syntax, structure and semantic heterogeneities of these models. For that, we used AgreementMakerLight, an ontology matching system, to evaluate the matching of EA models based on the ArchiMate and BPMN languages.

Keywords: Enterprise architecture · Ontology · Ontology matching · ArchiMate · BPMN

1 Introduction

Enterprise architecture (EA) is defined by Lankhorst as "a coherent whole of principles, methods, and models that are used in the design and realization of an enterprise's organizational structure, business processes, information systems, and infrastructure" [1]. In recent years, a variety of Enterprise Architecture languages, has been established to manage the scale and complexity of this domain. According to UK research firm Ovum [2] hybrid enterprise architecture frameworks are in the majority. It also reveals that 66 % of companies adopt or customize architectures using two or more frameworks. A framework should not be rigid and inflexible, but a living entity that evolves with the enterprise, retaining relevance for all stakeholders. Current model-based enterprise architecture techniques have limitations in integrating multiple descriptions languages due to the lack of suitable extension mechanisms. The integration approach here

© Springer International Publishing Switzerland 2016
W. Abramowicz et al. (Eds.): BIS 2016, LNBIP 255, pp. 315–326, 2016.
DOI: 10.1007/978-3-319-39426-8_25

means to analyse the models and to create links between these different models at structural and semantic levels, without changing the actual meta-model of the language along with the expressiveness of the language.

EA models, specially the ones dealing with the process perspective, are growing quite complex and heterogeneous at syntax, structure and semantic level due to the variety of process modelling languages. This issue causes difficulties of model interoperability [3]. The increasing volume size of process model repositories in the industry and also the need for automated processing techniques had led to the development of a variety of process model matching analyses [4], which are concerned with supporting the creation of an alignment between process models, i.e., the identification of correspondences between their activities. Examples of the application of process model matching include the validation of a technical implementation of a business process against a business-centred specification model [5], delta-analysis of process implementations [6] and a reference model and harmonization of process variants [7].

Model analysis, as a task, can be seen as the application of property assessment criteria to enterprise architecture models [8]. This includes heterogeneity analysis [9,10] which "identifies similar entities implemented in variants or on different platforms That should be reconsidered for standardization". This research work focuses on heterogeneity analysis of EA models with regards to process model similarity matching and aims at investigating the potential benefits and limitations of ontologies matching techniques for enterprise architecture model analysis.

This work is organized as follows: In Sect. 2 we start by characterizing the syntax, structure and semantic heterogeneities problems of EA models on two different levels, i.e., model and meta-model level (because heterogeneities may occur on different modelling levels). Section 3 described the implementation of the proposal and presents a systematic method to match the EA models, which we evaluate in three specific scenarios based on two popular EA languages. Next, Sect. 4 discuses the more interesting matching results obtained by the ontology matching system. Finally, Sect. 5 concludes the paper and provides directions for future work.

2   Heterogeneity in Enterprise Architecture

The heterogeneity of the enterprise architecture modelling languages can cause structural and semantic conflicts that hinder integration [11]. Since meta-models define the modelling concepts that can be used to describe models, these issues are also transposed to EA architectural representations using multiple modelling languages.

We can classify several type of heterogeneity in enterprise architecture modelling languages [12,13]. Structural heterogeneity represents the difference schematic, different meta-models and models, while, semantic heterogeneity represents the differences in the meaning of the considered meta-model and models concepts [14–16].

Haslhofer and Naiman [17, 18] have presented a survey about interoperability of metadata along with some of the heterogeneity issues that can occur. Transference of their heterogeneity classification to EA model transformation (thus interoperability) is possible [13]. Different heterogeneity types can have an impact at the structural and/or semantic level:

- **Element description.** *Naming Conflicts* are caused by synonym and homonym terms. In EA modelling languages, such heterogeneities can occur on both meta-model and model level. Also, *Constraints Conflicts* are caused by different modelling of constraints. These constraints are defined at the meta-model level. This kind of mismatches can be equal to the lexical mismatches in ontologies [19].
- **Domain representation.** *Abstraction Level Incompatibilities* arise when the same objects are modelled at different granularity levels. Such heterogeneity can occur on both meta-model and model level. This can correspond with Semantic heterogeneities as well. This kind of mismatches can be equal to one of the types of semantic mismatches called granularity in ontologies [19].
- **Domain Coverage** problems occur when elements are modelled in one model but not in the other. This can be seen both as a consequence of *Abstraction Level Discrepancies* or *Multilateral Correspondences*. It occurs, for instance, when multiple model elements have only one corresponding element in the target model, this is also known as construct overload [20]. This is associated with a loss of semantics. *Metalevel Discrepancy* occurs when model elements are assigned to different types of different meta-models or models. For example, an object of a certain class may model as an attribute or a relation in a different EA language [13]. This kind of mismatches can be equal to one of the types of semantic mismatches called coverage in ontologies [19].
- **Domain Conflicts.** These occur when domains overlap, subsume, or aggregate others, or when domains are incompatible. Such heterogeneity can occur on both meta-model and model level; this issue is also known as coverage conflict [21]. This kind of mismatches can be equal to two types of semantic mismatches called coverage and perspective in ontologies [19].

The notion behind the application of Ontology Matching techniques to handle EA heterogeneity is that since EA languages share many of the characteristics of ontologies, then matching techniques that can create meaningful liks between ontologies, can also be used to create them between EA models and meta-models. A categorization of ontology matching techniques to handle mismatches(heterogeneities) between ontologies based on the review of the following sources [22–24] supports this notion. This of course results in two challenges: (1) the encoding of EA meta-models and models using ontology languages and (2) the matching of these ontologies and their individuals.

In order to address the first challenge, we have encoded two popular EA modelling languages and some of their models in OWL-DL representation [25]. To address the second challenge we have conducted a study based on four representative case studies reported below.

3   Experimental Design

As case studies, we have selected three EA model matching tasks based on the ArchiMate and BPMN languages that showcase the heterogeneity challenges. To support the matching tasks we have used AgreementMakerLight, an ontology matching system that is extensible and implements several state of the art ontology matching algorithms. The evaluation of the produced alignments was manually evaluated.

3.1   The ArchiMate and BPMN Meta-Models

ArchiMate is an open and independent language for enterprise architecture that is supported by different tool vendors and consulting firms [26]. The ArchiMate framework organizes its meta-model in a three by three matrix. The rows capture the enterprise domain layers (business, application, and technology), and the columns capture cross layer aspects (active structure, behaviour and passive structure). We have specified the ArchiMate meta-model in OWL-DL. BPMN describes business processes using a Business Process Diagram (BPD), i.e., an annotated graph whose nodes explicitly represent activities, control flows, data, and auxiliary information about the process. Examples of BPMN elements are: Event, Activity Gateway and Sequence Flow. Properties of basic elements concern both the usage of the BPMN elements to compose the business process diagrams, and the behaviour of the elements during the execution of a process. The ontology created based on BPMN [27] is structured according to the description of the complete set of BPMN Element Attributes and Types contained in. The ontology currently consists of 189 Classes and 666 Class Axioms, 131 Object Properties and 262 Object Property Axioms, and 57 Data Properties and 114 Data Properties Axioms. Models were encoded as ontology individuals in both cases. Transformation Application as shown in Fig. 4 was used to convert each model into an ontological representation. The Transformation Application uses data to create and populate ontologies, independently from the schema used for organizing source data. Independence is achieved by resorting to the use of a mappings specification schema. This schema defines mappings to establish relations between data elements and the various ontology classes, properties and annotations. Those relations are then used to create and populate an ontology with individuals (instances), thus representing the original data in the form of an OWL ontology.

3.2   Case Studies

We have selected three case studies that demonstrate the heterogeneity challenges caused by element description conflicts and domain representation conflicts at the model level. Cases 1 and 2, showcase heterogeneities between models encoded in different languages, whereas Cases 3 illustrate heterogeneities between models using the same language.

- **Cases 1 and 2:** In cases 1 (Fig. 1) and 2 (Fig. 2) there are *Abstraction Level Incompatibilities* between an ArchiMate model and a BPMN model that model

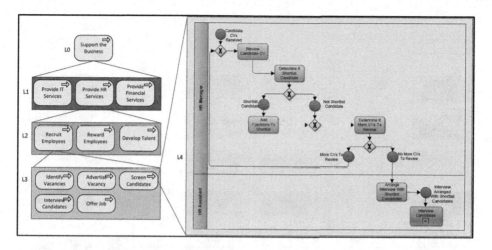

Fig. 1. Case 1: abstraction level incompatibilities heterogeneity between ArchiMate (left) and BPMN (right) models [28]

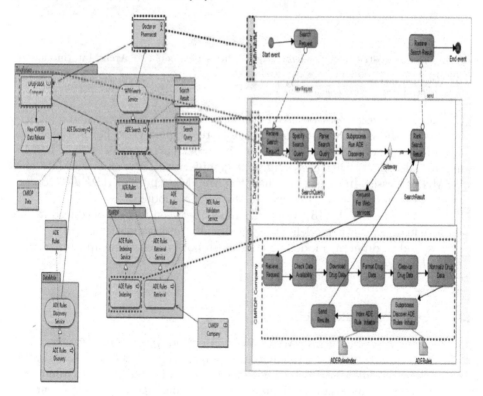

Fig. 2. Case 2: abstraction level incompatibilities heterogeneity between ArchiMate (left) and BPMN (right) (mappings correspond to dashed lines and boxes

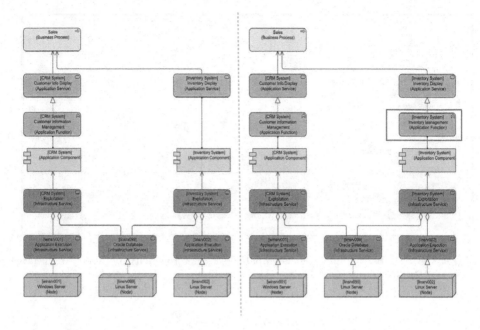

Fig. 3. Case 3: abstraction level incompatibilities between two ArchiMate models of the same situation modelled by two architects

the same Business Processes. ArchiMate processes are represented at a higher-level while the BPMN model provides a more detailed view. For instance, the ArchiMate process *Screen Candidates* in case 1 encompasses several BPMN individuals, illustrating these different granularities between ArchiMate and BPMN processes. In case 2, note how the ArchiMate process *ADE Rule Index-ing* corresponds to an entire layer in the BPMN model.

- **Cases 3** In case 3 (Fig. 3) we have conflicts between models produced in the same language. Both models are syntactically and semantically correct but correspond to different modelling choices. In both cases, there are *Abstraction Level Incompatibilities*.

3.3 AgreementMakerLight

The AgreementMakerLight (AML) [29] is an ontology matching system which has been optimized to handle the matching of larger ontologies. The AML ontology matching module was designed with flexibility and extensibility in mind, and thus allows for the inclusion of virtually any matching algorithm. AML contains several matching algorithms based both on lexical and structural properties and also supports the use of external resources and alignment repair. These features have allowed AML to achieve top results in several OAEI 2013, 2014 and 2015 tracks [30,31].

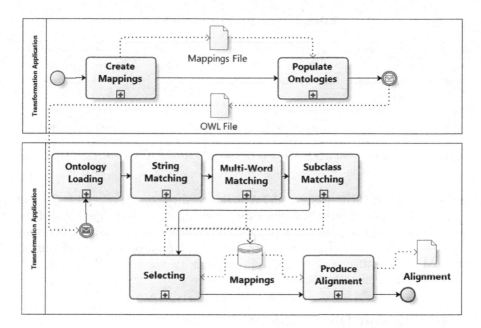

Fig. 4. Transformation application and AML workflow

The modularity and extensiblity of the AML framework made it an appropriate choice to handle the matching of our EA models. However, AML is not adapted to the matching of individuals, so the necessary changes were carried out to ensure individuals are properly loaded by the system. Furthermore, the current AML implementation only produces equivalence mappings, and since we were interested both in equivalence and subclass mappings, a novel subclass matcher was implemented. The business process that illustrates the pre-processing and subsequent use of AML (see Fig. 4) comprises of two sub business processes: transformation application and AML.Our pipeline for EA model mapping is based on a combination of AML's equivalence matchers and the novel subclass matcher. In a first step, we run the String Matcher, which implements a variety of similarity metrics and produces an all vs. all matrix of similarities. Running this matcher with a high similarity threshold coupled with a selection step to ensure 1-to-1 cardinality produces the set of equivalence mappings. Then, in a second step we run the Multi-Word Matcher, which is a fast word-based string similarity algorithm. For this matcher, we add to the set of words describing an entity, all of their WordNet synonyms. We use a lower threshold for this step since we are interested in capturing partial word-based matches. The thresholds were empirically defined. Next, These partial word-based matches are the input to our novel Subclass matcher. The Subclass Matcher works under the following

Table 1. Case 1 matching results

ArchiMate	BPMN	Relation	Measure	Evaluation
ScreenCandidates	AddCandidateToShortlist	≻	0.4498	correct
ScreenCandidates	NotShortlistCandidate	≻	0.4498	correct
ScreenCandidates	ShortlistCandidate	≻	0.6583	correct
ScreenCandidates	ReviewCandidateCV	≻	0.4144	correct
RecruitEmployees	InterviewArrangedWithShortlistCandidates	≻	0.3393	correct*
RecruitEmployees	CandidateCVsReceived	≻	0.3954	correct*

*mappings that are semantically correct but subsumed by more specific ones

premise: if a given entity from model A is matched to several related entities in model B, then we can infer that the entity from model A is a superclass of the related model B classes. Here, we consider two entities to be related when they co-exist in the same axiom. AgreementMakerLight then outputs an alignment composed of equivalence and subclass mappings that are scored from 0 to 1, according to their confidence.

4   Results and Discussion

Tables 1, 2 and 3 describe the more interesting matching results obtained by the AML strategy. For the most part, we have omitted from these tables the cases where individual's labels were identical and, therefore, straightforward to match. In case 1, we were able to generate several subclass mappings between *Screen Candidate* and BPMN individuals capturing eight out of a total of 12 correct mappings, as well as two mappings between *Recruit Employees* and BPMN individuals, which had already been captured by *Screen Candidate* subclasses. Table 1 presents a subset of these mappings.

In case 2, for *ADERulesIndexing* we were able to generate four subclass mappings, of which three were correct. *DiscoverADERulesInitiator* and *Index-ADERuleInitiator* were mapped thanks to their partial word matching, whereas *NormalizeDrugData* was discovered via the Subclass matcher. However, we were unable to discover the six remaining subclass mappings, since they don't share words with the ArchiMate process and are too distant to the processes that do.

Table 2. Case 2 matching results

ArchiMate	BPMN	Relation	Measure	Evaluation
ADERulesIndexing	NormalizDrugData	≻	0.4135	correct
ADERulesIndexing	IndexADERuleInitiator	≻	0.4546	correct
ADERulesIndexing	Gateway?No	≻	0.4051	incorrect
ADERulesIndexing	DiscoverADERulesInitiator	≻	0.7509	correct
ADERulesDiscovery	Gateway?No	≻	0.4051	incorrect
ADERulesIndexingService	RequestForWebService	≻	0.414	incorrect

Table 3. Case 3 matching results

ArchiMate V1	ArchiMate V2	Relation	Measure	Evaluation
Sales	Sales	=	0.9801	correct
ApplicationExecution	ApplicationExecution	=	0.9801	correct
CustomerInfoDisplay	CustomerInfoDisplay	=	0.9801	correct
CustomerInformationManagement	InventoryManagement	$\succ$	0.5207	correct

In case 3, both models are very similar, with processes bearing the same names. However, there is an *Abstraction Level Incompatibility*, since, in version 2, a new process has been added *InventoryManagement* that adds more detail to the model. This individual is a subclass of *CustomerInformationManagement*, a mapping that was captured by AML. Table 3 presents a representative subset of the obtained results.

The three case studies and their matching using a combination of ontology matching algorithms illustrate the challenges and opportunities in their application to addressing EA heterogeneities. As expected, string and word based techniques are effective at capturing the mappings between equivalent individuals who share similar names. However, when equivalent individuals had dissimilar labels, for which WordNet extension did not produce any shared synonyms, the applied algorithms failed to produce the correct mappings. Regarding *Abstraction Level Incompatibilities*, the results were related to the complexity of the models. In simpler model matching tasks, such as cases 1 and 2, the Subclass Matcher approach had a good performance, identifying 75 % of the subclass mappings. However, in more complex tasks, such as case 2, performance is reduced. Since the evaluated approaches relied only on model information to perform matching, there was no practical difference between matching models using the same or different languages.

We consider that the main limitation of the employed matching techniques was their inability to explore a considerable portion of the information modelled in the ontologies. The applied algorithms make very little use of this, considering all axioms as unlabeled links between individuals. In order to extend the application of ontology matching techniques in the EA domain, ontology matching systems need to be able to explore this semantic richness by producing semantic matching approaches that go beyond current strategies which are mostly Word-Net based [32]. In recent years, ontology matching systems have had a growing interest regarding reasoning capabilities, and we propose that a combination of these strategies with pattern-based complex matching approaches [33] may provide improved solutions to the EA model integration challenge.

Moreover, an evaluation of an approach based on upper level ontologies can be done in the future. Finally, there is also a need for adequate benchmarking, which bearing in mind the complexity of the matching between EA models and the amount of data embedded in EA models is a challenge in itself. There have been some efforts in this area with the creation of the Process Model Matching

Contest in 2013 [34], where models using different languages (including BPMN) are matched and evaluated against reference alignments. In future work, we will extend our evaluations to include these references.

5 Conclusion

This paper discusses how multiple EA domains can be integrated by matching the underlying meta-models or models through the use of ontology matching techniques. This can help to support the analysis of EA architectural representations. The contributions of this paper are: (1) Heterogeneity in EA model representations: A classification of heterogeneity that exists in EA model representations was proposed; (2) A systematic method for the analysis of EA models: A systematic method to apply heterogeneity analysis for EA models was proposed; We discussed the heterogeneity of EA modelling languages, illustrated by examples from the ArchiMate and BPMN languages. Next, we investigated the suitability of employing a state of the art ontology matching system, Agreement-MakerLight, to support EA model matching. This was accomplished by matching four selected case studies and revealed that existing techniques have a good performance when applied to less complex models. We argue that a combination of reasoning capabilities and pattern-based matching would be better suited to explore the semantic richness of EA models adequately. Improved matching performance would greatly benefit the analysis of EA architectural representations, by allowing a more thorough comparison of languages and models.

Acknowledgements. The authors are grateful to Marco Rospocher, for making the OWL representation of BPMN available. This work was supported by national funds through Fundao para a Cincia e a Tecnologia (FCT) with reference UID/CEC/50021/2013 and UID/CEC/00408/2013.

References

1. Lankhorst, M.: Enterprise Architecture at Work: Modelling, Communication and Analysis. The Enterprise Engineering Series. Springer, Heidelberg (2013)
2. Blowers, M.: Hybrid enterprise architecture frameworks are in the majority (2012)
3. Lin, Y.: Semantic annotation for process models: facilitating process knowledge management via semantic interoperability (2008)
4. Van Dongen, B., Dijkman, R., Mendling, J.: Measuring similarity between business process models. In: Bubenko, J., Krogstie, J., Pastor, O., Pernici, B., Rolland, C. (eds.) Seminal Contributions to Information Systems Engineering, pp. 405–419. Springer, Heidelberg (2013)
5. Castelo Branco, M., Troya, J., Czarnecki, K., Küster, J., Völzer, H.: Matching business process workflows across abstraction levels. In: France, R.B., Kazmeier, J., Breu, R., Atkinson, C. (eds.) MODELS 2012. LNCS, vol. 7590, pp. 626–641. Springer, Heidelberg (2012)

6. Küster, J.M., Koehler, J., Ryndina, K.: Improving business process models with reference models in business-driven development. In: Eder, J., Dustdar, S. (eds.) BPM Workshops 2006. LNCS, vol. 4103, pp. 35–44. Springer, Heidelberg (2006)
7. Weidlich, M., Mendling, J., Weske, M.: A foundational approach for managing process variability. In: Mouratidis, H., Rolland, C. (eds.) CAiSE 2011. LNCS, vol. 6741, pp. 267–282. Springer, Heidelberg (2011)
8. Närman, P., Johnson, P., Nordström, L.: Enterprise architecture: a framework supporting system quality analysis. In: 11th IEEE International Enterprise Distributed Object Computing Conference, EDOC 2007, p. 130. IEEE (2007)
9. Bucher, T., Fischer, R., Kurpjuweit, S., Winter, R.: Analysis and application scenarios of enterprise architecture: an exploratory study. In: Null, p. 28. IEEE (2006)
10. Lange, M., Mendling, J.: An experts' perspective on enterprise architecture goals, framework adoption and benefit assessment. In: 2011 15th IEEE International Enterprise Distributed Object Computing Conference Workshops (EDOCW), pp. 304–313. IEEE (2011)
11. Kühn, H., Bayer, F., Junginger, S., Karagiannis, D.: Enterprise model integration. In: Bauknecht, K., Tjoa, A.M., Quirchmayr, G. (eds.) EC-Web 2003. LNCS, vol. 2738, pp. 379–392. Springer, Heidelberg (2003)
12. Zivkovic, S., Kuhn, H., Karagiannis, D.: Facilitate modelling using method integration: an approach using mappings and integration rules (2007)
13. Lantow, B.: On the heterogeneity of enterprise models: archimate and troux semantics. In: 2014 IEEE 18th International Enterprise Distributed Object Computing Conference Workshops and Demonstrations (EDOCW), pp. 67–71. IEEE (2014)
14. Alexiev, V., Breu, M., de Bruijn, J.: Information integration with ontologies: experiences from an industrial showcase (2005)
15. Doan, A., Noy, N.F., Halevy, A.Y.: Introduction to the special issue on semantic integration. ACM SIGMOD Rec. 33(4), 11–13 (2004)
16. Klein, M.: Combining and relating ontologies: an analysis of problems and solutions. In: IJCAI-2001 Workshop on Ontologies and Information Sharing, pp. 53–62 (2001)
17. Haslhofer, B., Klas, W.: A survey of techniques for achieving metadata interoperability. ACM Comput. Surv. (CSUR) 42(2), 7 (2010)
18. Naiman, C.F., Ouksel, A.M.: A classification of semantic conflicts in heterogeneous database systems. J. Organ. Comput. Electron. Commer. 5(2), 167–193 (1995)
19. Euzenat, J., Shvaiko, P.: Ontology Matching, vol. 18. Springer, Heidelberg (2007)
20. Rosemann, M., Green, P.: Developing a meta model for the Bunge-Wand-Weber ontological constructs. Inf. Syst. 27(2), 75–91 (2002)
21. Bouquet, P., Euzenat, J., Franconi, E., Serafini, L., Stamou, G., Tessaris, S.: D2. 2.1 specification of a common framework for characterizing alignment (2004)
22. Bellahsene, Z., Bonifati, A., Rahm, E.: Schema Matching and Mapping. Data-Centric Systems and Applications, vol. 20. Springer, Heidelberg (2011)
23. Shvaiko, P., Euzenat, J.: Ontology matching: state of the art and future challenges. IEEE Trans. Knowl. Data Eng. 25(1), 158–176 (2013)
24. Otero-Cerdeira, L., Rodríguez-Martínez, F.J., Gómez-Rodríguez, A.: Ontology matching: a literature review. Expert Syst. Appl. 42(2), 949–971 (2015)
25. Bakhshandeh, M., Antunes, G., Mayer, R., Borbinha, J., Caetano, A.: A modular ontology for the enterprise architecture domain. In: 2013 17th IEEE International Enterprise Distributed Object Computing Conference Workshops (EDOCW), pp. 5–12. IEEE (2013)
26. Iacob, M., Jonkers, H., Lankhorst, M., Proper, E., Quartel, D.A.: Archimate 2.0 specification (2012)

27. Di Francescomarino, C., Ghidini, C., Rospocher, M., Serafini, L., Tonella, P.: Reasoning on semantically annotated processes. In: Bouguettaya, A., Krueger, I., Margaria, T. (eds.) ICSOC 2008. LNCS, vol. 5364, pp. 132–146. Springer, Heidelberg (2008)
28. van den Berg, M.: Archimate, BPMN and UML: an approach to harmonizing the notations. Orbus, software, white paper (2012)
29. Faria, D., Pesquita, C., Santos, E., Cruz, I.F., Couto, F.M.: AgreementMakerLight results for OAEI 2013. In: OM, pp. 101–108 (2013)
30. Dragisic, Z., Eckert, K., Euzenat, J., Faria, D., Ferrara, A., Granada, R., Ivanova, V., Jiménez-Ruiz, E., Kempf, A.O., Lambrix, P., et al.: Results of the ontology alignment evaluation initiative 2014. In: Proceedings of the 9th International Workshop on Ontology Matching Collocated with the 13th International Semantic Web Conference (ISWC 2014) (2014)
31. Cheatham, M., Dragisic, Z., Euzenat, J., Faria, D., Ferrara, A., Flouris, G., Fundulaki, I., Granada, R., Ivanova, V., Jiménez-Ruiz, E., et al.: Results of the ontology alignment evaluation initiative 2015. In: 10th ISWC Workshop on Ontology Matching (OM), pp. 60–115, No Commercial Editor (2015)
32. Giunchiglia, F., Autayeu, A., Pane, J.: S-Match: an open source framework for matching lightweight ontologies (2010)
33. Ritze, D., Meilicke, C., Sváb-Zamazal, O., Stuckenschmidt, H.: A pattern-based ontology matching approach for detecting complex correspondences. In: ISWC Workshop on Ontology Matching, Chantilly (VA US), pp. 25–36. Citeseer (2009)
34. Cayoglu, U., Dijkman, R., Dumas, M., Fettke, P., Garcıa-Banuelos, L., Hake, P., Klinkmüller, C., Leopold, H., Ludwig, A., Loos, P., et al.: The process model matching contest 2013. In: 4th International Workshop on Process Model Collections: Management and Reuse, PMC-MR (2013)

Service Science

Service Self-customization in a Network Context: Requirements on the Functionality of a System for Service Self-customization

Doreen Mammitzsch[✉] and Bogdan Franczyk

Information Systems Institute, Leipzig University, Leipzig, Germany
{mammitzsch,franczyk}@wifa.uni-leipzig.de

Abstract. Self-customization of services is an approach, where customers configure a service to their individual preferences by assistance of a system for self-customization. This paper concentrates on the self-customization of business services in a multi actor environment, where different service providers as part of a service network provide service-modules, which are selected and combined by the customer. Existing concepts as well as the exemplar of a service value adding system as given by a Fourth Party Logistics provider are used to define requirements on the functionality of a service self-customization system. The determined functionality is merged and presented in a model.

Keywords: Self-customization of services · Mass-customization · Services · Systems for mass-customization

1 Introduction

Providing individual services to customers at low costs of production can be seen as a main challenge for a service provider to be successful in the market. Mass-customization of services is a strategy, that targets the resolution of the discrepancy between customization and the efficiency, that equals a mass provisioning of services [1]. Cost benefits can be achieved by standardizing partial services whereas customization take place when partial services are individually combined to a total solution offering [2].

As the case of a logistics buying process [3] illustrates, the specification of individual service requirements and the following contracting phase are time-consuming. The customization of parameters, resources and activities take up to several months and bind human resources. In contrast to a small sales market and long lasting personalization processes in a make-to-order strategy, the concept of mass-customization has the capability to reduce the time of requirements analysis and service specification as well as the moment of the service provision [1]. Additionally, the service provider addresses a larger sales market [4]. Therefore the configuration process should be automated to reduce the transactional costs between customer and provider during the customization process [5]. By using information systems in the process of customization (configuration systems for mass-customization [5]), the service provider is able to serve a larger number of customers more efficiently [6] at lower transactional costs [1].

© Springer International Publishing Switzerland 2016
W. Abramowicz et al. (Eds.): BIS 2016, LNBIP 255, pp. 329–340, 2016.
DOI: 10.1007/978-3-319-39426-8_26

Systems for mass-customization of services can be used by salespeople of the service provider in the context of a consulting service. Another strategy is to provide customers with an interface that allows them to customize offerings to their own preferences [7]. Concepts of systems for mass-customization in the related literature focus on customizing physical products by salespeople or customers. There is a lack of concepts that focus on the self-customization of business services in a service-network context. In order to identify the characteristics, roles and tasks of a service-network, the business model of a Fourth Party Logistics Provider (4PL) is chosen as a single case study. 4PL-providing is characterized by the cooperation between diverse suppliers of logistics services. All suppliers contribute their specific service capability. The 4PL acts as an intermediary, who combines the individual service capabilities into a customer specific overall service. This scenario is taken as the basis for the requirements analysis to define the functionality of a **system for service self-customization**, which can be deployed by an intermediary service provider. The objective is to establish a self-customization process, which empowers the customer to create the required services. The purpose of this paper is to analyze the requirements on a service self-customization tool in a business-network context. Furthermore the paper aims at developing an appropriate concept of functionalities.

It is structured as follows: After explicating the meaning of mass- and self-customization of services a scenario of a 4PL-service-network is given as an example to illustrate the actors, roles and tasks within such a service-network. Related literature is reviewed and taken as the basis for the definition of requirements on a service self-customization system. Finally the identified requirements are translated into a model of functionality.

The contribution of this paper can be brought into relation to the discipline of mass-customization of services, especially systems for mass-customization, because it aims at investigating design principles for service self-customization tools. To design a service self-customization tool in a multi actor environment, there are several aspects, which need to be brought up to build a basis for a consistent solution. The scientific framework is made up from the science of services with the main elements services, mass-customization of services and service self-customization. Farther relevant are service modularization and description.

2   Self-customization of Business Services Within a Mass-Customization Strategy

2.1   The Concept of Self-customization of Services

Self-customization is an approach within the mass-customization strategy, where the customers configure the product or service to their specific preferences [8]. The customer is integrated in the service process by delegating him those items of work, which are necessary to specify the service he really wants. Customer and service provider may consequentially cut time and transactional costs [1, 9]. Self-customization can be realized by the use of ICT [1]. Systems for self-customization assist the customer to translate their preferences in a concrete service [7]. A precondition is the existence of a modular service portfolio. This is addressed by different authors with different concepts. One concept is the platform-based approach, which is established in the manufacturing industry and transferred to the service sector [10–13]. Further concepts describe modularization methods, such as the modularization of services in the context of service engineering [14–16].

Service-modules are standardized, whereas the overall service, combined from several service-modules, is customer specific [2]. The process of mass-customization of services can be specified in the context of self-customization as follows [2]:

1. Definition and provision of service-modules by service-providers,
2. Selection of service-modules and combination into an overall service by the customer.

Within a service-network context, service provision is based on the cooperation of multiple companies. To combine several service-modules in a self-customization procedure means combining the service providing companies as well. The self-customization process is thus complemented by another step: configuration of service-providers/ construction of the service-system. This step does comprise more than the creation of individual services – it includes the setting up of the corresponding value-adding system, which integrates several service-providers.

2.2 Services for Self-customization

Service terms (e.g. "service", "e-service", "business service") often address related concepts from different domains such as computer science, information science and business science [17]. Usually there is a lower degree of consensus among those who use these terms. The result is a collection of manifold service types and definitions. Within business and especially marketing oriented literature, some agreement as to what characterizes services exist [18]. The main common and recurring elements of different service definitions, which represent the service understanding of this paper, are listed below [19–35]:

– Activity or series of activities or process
– Offer/provide benefits or solutions
– Interaction of economic entities, mainly the provider and the customer
– Involved elements: persons (knowledge and skills), physical resources (also facilities), goods, systems (also ICT and internet-related systems)

Services can be classified by means of many criteria [36]. One criterion is the type of buyer: private consumers vs. business customers. The latter require business services, which are delivered and bought by organizations [37]. The focus of this paper is on business services. Services can be further categorized by the integration of external factors like persons, physical goods or information [35]. Services for information and physical processing suit best for the use of self-customization systems and are thus relevant. The level of automation indicates the extent to which people are substituted by machines or information systems in executing services [2]. Services can be performed manually, automated or semi automated. The more parts of a service can be processed by information systems the more the service is suitable for self-customization [38]. The minimum condition to qualify services for self-customization is a formalized service description so that service-modules can be saved, identified, compared and composed in a self-customization procedure.

A final criterion to differentiate services is their time aspect of consumption. Services can be consumed either over a short or over a longer period of time, where tasks are executed continuously [37]. Long lasting services in particular put special demands on the functionality of a mass-customization system. The continuous service delivery is accompanied by a steady integration of information and physical objects of the customer and has to be monitored over an extended period of time.

Services can be characterized by some criteria which are equally relevant for design decisions of a system for service self-customization. The main criteria, mentioned in literature and helpful to extract the characteristics of services, are immateriality, integration of external factors, unable to store, simultaneity of services rendered and consumed [2]. As services cannot be stored, the service provider does not have the option to decouple assembly and distribution from production by building up stocks [2]. This is of major importance for the mass-customization process. Whilst physical product components can be produced in advance and finally assembled after a customer's order, service-modules can only exist as a concept. Service delivery and consumption are identically and no value creating product is being transferred between provider and customer as it is customary in the manufacturing sector.

2.3   Systems for Self-customization

Within academic research the concept of self-customization of services is barely picked out as a central theme. Configuration methods as a subdomain of self-customization are addressed by several authors. Configuration means the design of a new artifact from preexisting components by consideration of certain rules. This happens by the specification of requirements on a desired result and the automated search for appropriate components. An overview of configuration procedures is given by [39]. Two widely used self-customization methods in the marketplace are by-attribute and by-alternative customization procedures [7]. By-attribute customization means to determine product attributes to individual preferences. The appropriate components are chosen and virtually assembled in order to show the result to the customer prior to his order. If services are subject of such a customization procedure, service-modules have to be available well described, e.g. by functional and non-functional properties. This enables their identification and composition according to a customer's preferences [40]. The customization procedure as well as the structured description of service-modules is element of a system for service self-customization. A comprehensive model of a service self-customization system´s functionality including these partial aspects among others is not to be found in current literature.

Only few authors give a proposition of the functionality of systems for mass-customization [9, 41]. Their focus is either on product-customization or kept generic, without a particular consideration of self-customization of services. Nonetheless they provide design principles that can be adapted to formulate the features of a system for service self-customization. Requirements on the functionality of a mass-customization system with focus on products are given below [41]:

- Creation of product models (model of components/list of parts)
- Requirements analysis
- Ascertainment of manufacturer/supplier capabilities
- Selection of manufacturers/suppliers
- Enable product specification
- Assignment of actors
- Formation of the value added system (instantiation)
- Order placement (involves all actors in the value added system)
- Exchange of product specifications
- Check of the manufactured object

These requirements were determined in the context of a value added system, where the designed products are manufactured by several companies. They give some input on the design of a service self-customization system and are therefore adapted and supplemented by further requirements in the context of a service-network.

3   A System for Self-customization of Services

3.1   Use Case Fourth Party Logistics Service Providing and the Implications to the Design of a Service Self-customization System

In this chapter the case of a fourth party logistics provider (4PL) is chosen to present a service business, which is faced with individual customer requirements as well as competition. A study of the third party logistics (3PL) market [42] revealed, that the flexibility of to accommodate customers' needs and the ability to achieve cost and service objectives are perceived as key success factors of a contract logistics provider. In sum these requirements also apply to the 4PL market. The difference between 4PL and 3PL providing is in integrating several sub logistics providers to offer a major spectrum of value-added services for serving whole customer supply chains. Compared to the 3PL, the 4PL is an enhanced business model where the logistics provider acts as an integrator or general contractor [43]. The 4PL service business is characterized by the following definition (adapted and slightly modified from [44–46]): *A 4PL provider has the ability to fulfill all intra- and cross-company logistics functions concerning planning, execution and monitoring. Furthermore he provides additional services like consulting, IT related and financial services. As an integrator he combines and manages the resources, capabilities, and technology of its own organization with those of other service providers to create a customized complex logistics service. The degree of in-house service delivery can vary from a total self-accomplishment of tasks with proprietary resources to complete external processing of service tasks.*

Assuming that a 4PL intends to serve more than one customer, he is confronted with the issue of acting for many individual customers without deficit in quality and attention for each customer [44]. The strategy is to link the efficiency of a mass service provider with the adaptability of an individual service provider [44]. The Mass-customization strategy combines both concepts. To follow this strategy is an option for the 4PL to cope with the market conditions/requirements.

Information flow management is a further basic challenge of the 4PL acting as an integrator, since he has to coordinate multiple clients and logistics service providers. ICT offers the possibility to support the information flows relating to logistics services as well as to coordination and communication issues. ICT also enables the expansion and customization of logistics service offerings [47].

By the use of modularity in services (as an instrument of the mass customization strategy), the 4PL is able to shorten the contracting phase. Well defined service-modules can be easily identified and combined to an individual overall service with the aid of information systems/service-configuration systems.

The following key words and their meaning should be noted, because they are taken as the basis for the role model of a value-added system pursuing the strategy of self-customization of services:

- Service delivery by several/multiple service-providing companies
- Bundled service-proposition by a company acting as an integrator with or without own assets/resources (neutral)
- Modular service architecture/offering as a precondition for self-customization and an instrument of mass-customization strategy
- Service customers are manufacturing companies requesting and buying business services

The following main characteristics derived from the value creation constellation as described previously, are necessary to determine the architecture of a service self-customization system:

- Value added network, consisting of different actors and transactions between them; value creation by services; actors are legally and economically independent
- For each customer service, which is the result of the service-customization procedure, a specific value added network is created, kept up and dissolved after service delivery
- An intermediary company bundles the service offers and sells them to the customer; the intermediary analyses the requirements and create a customized service; by use of the tool, these activities are transferred to the customer; the intermediary also monitor the service delivery

3.2 Model of a Self-customization Service System

From the previously mentioned key words three main roles can be derived, as shown in Fig. 1. Tasks can be assigned to each role as well as preconditions and requirements to offer and buy services in a self-customization valued added service system. In the following the identified roles are shortly described.

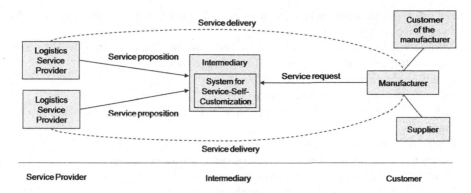

Fig. 1. Model of roles in a multi-actor service-system

In order to let the customer compose single services to an overall customer service, the *service providers'* service architecture needs to be modular. They also have to define parameters, which can be personalized by a customer in the self-customization process (e.g. conditions to store perishable products). Additional to the availability of own resources, the service provider needs to know the capacity of each resource. Furthermore, flows of activity may be limited in physical, staff or time respects. To sell services to a customer, the provider must know his costs and determine the amount he intends to charge. The basic tasks of the provider are making service offerings and deliver these services, which are requested by the customer.

The *customer* as the buyer of the services must be able to identify his needs and how to meet them. His main task is to get in touch with the company, which offers appropriately bundled services, and to identify the single services that best fit the needs and compose them to an overall service. This may be supported by a system for self-customization, which the intermediary provides. The company, acting as an *intermediary*, bundles the services of several service providers and offers them centrally to customers.

3.3 Model of a Service Self-customization Systems Functionality

The requirements on a system for service self-customization arise from the previously mentioned roles and their basic tasks. Additionally some input was given by related literature (Sect. 2.3). A subset of the requirements and functionality, which is usually discussed exclusively in the context of mass-customization systems for goods, can be adapted for the model presented here. The following demands are made at least on a system for service self-customization:

- Provide a service provider interface to accumulate all functions that are related to interactions of the provider with the system, e.g. enter, edit, administrate service offerings or give an overview of the service-configurations the provider is involved in
- Facilitate the definition of service-modules, basic rules and restrictions for the composition of services
- Save rules and service-modules in a database/service catalogue

- Use of a configuration method to determine the requirements of the customer with an appropriate visualization
- Allow for searching service-modules as well as for the automated retrieval of potentially suitable modules based on the identified requirements
- Combine identified service-modules
- Presentation of the composed service
- Edit service-modules in the customization process in order to personalize/modify them
- Check for capacity, time, logical and other constraints
- Determine the final price as well as the estimated time of delivery
- Identify the final suppliers for each service-module
- Transfer the service specification to each supplier
- Support the process of ordering the customized service
- Saving user profiles and each service-composition the customer assembled
- Monitoring of the status of the service-delivery

These requirements are translated into individual functionality of a service self-customization system (Fig. 2).

Two stages of functionality are distinguished: the front stage and the back stage. The front stage functionality is represented by interactions of a service customer and service provider with the system. Therefore it is divided into a customer as well as a provider interface with user specific functionality. The back stage consists of functionality, which is not directly perceived by the customer or provider. Automated processes use and transfer data, which is received and visualized by the frontend functionality in order to allow for user interaction. By use of a *service-provider interface*, service-modules can be created. This happens in form of formalized descriptions with the use of attributes, which characterize a service-module. To differentiate the steps of a service-module´s life cycle, the status "design", "implemented" and "archived" are given. A provider has the possibility to change or update a service-module. Only in design phase service-modules are able to delete. If a service-module is part of a current service-composition, but should no longer been offered, it can be archived. To enhance the valid composition of service-modules the provider can add and update constraints and rules. After the composition of a service and its personalization by the customer, the corresponding service-modules (as part of the composition) are sent to the provider for validation. If the service-modules are highly standardized and well described and if the capacities of the provider to fulfill the service are up to date, the validation can be done automatically. The providers are informed about the service-modules to be approved. After the approval of every service-module by the providers, the service-composition is offered to the customer.

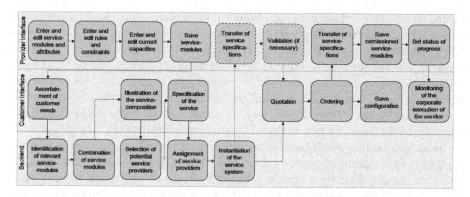

Fig. 2. Model of functionality of a service self-customization system

The service-composition is generally hidden to the provider. He only sees the modules relating to him and the data, which is necessary to manage the transfer of objects or information between the cooperating actors. With the help of a monitoring component, the providers can set the status of the service-delivery. For example they inform the customer about the actual start of the service-delivery and its progress. Maybe some interrupts or delays can be transmitted.

By assistance of a *customer interface,* needs of the customer are ascertained and used to identify and select relevant service modules. This procedure is realized by a configuration method (e.g. by attribute configuration). After selecting and composing the relevant service modules, the related service providers are selected too. Since a service-module can be offered by different providers and the customer has the possibility to change his preferences and therefore to change the service composition, the pool of identified providers is temporarily. By a presentation of the composed service, the customer has an overview of the progress of the customization. To personalize the service and thereby all corresponding service-modules, attributes are requested, which the customer has to edit. After the collection of all necessary values, suitable providers can be assigned. Therefore the values are checked against constraints in time, capacity and region. Every service-composition informs the customer about the estimated price, start of delivery and delivery time. The customer is able to change attribute values to test alternative scenarios if possible. If the offered service-modules are highly standardized, the validation of the composed service and all corresponding service-modules happens within the composition process by means of stored rules. Otherwise it takes place in a validation process, where the providers have to give a feedback. After the automated validation or approval by the providers, an offer is generated and the customer is able to place an order.

4 Conclusion

In this paper main requirements on a service self-customization system are identified. Self-customization is a concept in the mass-customization strategy, which aims at transferring tasks of product or service customization to the customer. This procedure is enabled by assistance of information systems. In current literature there are few contributions focusing on the development of self-customization systems. Most of them relate to the customization of goods rather than on business services and focus only one partial aspect of self-customization. There is a lack of presenting functionality for the self-customization of services in a multi-actor environment. Hence, a concept of functionality is introduced, which seeks to fulfill the identified requirements on a service self-customization system. The requirements are deduced from a role model, which was developed on the basis of a Fourth Party Logistics providing network. The 4PL provider acts as an intermediary, who bundles the different service offerings of various service suppliers and provides the customers an overall solution. The service self-customization system can be provided by a company like this to support and fasten the process of service specification. If customers bring out steps of the service design by themselves, transactional costs between customer and provider, which occur during the customization process, can be reduced.

The present model of functionality serves as the basis for further work. The focus will be on a differentiated consideration of the customer and provider interface. Single functions assigned to each interface are to be subdivided into precise partial functions of the system in order to describe them formally. The result will be a model of a system for service self-customization, which can be transferred into specific use cases in the service industry.

References

1. Büttgen, M.: Mass customization im Dienstleistungsbereich - Theoretische Fundierung und praktische Umsetzung eines Konzeptes zur Erlangung hybrider Wettbewerbsvorteile. In: Mühlbacher, H., Thelen, E. (eds.) Neue Entwicklungen im Dienstleistungsmarketing, pp. 257–285. Deutscher Universitätsverlag, Wiesbaden (2002)
2. Burr, W., Stephan, M.: Dienstleistungsmanagement. Kohlhammer, Stuttgart (2006)
3. Selviaridis, K., Spring, M.: The dynamics of business service exchanges: insights from logistics outsourcing. J. Purchasing Supply Chain Manag. **16**, 171–184 (2010)
4. Piller, F.T.: Mass customization. In: Albers, S., Herrmann, A. (eds.) Handbuch Produktmanagement, pp. 929–955. Springer, Berlin (2003)
5. Rogoll, T.A., Piller, F.: Konfigurationssysteme für Mass Customization und Variantenproduktion. ThinkConsult Marktstudie, München (2003)
6. Thomke, S., von Hippel, E.: Customers as innovators: a new way to create value. Harvard Bus. Rev. **80**, 74–81 (2002)
7. Valenzuela, A., Dhar, R., Zettelmeyer, F.: Contingent response to self-customization procedures: implications for decision satisfaction and choice. J. Mark. Res. **46**, 754–763 (2008). XLV (Article Postprint)
8. Reiß, M., Beck, T.C.: Mass customization - ein Weg zur wettbewerbsfähigen Fabrik. ZWF **89**(11), 570–573 (1994)
9. Piller, F.: Mass Customization. Ein wettbewerbsstrategisches Konzept im Informationszeitalter, vol. 4. Deutscher Universitätsverlag, Wiesbaden (2006)
10. Pekkarinen, S., Ulkuniemi, P.: Modularity in developing business services by platform approach. Int. J. Logistics Manag. **19**(1), 84–103 (2008)
11. Meyer, M.H., DeTore, A.: Perspective: creating a platform-based approach for developing new services. J. Prod. Innov. Manag. **18**, 188–204 (2001)

12. Stauss, B.: Plattformstrategie im Dienstleistungsbereich. In: Bullinger, H.-J., Scheer, A.-W. (eds.) Service Engineering, pp. 321–340. Springer, Heidelberg (2006)
13. Hillbrand, C., März, L.: Modellierung von Produktplattformen für Logistikdienstleistungen. In: Thomas, O., Nüttgens, M. (eds.) Dienstleistungsmodellierung, pp. 17–34. Physica Verlag, Heidelberg (2009)
14. Corsten, H., Gössinger, R.: Modularisierung von Dienstleistungen. Untersucht am Beispiel von Logistikdienstleistungen. In: Gouthier, M.H.J., et al. (eds.) Service Excellence als Impulsgeber, pp. 163–185. Gabler Verlag, Wiesbaden (2007)
15. Böhmann, T., Krcmar, H.: Modulare Servicearchitekturen. In: Bullinger, H.-J., Scheer, A.-W. (eds.) Service Engineering, pp. 377–401. Springer, Heidelberg (2006)
16. Böhmann, T.: Modularisierung von IT-Dienstleistungen: Eine Methode für das Service Engineering. DUV, Wiesbaden (2004)
17. Baida, Z., Gordijn, J., Omelayenko, B.: A shared service terminology for online service provisioning. In: ICEC 2004 Sixth International Conference on Electronic Commerce. ACM (2004)
18. Edvardsson, B., Gustafsson, A., Roos, I.: Service portraits in service research: a critical review. Int. J. Serv. Ind. Manag. 16(1), 107–121 (2005)
19. Hill, T.P.: On goods and services. Rev. Income Wealth 23(4), 315–338 (1977)
20. Lehtinen, U.: On defining service. In: Proceedings of the XIIth Annual conference of the European Marketing Academy (1984)
21. Grönroos, C.: Services Management and Marketing: Managing the Moments of Truth in Service Competition. New Lexington Press, San Francisco (1990)
22. Grönroos, C.: Service Management and Marketing: A Customer Relationship Management Approach, vol. 2. Wiley, Chichester (2000)
23. Gustafsson, A., Johnson, M.: Competing in the Service Economy. Jossey-Bass, San Francisco (2003)
24. Vargo, S.L., Lusch, R.F.: Evolving to a new dominant logic for marketing. J. Mark. 68, 1–17 (2004)
25. Pine, B.J., Gilmore, J.H.: The Experience Economy: Work is Theatre and Every Business a Stage. Harvard Business School Press, Cambridge (1999)
26. Alter, S.: Integrating sociotechnical and technical views of e-services. e-Service J. 7(1), 15–42 (2010)
27. Lovelock, C.H., Gummesson, E.: Whither services marketing. J. Serv. Res. 7(1), 20–41 (2004)
28. Lehtinen, U., Lehtinen, J.R.: Two approaches to service quality dimensions. Serv. Ind. J. 11(3), 287–303 (1991)
29. Grönroos, C.: Service logic revisited: who creates value? And who co-creates? Eur. Bus. Rev. 20(4), 298–314 (2008)
30. Chesbrough, H., Spohrer, J.: A research manifesto for service science. Commun. ACM 49(7), 33–40 (2006)
31. Fitzsimmons, J.A., Fitzsimmons, M.J.: Service Management: Operations, Strategy, and Information Technology, vol. 3. McGraw-Hill, New York (2001)
32. Sampson, S.E.: Understanding Service Businesses: Applying Principles of Unified Systems Theory, vol. 2. Wiley, New York (2001)
33. Tapscott, D., Ticoll, D.: The Naked Corporation: How the Age of Transparency Will Revolutionize Business. Free Press, New York (2003)
34. Colecchia, A., et al.: A New Economy: The Changing Role of Innovation and Information Technology in Growth. OECD, Paris (2002)

35. Kleinaltenkamp, M., Hellwig, A.: Innovationen durch Kundenintegration bei unternehmensbezogenen Dienstleistungen. In: Gouthier, M.H.J., et al. (eds.) Service Excellence als Impulsgeber. Gabler, Wiesbaden (2007)
36. Corsten, H.: Dienstleistungsmanagement, vol. 3. Oldenbourg Wissenschaftsverlag, München (1997)
37. Axelsson, B., Wynstra, F.: Buying Business Services. Wiley, Chichester (2002)
38. Reichwald, R., Piller, F., Meier, R.: E-Service Customization — Strategien zur effizienten Individualisierung von Dienstleistungen. In: Bruhn, M., Stauss, B. (eds.) Electronic Services: Dienstleistungsmanagement Jahrbuch 2002, pp. 225–241. Gabler Verlag, Wiesbaden (2002)
39. Hümmer, W.: Vertragsverhandlungen um konfigurierbare Produkte im elektronischen Handel (2004). https://opus4.kobv.de/opus4-fau/files/82/Dissertation+Hümmer.pdf. Accessed 22 Dec 2015
40. Böttcher, M., Klingner, S.: Providing a method for composing modular B2B services. J. Bus. Ind. Mark. **26**(5), 320–331 (2011)
41. Dietrich, A.J.: Informationssysteme für Mass Customization: Institutionenökonomische Analyse und Architekturentwicklung. Deutscher Universitäts-Verlag, Wiesbaden (2007)
42. Langley Jr., C.J.: The State of Logistics Outsourcing. 2010 Third-Party Logistics. Results and Findings of the 15th Annual Study. Capgemini, Atlanta (2010)
43. Vivaldini, M., Pires, S.R.I., Souza, F.B.D.: Collaboration and competition between 4PL and 3PL: a study of a fast-food supply chain. Flagship Res. J. Int. Conf. Prod. Oper. Manag. Soc. **1**(2), 17–29 (2008)
44. Nissen, V., Bothe, M.: Fourth Party Logistics - ein Überblick. Logistik Manag. **4**(1), 16–26 (2002)
45. Bauknight, D., Bade, D.: Fourth party logistics-breakthrough performance in supply chain outsourcing. Supply Chain Manag. Rev., Global Suppl. **2**(3) (1998)
46. Schmitt, A.: 4PL-Providing als strategische Option für Kontraktlogistikdienstleister. DUV, Wiesbaden (2006)
47. Kutlu, S.: Fourth Party Logistics: The Future of Supply Chain Outsourcing? Best Global Publishing, Brentwood (2007)

Risk-Aware Pricing of B2B Services: Approach, Realization and Application to a Payments Transaction Processing Service

Michael Fischbach[1](✉) and Rainer Alt[2]

[1] Business Engineering Institute St. Gallen AG,
Lukasstrasse 4, 9008 St. Gallen, Switzerland
michael.fischbach@bei-sg.ch
[2] University of Leipzig, Grimmaische Strasse 12,
04109 Leipzig, Germany
rainer.alt@uni-leipzig.de

Abstract. This paper proposes a risk-aware B2B service pricing approach. The approach characterizes relevant cost positions according to their quantity and adaptiveness towards changes in the quantities sold. Based on the achieved transparency about the cost structure, cost niveau and cost adaptiveness, the approach allows to configure a risk-aware pricing scheme with an arbitrary number of different price components. It also allows for several different pricing schemes and provides analysis functionality for comparing these schemes with respect to risk and return criteria. The approach contributes to the domain of service science, which historically has not been discussing risk-based pricing approach in-depth.

Keywords: Service-orientation · B2B pricing · Risk-aware pricing · Finance industry

1 Introduction

For some time now, trends towards the industrialization of the financial services industry (Lamberti [8]) or the "lean bank" [1] are discussed. Financial services companies consolidate and specialize on certain sub-processes within the value chain, i.e. they concentrate on their core competencies [5]. This consolidation and specialization results in the formation of value creation networks [14]: companies engage in more or less stable co-operations in order to source those services which they are no longer to produce themselves. Thus, normally a multitude of different companies is engaged in the production of financial services offered to the end-customer.

Generally, three types of banks prevail: product-, distribution-, and transaction-banks. Distribution banks focus on the interface to the customer, their core competency lies in the choosing and selling of banking services to the end customer. This type of bank mostly corresponds with most peoples common understanding of a bank. However, in order to be able to offer banking services to the customer, distribution banks have to source these services from specialized providers, the product banks. Product

© Springer International Publishing Switzerland 2016
W. Abramowicz et al. (Eds.): BIS 2016, LNBIP 255, pp. 341–355, 2016.
DOI: 10.1007/978-3-319-39426-8_27

banks engineer financial products and deliver them to the sales banks, which in turn put their label on it and sell it to the customer. The usage of many banking services causes a whole bunch of activities to be performed until the desired outcome is realized. For instance, once the end customer submits a payment transaction via the banks e-banking interface, many activities follow until the payment is credited on the receiver's bank account. As these processing activities are usually very technology- and know-how intensive, specialized providers emerged, called transaction banks. Depending on the sourcing model, these providers take over a certain amount of activities from the sales bank. One rationale for the sales bank to source the processing of payments transactions from external providers is the possibility to not have to maintain costly IT-infrastructure such as equipment for the digitalization of payment slips. Especially smaller banks cannot afford such investments, because their transaction volume does not allow them to achieve substantial economies of scale.

A bundling of transactions volumes from several banks on the provider's side seems promising with respect to these issues. However, significant utility for the client banks is only achieved if the provider applies variable pricing schemes. This in turns poses a possibly substantial risk on the provider, as he risks not being able to cover his fixed costs due to fluctuating transaction volumes.

The trade-off between fulfilling customer's demand of variable pricing and provider's demand of reduced risk affords a systematic approach to the design of a viable pricing scheme. Today, pricing decisions in the backoffice service provision area mostly are made out of a gut feeling, without systematic analysis. As we show, this lack of sophistication is mainly due to the non-availability of applicable approaches. Consequently, we propose a simple yet purposeful approach to the creation of risk-aware pricing-schemes.

The research questions we follow are:

Q1. What requirements are posed on a risk-aware B2B service pricing approach for the financial industry?

Q2. To what extent do current pricing approaches for B2B services in the banking industry fulfil the identified requirements?

Q3. Based on the answer of RQ1, how could a risk-aware B2B service pricing approach for the financial industry look like?

By answering the research questions, this paper adds to the literature body of service science [4], as risk-based pricing has not been discussed extensively so far, is however crucial for network-based collaboration structures, as argued before.

Section 2 points out the chosen research methodology. Section 3 derives general and specific requirements on a risk-aware pricing scheme for B2B services in the banking industry *(Q1)*. Based on these requirements, existing approaches are assessed and shortcomings identified *(Q2)*, Sect. 4. An approach is presented that addresses most of the shortcomings *(Q3)*, Sect. 5. First, the basic functioning is described in a qualitative manner. Second, a formalized interpretation helps transferring the approach into a software application. A stringent example shows the practicability of the approach. Finally, Sect. 6 concludes and identifies possible future research directions.

2 Research Methodology

This paper refers to a multilateral research program (i.e. consortium research, see [10]) that started it's fourth phase in summer 2010 and investigates the topic of service-oriented design and valuation of banking services. The research team consists of academics from three universities in Germany and Switzerland as well as practitioners from 18 companies of various sizes and roles in the banking value chain (e.g. regional retail bank, international private bank, outsourcing provider, software provider). The companies contribute to the research by playing an active role in biannual steering committee meetings and quarterly workshops as well as bilateral projects to ultimately enhance the development of the envisioned methodologies and verify its applicability. Consequently the research program follows the paradigm of "emphasizing collaboration between researchers and practitioners" [3]. The chosen consortium research method is based on a process model for Design Science Research proposed by [11] and the corresponding guidelines proposed by [7] to ensure a rigorous link to existing research as well as the relevance of the generated artefacts. The basic principle of consortium research is the collaboration between academic institutions and companies, ensuring both an academic and a pure practice oriented view on the problems. Both parties are engaged in the definition of the problems and objectives as well as in the design, development, evaluation and diffusion of artefacts.

Besides the valuation of banking services the research program also focuses on the customer-oriented networking of the future bank. Following Design Science [7], the artefacts combine to a methodological approach towards an approach for customer- and service-oriented networking in the financial industry.

3 Requirements on an Approach for Risk-Aware Pricing of B2B Services in the Banking Industry

3.1 Requirements

Out of an analyses of three cases from the financial industry (a core banking software provider, a retail bank and a banking services provider), a pricing approach for B2B services in the banking industry has to meet several requirements:

R1: Complexity has to be kept on a rather low level to ensure practical acceptance. Especially in the domain of transaction banking services, which are mostly IT-intensive, complexity in terms of involved cost positions can be dramatically high. So the approach has to offer mechanisms that reduce complexity by offering the possibility to abstract at an acceptable level (i.e. apply a suitable granularity of the cost positions).

R2: The approach has to be straightforward and relatively easy to understand. Talks to decision-makers indicate that the time spent on pricing activities in the transaction banking area is rather limited. Therefore, given this circumstance, the approach has to be fast. While R1 relates to the data input procedure, this requirement focuses on the model itself.

R3: Accounting structures in the transaction banking domain are rather rudimentary, often not providing more information than distinguishing between fixed and variable costs. A pricing approach has to take this into account by not relying on sophisticated cost information.

R4: The approach has to determine the cost structure of the service that is to be priced, as this is a crucial step towards assessing the risk potential. Especially important is the distinction between fixed and variable costs. As stated earlier, the more variable the pricing scheme is, the more benefit it provides to the customers (i.e. the distribution bank). For instance, a completely variable pricing scheme (i.e. a price per transaction), leaves no risk for the customer, because as transactions are dropping the overall sum paid to the provider drops as well. However, the provider risks being left over with a substantial amount of uncovered fixed costs. On the contrary, if the provider would simply divide the total costs of providing the payments transactions processing service among his clients (i.e. a fixed price per client), all risk is transferred to the customer. In a competitive market, surely the latter constellation is not stable. However, the former pricing scheme would not be stable either, as the provider bears a significant amount of risk which would eventually drive him bankrupt. Therefore, the pricing approach has to offer a mechanism for approaching the optimum degree of price variability, which is nothing else than finding a risk-adequate pricing scheme. Thus, the approach has to provide a mechanism that allows the decision-maker to achieve a reasonable portion of risk-taking by considering the sensitivity of cost positions when deciding about the variability of the anticipated pricing scheme.

R5: Most B2B services in the financial services domain are priced by a multi-component pricing scheme. Usually, the pricing scheme consists of a mixture of basic price models, such as base prices, variable prices (linear, degressive, sliding-scale prices and so on). Consider the provision of a core banking software as a service. The pricing scheme could include the following price components: Component 1: a fixed license fee (fixed); Component 2: a fee per workstation (fixed); Component 3: a fee per branch (fixed); Component 4: a fee per ticket (variable); Component 5: a fee per hour of manual intervention (variable). Consequently, a B2B pricing approach in the banking domain has to be able to model a pricing scheme with more than one price component.

R6: Due to reasons laid out before, a pricing approach generally has to take into account multiple sources of information. The costs of providing the service constitute the (long-term) lower boundary of the price. Adding desired margins yields the minimum price that is to be achieved (cost-plus). However, even though this procedure ensures the calculatory profitability of the offering, decision-makers still take a substantial amount of risk by just using a cost plus mechanism: they risk their price not being accepted by the market, thus not being able to realize the anticipated transaction volume and consequently running into possibly large amounts of losses. Therefore, a viable risk-aware pricing approach for B2B services in the banking domain has to fulfill the following requirement: The approach has to take into account prices of competitors and customer's willingness to pay. Such an integrated view on the costs and on the market conditions adds further to the goal of achieving a risk-aware pricing scheme.

4 Existing Approaches

Pricing literature brought about a huge amount of pricing approaches. Basically, pricing these can be characterizes as either one-dimensional or multi-dimensional approaches [13]. One-dimensional approaches aim at designing one price for a specific product, without considering price differentiation with respect to time, region, customer etc. However, the term one-dimensional is a bit misleading, because it surrogates that the final price consists of only one component. This of course is not necessarily true. Especially in B2B relationships in the banking industry, multi-component prices involving for instance a base fee and a usage-depending component are rather common. Multi-dimensional approaches aim at creating different prices for the same good or service, depending on the location, time or kind of customer. The ultimate goal of these approaches is to maximize profits by optimally skimming customer's rent. An example is the airline industry, which extensively applied the yield management approach, in which flights are priced dependant on the capacity situation and the expected customer willingness to pay.

Although multi-dimensional approaches are punctually deployed in B2B sourcing relationships in the banking industry, as for example in the context of payment transaction delivery times, it is rather uncommon to find multi-dimensional pricing approaches in this domain. Several reasons account for this. First, B2B providers regional expansion is usually limited to a certain area; cross-border sourcing relationships are not common. Further, market transparency is rather low, meaning that providers have reliable information that would enable them to model multi-dimensional prices. For instance, price elasticities are not known. Due to these reasons, we subsequently focus our attention on one-dimensional pricing approaches. [9, 12] differentiate between demand, cost- and competition-oriented pricing mechanisms. [15] argues that this classification neglects the complexity of pricing decisions and also does not reflect the diversity of pricing mechanisms applied in practice [13]. Therefore, he proposes to take the information usage as the central differentiation criterion. The result is a classification scheme consisting of: one-sided fixed-, flexible intuitive- and simultaneous pricing mechanisms. One-sided fixed pricing mechanisms involve a single-stage processing of information, meaning that only one kind of information is considered, e.g. only cost- or competition-related information. An example is the cost-plus method. Based on the expected cost of service provision, the price is simply set by adding a desired margin. Often cost-plus pricing is defended as a heuristic allowing fair or reasonable pricing without being aware about the market conditions [6]. An orthogonal example is the approach of competition-based pricing. This method sets prices in relation to the prices of competitors. Assuming the exact same service level – prices are set cheaper or equal to the prices of competitors. One-sided fixed pricing approaches inevitably neglect important circumstances. In the case of cost-plus, competition-based criteria are neglected, while the competition-based approach does not consider cost-related issues. Therefore, in operational practice decision-makers usually consider more than one kind of information which they process in two ore more subsequent steps. In Wiltinger's terminology, such pricing approaches are designated as flexible-intuitive. With this approach, the first step is to gain a rough indication – based on some information source - of what the price could be and then to refine this indication by subsequently taking further information into account. The third category, namely simultaneous pricing approaches, differs from the second category with respect to the non-seriality of information processing. These approaches consider multiple sources of information simultaneously, as for instance controlling-, market- and target information. Inherent to this class of approaches is a comparison between multiple pricing schemes. Exemplary methods include direct costing, decision tree analysis, decision support systems and marginal-analytic methods [13]. Approaches in this category usually require IT support due to their (calculation) complexity.

The methodology presented in this article belongs to the simultaneous pricing approaches, thus considering multiple sources of information and providing the possibility to compare different pricing schemes with respect to their riskiness. However, it exhibits some elements of flexible-intuitive approaches as there is a certain degree of seriality in information processing.

In the following we examine current pricing mechanisms for their fulfillment of the requirements posed on a risk-aware B2B pricing approach. First and foremost it is worth mentioning that there is no dedicated pricing mechanism for B2B services in the banking industry to be found in literature. Thus, we focus on the approaches mentioned. We further only consider pricing approaches which are likely to be used in B2B pricing of banking services. For instance, due to its complexity yield management is not likely to be applied in this context in the coming years. In addition, "Decision Support Systems" was also not included because the variety of possible DSS systems is rather broad. Table 1 sums of the findings.

Table 1. Assessment of existing pricing approaches.

Approach	R1	R2	R3	R4	R5	R6
Cost-plus only	4	4	4	0	1	0
Competition-based only	3	3	n.a.	1	4	0
Direct costing	2	1	3	0	1	0
Decision tree analysis	0	0	4	0	0	2
Marginal analytical methods	0	0	1	0	1	3

0 no fulfillment 1 partial fulfillment 2 moderate fulfillment 3 good fulfillment 4 complete fulfillment

5   Risk-Aware Pricing of B2B Services in the Banking Industry

5.1   Case "Payments Transaction Processing"

To convey an understanding of how the following approach is applied in practice, we first outline a real-world example of a domestic payments processing service. A universal bank outsourced the transaction processing to a provider, who in turn is bundling transaction volumes from several banks, thus achieving economies of scale. Besides the depicted service, the provider also offers the services "international payments processing" and "securities transactions processing". The service under considerations amounts for 40 % of the companies' turnover. Because the theoretical number of possible sourcing model configurations (i.e. the process cut) is countless, we subsequently describe one scenario that we denote as "partial outsourcing". Based on a reference process "payments" by [2] seven distinct process steps can be distinguished. First, the customer places a payment order (either electronically or paper-based), which is received by the distribution bank. Second, in case of paper-based delivery the payment slip has to be digitized. Eventual manual corrections are performed (such as correcting the spelling of an address) and the instruction is checked for textual inconsistencies. Third, several other checks are performed in order to ensure that the customer is allowed to conduct the payments, e.g. limit- and regulatory checks. Once all checks are positive, the order is approved. The fifth step is the internal processing of the order, as for instance fees calculation, payment booking, the printing and delivery of customer outputs (such as account statements) and the archiving for regulatory compliance. After that, the transaction is processed in the interbank market. Simply put, this step is to transfer the payment from the sending bank to the receiving bank, often referred to as interbank clearing.

In our scenario we assume that the distribution bank outsources everything from digitization to interbank processing, except for customer output creation and archiving, which remains in-house.

Table 2 provides some further details about the assumed transaction volumes and characteristics of the provider. The stated figures adhere to real-world relations of typical mid-sized Swiss universal banks and respective providers (the figures have been validated by several consortium research partners in a dedicated workshop).

Table 2. Further assumptions about the provider, the service and the transaction volumes.

General assumption about the provider	
Employees:	150 FTE
Services offered:	Domestic payments transactions processing
	International payments transactions processing
	Securities transactions processing
General employee costs (p.a.):	3.0 FTE General Management (300'0000 per FTE)
	0.5 FTE Product Management (180'000 per FTE)
	0.5 FTE Service Management (180'000 per FTE)
	2.0 FTE HR (180'000 per FTE)
	2.0 FTE Accounting (180'000 per FTE)
	1.0 FTE Sales (180'000 per FTE)
Assumptions about the service "payments transactions processing"	
Customers:	5 mid-sized Swiss universal banks
Service turnover/company turnover (p.a.):	40 %
# domestic payments transactions (p.a.)	12.3 Mio. p.a.
- thereof paper based	4.92 mn. (80 % ESR, 20 % form-standardized)
Currency:	100 % CHF
FTE:	25 FTE for domestic payments processing (thereof 40 % hourly wage workers)
	40 TE for digitization

5.2   Description of the Approach

We propose an iterative three-step approach. First, all relevant costs are defined, characterized and quantified. Second, price schemes are designed. This is the point where an integrated view on costs and prices is taken in order to achieve a risk-adequate pricing. Finally, certain ratios are calculated which allow to assess and compare the different pricing schemes. Based on the result of the ratios, eventually various adaptions have to be made in the preceding steps to achieve a pricing that is in conformity to risk guidelines or market price acceptance. In the following, each step is described in detailed and illustrated by means of the introduced exemplary service "payments transaction processing".

First, all relevant cost positions have to be named and listed *(1.1)*. To keep complexity low, only cost positions that have a substantial influence are to be included. All others could possibly be considered by including a lump sum. Next, the determined cost positions are characterized as either fixed or variable *(1.2)*. All sales figures that are cost relevant have to be estimated, i.e. all figures that have an influence on the size of the cost positions *(1.3)*. In our example these are the expected transaction volumes (5 customers × 12'300'000 transactions per customer = 61'500'000 transactions) and the number of digitizations (5 customers × 4'920'000 = 24'600'000, 80 % of which are ESR). Based on these figures, the cost positions identified in 1.1 are quantified. The next step is where our proceeding offers substantial decision support. For each fixed cost position, sensitivities are estimated *(1.4)*. Sensitivities quantify the ability of a cost position to adjust to changed demand situations, i.e. the degree of variability. Variable cost positions by definition have a sensitivity of 1. However, most fixed cost positions do not have a sensitivity of 0 (which would mean that they are completely fixed). This assumption might hold for positions such as buildings or substantial IT infrastructure. However, when employee costs are considered the picture is slightly different: assume a decision-maker runs through the procedure we propose (i.e. calculating all relevant costs, designing pricing schemes, comparing pricing schemes and finally take the final decision about prices). Further assume that due to market changes the anticipated transaction volumes cannot be reached. In this case the decision-maker could lay off some employees that are no longer needed. However, due to a mandatory period of notice, let's say six months, the termination only becomes effective after this time. Thus, half of the employee's salary has to be paid, meaning that the other half can be saved. Therefore, this cost position would have a sensitivity of 0.5. Sensitivities serve two distinct purposes: first, they can be utilized for semi-automatic scenario simulations, a topic we touch upon later; second, thinking about sensitivities gives a sense for the risk that is inherent in each cost position, which in turn will (and should) have an impact on the pricing decision.

At this point, all cost positions are defined and quantified. Following the cost-plus approach, the next step is to add the desired margin *(1.5)*. After completion of (macro-) step 1 the situation in our example could look as in Table 3.

To sum it up, (macro-)step 1 provides the following decision-crucial information: a list of all relevant cost positions, their respective quantities and sensitivities as well as the total required turnover.

Table 3. Estimated cost situation for the exemplary service after completion of (macro-) step one.

Cost position	Unit	Cost/unit	Quantity	Total	S.
Fixed costs					
Employee costs for the team "payments", permanent staff	FTE	110'000	15	1'650'000	00.5
Material expenses (IT infrastructure, workplace, workstation etc.)	FTE	30'000	15	450'000	00.2
Variable costs					
Employee costs for the team "payments", temporary staff	Hours	55	19'320	1'062'600	1
Material expenses (IT infrastructure, workplace, workstations etc.)	FTE	10'000	10	100'000	1
Swiss Interbank Clearing costs	TRX	0.03	61.5 mn	1'845'000	1
Digitalization per payment slip ESR	Pymt. slips	0.12	19.68 mn	2'361'600	1
Digitalization per payment slip form-standardized	Pymt. slips	0.52	4.92 mn	2'558'400	1
Indirect costs					
Surcharges (overhead costs etc.)	4.36 % on direct costs			800'217	1
Risk surcharge	1.00 % on direct costs			100'276	1
Margin	4.00 % on total costs			437'124	1
Total required turnover				11'365'217	

The next step is to design different pricing schemes. Suppose the provider is risk-averse. Thus, the pricing scheme should reflect this attitude by including a fixed price component. Having in mind that our service consists of basically two main components, digitalization and processing, we decide to additionally include two further price components. As our customers demand a pricing scheme as variable as possible, these two components are variable *(2.1)*. Next, for each component, decide on the key according to which the costs are distributed among the customers. Take over the corresponding estimated sales figures from 1.1. For the fixed component the decision is straightforward: the corresponding key is the number of customers (i.e. five), because the fixed price is equally shared among all prospective customers. As indicated before, the second and third price components base on the keys "number of transactions" and "number of digitizations", with the corresponding figures estimated in 1.3 *(2.2)*. The next step enables the decision-maker to design the pricing scheme under the aspect of risk-adequacy *(2.3)*: for each cost position identified in (macro-)step one, he decides which portion of the respective cost position he wants to earn with each of the price components. Consider our example: the fixed employee cost position is classified as a fixed cost. Consequently, a risk averse provider should allocate 100 % of this position to the fixed price component in order to minimize the risk of not covering his fixed costs in case the transaction numbers tumble. However, in case the market does not accept such a fixed amount, the provider will not be able to enforce the desired allocation. Hence, he will eventually decide to allocate only 60 % of this fixed cost position to the fixed price component, and allocate the remainder to the variable price components. Similar considerations are applied for all other cost positions. For instance, the digitization costs are completely allocated to the third price component to achieve a usage-based charging (it certainly would not be fair to allocate these costs to the second component, as a transaction does not necessarily cause a digitization). Naturally, for each cost position the percentage distributions have to sum up to 100 %. This step enables the decision-maker to design a risk-aware pricing scheme, based on his risk appetite and restraints imposed by customer acceptance. After allocating each cost position in the described way, the next step is to sum up the allocated costs per cost component and divide the sums by the respective key values (e.g. 61'500'000 transactions for the second price component) *(2.4)*. The result is a price per price component that has to be earned on average in order to achieve the earning goals. These prices constitute the basis for further decision. In that respect it provides substantial decision support. Next, for each price component, decide on the price model to be used (linear, degressive, sliding-scale price etc.) *(2.5)*. Based on the decision support that the preceding step provides, decide on the definitive price per price component *(2.6)*. No matter which price model was chosen, the only important thing is that the average price per component is at least equal to the price obtained in 2.4. However, if the market allows for higher prices, the decision-maker is free to raise the suggested prices. The result of (macro-)step two is a complete price scheme. In case more than one price scheme is desired (e.g. a more risk-affine price scheme without a fixed price component) (macro-)step two is repeated. Figure 1 shows a possible pricing scheme for the exemplary payments transactions processing service.

The final step of our proposed approach is an assessment of the constructed pricing scheme(s). Basically, owing to subjective preferences each user may have his/her favorite analyses, graphs and tables. However, subsequently we selectively propose some ratios and figures that are useful in assessing the results of our approach: A. Turnover per price scheme; B. Revenue per price scheme; C. % of variable costs/% of variable turnover.

A and B. If the decision maker always exactly adopts the prices suggested by the approach, each pricing scheme is going to yield the same turnover and subsequently the same revenue. However, if for some reasons such as market intransparency, one price scheme enables the provider to charge overall higher prices than other price schemes would enable him to do, the former price scheme is superior to the others with respect to turnover and revenue.

C. The risk of a price scheme in the presented approach is defined as the relation between the total sum of variable cost positions and variable turnover. The higher this ratio, the lower the inherent risk. A redefinition of this ratio in terms of fixed costs makes matters clear: the higher the fixed costs in relation to the sum of the fixed price components, the higher the provider's risk of not covering his fixed costs once the transactions and/or number of digitizations decrease. For reasons of simplicity this

Price component			Base fee	TRX processing	Digitization
Type			fix	variabel	variabel
Key			Customers	Transactions	Receipts
Price model			Linear	Linear	Linear

Planmengen			Number of Customers	Number of Transactions	Number of Receipts
Customer 1			1	12'300'000	4'920'000
Customer 2			1	12'300'000	4'920'000
Customer 3			1	12'300'000	4'920'000
Customer 4			1	12'300'000	4'920'000
Customer 5			1	12'300'000	4'920'000
Gesamtmenge			5	61'500'000	24'600'000

Fixed costs	Sensitivity	Sum	%	absolute	%	absolute	%	absolute
Personnel costs team payments	0.5	1'650'000 100%	60%	990'000	40%	660'000	0%	0
IT and workplace costs	0.2	450'000 100%	80%	360'000	20%	90'000	0%	0
TOTAL fixed costs				1'350'000		750'000		0
Variable costs								0
Personnel costs team payments (variable)	1	1'062'600 100%	0%	0	100%	1'062'600	0%	0
IT and workplace costs (variable)	1	100'000 100%	0%	0	100%	100'000	0%	0
SIC-Costs	1	1'845'000 100%	0%	0	100%	1'845'000	0%	0
Digitization ESR	1	2'361'600 100%	0%	0	0%	0	100%	2'361'600
Digitization Standard	1	2'558'400 100%	0%	0	0%	0	100%	2'558'400
TOTAL variable costs				0		3'007'600		4'920'000
TOTAL direct costs				1'350'000		3757'600		4'920'000
Indirect costs		900'493 100%			50%	450'247	50%	450'247
TOTAL indirect costs				0		450'247		450'247
TOTAL costs				1'350'000		4'207'847		5'370'247
Margins								
Margin		437'124 100%	30%	131'137	50%	218'562	20%	87'425
TOTAL Margins				131'137		218'562		87'425
TOTAL required turnover		11'365'217		1'481'137		4'426'408		6'457'671
Calculated average to be achieved per key unit				296'227.42		0.07		0.22

Linear price model					
Price per key unit			296227.42	0.07	0.22
TOTAL effective turnover		11'365'217	1'481'137	4'426'408	5'457'671
Over-/Under-coverage		0	0	0	0

Share of fixed turnover	1'481'137	13%
Share of variable turnover	9'884'080	87%

Fig. 1. A possible pricing scheme.

seemingly straightforward calculation of the ratio neglects the effect of sensitive fixed costs. Principally, a whole bunch of other ratios and figures could be applied in the assessment of the different pricing schemes, especially ratios concerned with the results of different scenario simulations, a topic we will briefly touch upon later. Figure 2 gives a complete overview of our proposed approach.

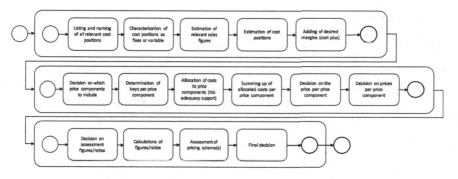

Fig. 2. Summary of the proposed approach.

Going Back the Spiral

Suppose it becomes apparent that the determined price scheme is not accepted by the market. Based on what the market is willing to pay the price scheme and/or the underlying cost positions have to be checked for whether there are possibilities to lower costs or to reconfigure the pricing scheme in order to meet market's requirements. Such a proceeding is referred to as target pricing. Although the proposed procedure does not yield definitive suggestion on how to lower costs and/or rearrange pricing schemes, it yet creates some level of transparency, which is the basis for profound target pricing decisions.

Formalization of the Approach

This section briefly shows how the quantitative parts of the proposed approach can be formalized so that a transfer into an IT-enabled solution becomes feasible. We do not aim at giving a complete implementation guide, but rather want to show the basic quantitative relationship within the approach.

First, for each of the $m = 1\ldots M$ cost positions the total costs cp_m, given the estimated sales figures, are calculated. These consist of the cost per unit uc_m multiplied by the number of units u_m plus eventually an absolute abs_m:

$$cp_m = uc_m * u_m + abs_m$$

Next, for each cost position $m = 1\ldots M$, the cost sum cp_m is distributed among all N price components pc_n, where $n = 1\ldots N$. The percentage of cost position cp_m that is allocated to price component pc_n is denoted as $perc_{mn}$. These percentages are put in by the user, they are the main instrument for risk-aware pricing within our approach. The double sum

$$\sum_{m=1}^{M}\sum_{n=1}^{N} perc_{mn} * cp_m$$

has to be equal to the sum of all cost positions. This is true if and only if $\sum_{n=1}^{N} perc_{mn} = 1 \forall m$, or – stated different – if each cost position is completely allocated to the price components.

The sum $\sum_{m=1}^{M} perc_{mn} * cp_m$ yields the calculatory turnover pc_n that has to be earned by price component n. For each price component, the calculatory turnover pc_n has to be equally split among the quantity of the underlying key unit key_n. The result is the calculatory required turnover per underlying key unit per price component (denoted as $price^{calc}$):

$$price_n^{calc} = \left(\sum_{m=1}^{M} perc_{mn} * cp_m\right)/key_n$$

in the case of price component n. In the illustrated example, the key of the second price component is the number of transactions, 61.5 mn. Thus, $key_n = 61'500'000$.

The figures $price_n^{calc}$ $(n = 1\ldots N)$ provide substantial decision support in that they lay the basis for the final decision about the definitive price per price component, $price_n^{definitive}$. Table 4 indicates the valid value range for each of the used variables and their respective meaning.

Table 4. Valid value ranges and summary of variable declarations.

Variable	Value range	Meaning
m	$\mathbb{N}/0$	Index of cost positions
cp	$\mathbb{R}^+$	Value of a cost position
u	$\mathbb{N}/0$	Number of units
abs	$\mathbb{R}^+$	Absolute costs
n	$\mathbb{N}/0$	Index of price components
uc	$\mathbb{R}^+$	Costs per unit u
pc	$\mathbb{N}/0$	Costs allocated to a price component
$perc$	$\geq 0 \wedge \leq 1$	Allocation percentage
key	$\mathbb{N}/0$	Quantity of resp. key units
$price^{calc}$	$\mathbb{R}^+$	Calculatory price
$price^{definitive}$	$\mathbb{R}^+$	Definitive (set) price

6 Evaluation and Conclusions

This contribution proposes an approach for risk-adequate B2B service pricing. It is derived from existing approaches and practical experience. At the example of a payments transactions processing service the methodology has been applied.

With respect to requirements fulfilment, the following findings arise (Table 5):

- Reportedly, the approach is straightforward and easy to apply in practice (R1&R2)
- The approach only requires rudimentary cost information, which should be available in almost every company. It analyzes the cost structure, which is crucial to assess the inherent risk (R3&R4)
- The pricing method includes multiple price components, which is crucial for practical applicability (R5)

By fulfilling these requirements, it adds to the scientific body of knowledge a new multi-dimensional, cost-based pricing model.

Table 5. Requirements fulfilment.

Approach	R1	R2	R3	R4	R5	R6
Risk-aware pricing approach	3	4	3	4	4	2

0 no fulfillment 1 partial fulfillment 2 moderate fulfillment
3 good fulfillment 4 complete fulfillment

The following benefits have been reported by the two consortium members (banking software vendors), who have applied it within their organization:

- Better management support because of transparency of the price-finding process
- Faster pricing decisions
- Better understanding of the inter-relationships between prices, costs, margins and risks
- Better risk control and risk prediction

Further research potential especially prevails in one direction: while the approach might be used to analyze competitor's pricing actions, it has not yet been validated in this respect. Eventually, the model needs some extension in order to directly enable the user to compare multiple pricing schemes and pricing compositions. Further, the artifact needs to be more closely tied to existing results from the domain of service science. In this respect, a central future research need is to investigate whether the risk-based pricing approach equally works for all types of services (i.e. business- and technical services). Another - rather practice-oriented - opportunity is to evaluate possible production environments for the approach. These range from dedicated implementation to implementing the approach directly into core-banking systems or pricing engines.

References

1. Allweyer, T., Besthorn, T., et al.: IT-Outsourcing: Zwischen Hungerkur und Nouvelle Cuisine, Digitale Ökonomie und struktureller Wandel. Deutsche Bank Research, Frankfurt (2004)
2. Alt, R., Bernet, B., et al.: Transformation von Banken: Praxis des In- und Outsourcings auf dem Weg zur Bank 2015. Springer, Heidelberg (2009)
3. Avison, D.E., Lau, F., Myers, M., Nielsen, P.A.: Action research. Commun. ACM **47**(1), 94–97 (1999)
4. Bardhan, I.R., Demirkan, H., Kannan, P.K., Kauffman, R.J., Sougstad, R.: An interdisciplinary perspective on IT services management and service science. J. Manag. Inf. Syst. **26**(4), 13–64 (2010)
5. Gottfredson, M., Puryear, R., et al.: Strategic sourcing: from periphery to the core. Harvard Bus. Rev. **83**(1), 132–139 (2005)
6. Hanson, W.: The dynamics of cost-plus pricing. Manag. Decis. Econ. **13**, 149–161 (1992)
7. Hevner, A.R., March, S.T., et al.: Design science in information systems research. MIS Q. **28**(1), 75–105 (2004)
8. Lamberti, H.-J.: Industrialisierung des Bankgeschäfts. Die Bank **6**, 370–375 (2004)
9. Meffert, H., Burmann, C., Kirchgeorg, M.: Marketing: Grundlagen marktorientierter Unternehmensführung: Konzepte, Instrumente, Praxisbeispiele. Gabler, Wiesbaden (2008)
10. Oesterle, H., Otto, B.: Consortium research: a method for relevant IS research. Bus. Inf. Syst. Eng. **5**, 1–24 (2010)
11. Peffers, K., Tuunanen, T., et al.: A design science research methodology for information systems research. J. Manag. Inf. Syst. **24**(3), 45–77 (2008)

12. Sander, M.: Internationales Preismanagement: Eine Analyse preispolitischer Handlungsalternativen im internationalen Marketing unter besonderer Berücksichtigung der Preisfindung bei Marktinterdependenzen. Physica, Heidelberg (1997)
13. Simon, H., Fassnacht, M.: Preismanagement. Gabler, Wiesbaden (2009)
14. Sydow, J.: Management von Netzwerkorganisationen - Zum Stand der Forschung. Management von Netzwerkorganisationen. J. Sydow. 4, 387–472 (2006). (Wiesbaden, Gabler)
15. Wiltinger, K.: Preismanagement in der unternehmerischen Praxis: Probleme der organisatorischen Implementierung. Gabler, Wiesbaden (1998)

On the Maturity of Service Process Modeling and Analysis Approaches

Florian Bär[1,2(✉)], Kurt Sandkuhl[1], and Rainer Schmidt[2]

[1] University of Rostock, Albert-Einstein-Str. 22, 18059 Rostock, Germany
{Florian.Baer,Kurt.Sandkuhl}@uni-rostock.de
[2] Munich University of Applied Sciences, Lothstr. 64, 80335 Munich, Germany
{Florian.Baer,Rainer.Schmidt}@hm.edu

Abstract. Prior research has provided a number of approaches for the specification and analysis of service processes. However, little is known about their level of maturity regarding considered dimensions and characteristics. The present study represents a first step towards filling this gap. Drawing upon recent formalizations and delineations of service, a model for assessing the maturity of service modeling and analysis tools is derived. As part of a systematic literature review, it is applied to a set of 47 service blueprinting techniques to determine their maturity. The study's findings indicate a high level of maturity regarding control flow and resource integration for most of the identified approaches. However, there are several shortcomings with respect to the input and output dimensions of the service process. The study proposes a set of research questions to stimulate future research and address the identified shortcomings.

Keywords: Service blueprinting · Service process · Specification and analysis · Maturity

1 Introduction

In recent years, the characterization of service has shifted from an intangible product focus to a more process oriented focus [1, 2]. A service process "[...] can be viewed as a chain or constellation of activities that allow the service to function effectively" [3, p. 68]. It represents a basis for advancements in service, such as service productivity and service quality [4, 5]. A systematic method for service process specification supports service innovation and improvement and allows service organizations to gain market success and growth [3, 5]. Scientific literature provides various formalizations and delineations regarding the service process [6, 7], which together define a set of dimensions and characteristics, from which service process specifications can be understood [4].

So far, research in the areas of service science, management and engineering (SSME) has introduced a multitude of approaches for the specification and analysis of the service process [8, 9]. However, little is known about the extent to which they allow for service process specifications considering proposed dimensions and characteristics, as evaluations have focused on ontological and conceptual comparisons [10, 11]. Thus,

© Springer International Publishing Switzerland 2016
W. Abramowicz et al. (Eds.): BIS 2016, LNBIP 255, pp. 356–367, 2016.
DOI: 10.1007/978-3-319-39426-8_28

the research question of *what is the level of maturity of existing approaches for modeling and analyzing service processes*, still remains unanswered.

To extend our understanding in this regard [4, 12], a systematic literature review following the principles of [13] is conducted in the article at hand. It aims for the identification of existing service process specification and analysis approaches and the assessment of their level of maturity. For the latter, a data extraction framework is developed, which is based on the combination of novel formalizations and delineations of the service process presented in [4]. On the basis of the review's results, research questions are formulated to stimulate future research on service production and delivery.

The remainder of the article is organized as follows. First, the data extraction framework for maturity assessment is described. In Sect. 3, we present the research design underlying our systematic literature review. The results of the study are depicted and discussed in 4. The final section concludes the article and discusses its implications for the scientific community.

2   Theoretical Foundations

In [4], extant literature on the production process in services is reviewed and aggregated to propose a generalized formalization and delineation of the service process. It defines the set of information, in terms of dimensions and characteristics, which has to be captured in the service process specification. For this research, we define the maturity of a service blueprinting approach, as the degree to which it allows for the specification and analysis of this information. Approaches supporting the whole set of dimensions and characteristics, offer the highest level of maturity (Fig. 1).

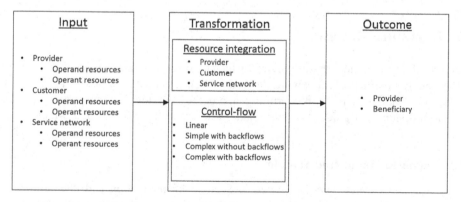

Fig. 1. Perspectives and dimensions of service processes (according to [4]).

The data extraction framework, which is used for the maturity assessment as part of the subsequent systematic literature review, is based on the findings of [4]. However, adding additional sub-dimensions, it also extends the suggested set of dimensions and characteristics to allow for a more detailed evaluation and analysis. In this section, the framework for data extraction is described.

Service processes have to be understood from three overall dimensions: input, transformation process and outcome [4]. Contrary to manufacturing-based production processes, "with service processes, [not only the service provider, but also] the customer provides significant inputs into the production process" [14, p. 16]. More generally, this is true for all of the actors involved in the service process. Input resources can be categorized into operand and operant resources [6, 15]. While former are defined as (tangible) resources on which actions are performed to generate certain effects, latter refer to (intangible) resources that act upon operand and/or other operant resources [6].

Transformation in service, is characterized as the alteration and conversion of a set of input resources into desired outcome [16]. According to Service-Dominant Logic [6, 15], actors within the service process share access to one another's input resources to integrate them and co-create reciprocal outcome (value). The complexity of such transformational processes is determined by the number of participating actors and the structure of the process flow. As service processes can span multiple customers and stakeholders, they are able to evolve into complex service networks [17, 18]. Representing routing options of customer orders, process flows can be simple, linear or complex [19].

Regarding the service process, it has to be distinguished between output and outcome. For a while, research in the service operations management domain has focused on output as a measure of service volume [4, 20]. However, academics have realized the inability of these measures to reflect received benefits from service [4]. Outcome, in this context, identifies the increase in an actor's well-being through service [20]. In service marketing literature, it is primarily referred to as value, which "[...] is always uniquely and phenomenologically determined by the beneficiary" [15, p. 7]. Conceptualizations of value are, among others, stakeholder satisfaction and means-end [21].

3   Literature Review

As part of a systematic literature review, which follows the guidelines recommended in [13], the data extraction framework derived in Sect. 2 is applied to assess the maturity level of identified service blueprinting approaches. In the following sections, the research design, that guides our systematic literature review, is depicted briefly.[1]

3.1   Population and Intervention

Based on the research question stated in Sect. 1, the population and intervention of interest are derived as major terms. Our population is that of service process specification and analysis approaches, and the intervention includes their application on either exemplary service processes, or within the context of empirical (qualitative and

---

[1] A more detailed description of the research design can be obtained from the principal author upon request and the following link: http://1drv.ms/1LDAjR2.

quantitative) inquiries. We are interested in those outcomes of the identified studies, which describe the constructs and capabilities of the proposed approaches and thus allow us to assess their maturity level.

3.2   Search Strategy and Process

Population, intervention and outcomes derived from the research question in Sect. 3.1, represent major terms for constructing our search term [13]. Alternative spellings and synonyms are determined for each of them. In addition, keywords of already reviewed literature are checked. Designed search strings are piloted and the results are documented and reflected within an iterative procedure. The final search term is: *"service blueprinting"*. It is appropriate for our systematic literature review, as the Service Blueprinting method [5] is widely adopted and well known to service researches [18, 22]. This is why many of these approaches can be seen as extensions to it, and it is discussed in nearly every topic-related article [23–25].

We organize our search process in two separate phases: an initial search phase and a secondary search phase. These two search phases are described in more detail in the following sections.

Initial Search Phase. In the initial search phase, the derived search string is applied to the meta search engine provided by the library of the Queensland University of Technology to identify candidate primary online sources. It includes electronic databases, such as Emerald Insight, ScienceDirect, SpringerLink, JSTOR, ABI/INFORM Complete and PsycARTICLES. A total number of 159 articles are identified as being relevant out of 246 hits in the meta search engine. For the initial search process, peer-reviewed journal articles publicized in English language to the date of 2015-09-01 are included. In this phase, only the studies' titles, abstracts and keywords are reviewed. Identified articles are published in 87 different journals, which serve as our primary online sources in the secondary search phase.

Secondary Search Phase. As part of the secondary search phase, identified online databases are searched for existing service blueprinting approaches. Therefore, to each primary online source, the constructed search string is applied. Based on the final set of articles perceived as relevant to our study, a forward and backward search is performed to find any additional research of interest [26]. Our initial sample consists of 278 studies from which we exclude the ones that do not match our inclusion criteria. As a result, a total number of 47 service process modeling and analysis approaches are identified as being relevant (see Table 1). Information of each primary study is extracted according to the framework derived in Sect. 2.

4   Descriptive Data Synthesis and Discussion

In Table 1, information extracted from the primary studies is tabulated in a manner highlighting dimensions and characteristics considered by identified approaches for service process specification and analysis. It therefore represents the structure of our data extraction framework derived in Sect. 2. In this section, the results of the systematic literature review are discussed and thereupon a set of research questions is proposed to stimulate future research on service production and delivery.

4.1   General Discussion

As revealed in Table 1, none of the identified approaches offers the highest possible level of maturity. Though, both of the approaches in [27, 28] take into account most of the proposed dimensions and characteristics, they do not allow for the specification of complex process flows. Therefore, future research might enhance their meta models to introduce additional constructs for control flow. Nevertheless, it is this control flow dimension in respect to which the majority (31 out of 47) of identified approaches has to be considered as highly mature. This is also true for the dimension of resource integration. While, in total, 32 of the identified approaches can be used to specify and analyze the resource integration procedures of at least two process actors (service provider and customer), 13 of them also provide support for the representation of service networks. As a result, 15 tools focus on only a single stakeholder perspective, that is either the service provider or customer.

Table 1. Dimensions and characteristics considered by the identified approaches.

Tool	Input			Resource integration			Ctrl.-Flw.	Outcome	
	Prov.	Cust.	Net.	Prov.	Cust.	Net.		Prov.	Ben.
[29]	–	–	–	–	X	–	Cplx.	–	Sq.
[30]	Od.	–	–	X	–	–	Cplx.	Profit.	Sq.(t)
[31]	Od.	Od.	–	X	X	–	Cplx.	Profit.	Sq.(t)
[32]	Od./Ot.	–	–	X	X	–	Cplx.	Effic.	–
[22]	Ot.	Ot.	Ot.	X	X	X	Cplx.	Profit.	Sq.(t)
[8]	Od./Ot.	Od.	–	X	X	–	Cplx.	Effic.	Sq.(t)
[33]	–	Od./Ot.		–	X	–	–	–	Sat.
[34]	–	–	–	–	X	–	Spl.	–	M.E.
[35]	–	–	–	X	X	–	Spl.	–	–
[23]	Od./Ot.	Od.	–	X	X	–	Cplx.	–	Sat.
[36]	Od.	Od.	–	X	X	–	Cplx.	Profit.	Sq.(t)
[37]	Od./Ot.	–	–	X	X	–	Cplx.	Profit.	Sq.(t)
[38]	Od.	–	–	X	X	–	Cplx.	Profit.	Sq.(t)
[39]	–	–	–	X	X	X	–	–	–
[40]	Od./Ot.	Od./Ot.	Od./Ot	X	X	X	Cplx.	–	–
[41]	Od.	–	–	X	X	–	Cplx.	Profit.	Sq.(t)
[42]	Od./Ot.	Od./Ot.	–	–	–	–	–	N.B.	N.B.
[27]	Od./Ot.	Od./Ot.	Od./Ot	X	X	X	Spl.	Res.	Res.

(*Continued*)

Table 1. (*Continued*)

Tool	Input			Resource integration			Ctrl.-Flw.	Outcome	
	Prov.	Cust.	Net.	Prov.	Cust.	Net.		Prov.	Ben.
[5]	Od.	–	–	X	–	–	Cplx.	Profit.	Sq.(t)
[43]	Od.	Od.	Od.	X	X	X	Cplx.	–	Sq.(t)
[44]	Od.	–	–	X	–	–	Cplx.	i.Sq.	Sq.(t)
[45]	Od.	–	–	X	X	X	Spl.	–	Sat.
[46]	–	–	–	X	X	–	Spl.	Effic.	Sq.(t)
[47]	Od.	–	–	X	X	–	Cplx.	Profit.	Sq.(t)
[48]	Od.	Od.	Od.	X	X	X	–	–	Sq.
[24]	Od./Ot.	–	–	X	–	–	Cplx.	i.Sq.	Sq.(t)
[49]	Od./Ot.	–	–	–	X	–	Cplx.	–	Sat.
[28]	Od./Ot.	Od./Ot.	Od./Ot	X	X	X	Spl.	Pb.V.	Pr.V.
[50]	Od.	Od.	Od.	X	X	X	Cplx.	–	–
[25]	Od.	Od.	–	X	X	–	Spl.	–	–
[51]	Od./Ot.	Od.	–	X	X	–	Cplx.	–	Sq.(t)
[52]	–	–	–	–	X	–	Cplx.	–	Sq.
[53]	Od.	–	–	–	X	–	Cplx.	Effic.	Sq.(t)
[54]	Od./Ot.	–	–	X	–	–	Spl.	Profit.	–
[3]	Od.	–	–	X	X	–	Cplx.	Profit.	Sq.(t)
[55]	–	–	–	X	X	–	Cplx.	Profit.	Sq.(t)
[56]	Od.	–	–	X	X	–	Cplx.	–	E.V.
[9]	Od.	–	–	X	X	–	Cplx.	Profit.	Sat.
[57]	Od.	–	–	X	X	X	Cplx.	Profit.	Sq.
[58]	Od./Ot.	Od./Ot.	Od./Ot	X	X	X	–	M.E.	M.E.
[59]	–	–	–	–	X	–	Spl.	–	Sat.
[60]	–	–	–	X	X	–	Cplx.	–	Sq.
[61]	–	–	–	X	X	–	Cplx.	Effic.	Sq.(t)
[17]	–	–	–	X	X	X	–	–	–
[18]	Od.	Od.	Od.	X	X	X	Cplx.	Profit.	N.B.
[62]	Od./Ot.	Od.	–	–	X	–	Cplx.	–	Sq.
[63]	–	–	–	–	X	–	Spl.	–	Sat.

Notes: (Prov.) = Provider   dimension;   (Cust.) = Customer   dimension;
(Net.) = Service network dimension; (Ctrl-Flw.) = Control flow dimension;
(Ben.) = Service beneficiary dimension; (X) = Dimension is considered by
approach; (-) = Dimension is not considered by approach; (Od.) = Operand
resource; (Ot.) = Operant resource; (Cplx.) = Complex process flow with or
without backflows; (Spl.) = Simple process flow with or without backflows;
(Profit.) = Profitability;   (Sq.) = Perceived   service   quality;   (Sq.(t)) = Time
dimension of perceived service quality; (Effic.) = Service process efficiency;
(M.E.) = Value as means-end; (Sat.) = Perceived satisfaction; (N.B.) = Value as
net benefit; (Res.) = Results of the service process; (iSq.) = Time and cost
dimension of perceived internal service quality; (Pr.V) = Perceived private
value; (Pb.V.) = Perceived public value; (E.V.) = Perceived emotional value.

To obtain insights on the maturity regarding outcome and input of an individual approach, a more detailed discussion on the interrelations of these two dimensions and possible conceptualizations of service outcome is required.

4.2   Discussion on Outcome Conceptualizations

Beneficiary Perspective. From the service beneficiary perspective, conceptualizations of service outcome can be classified into two primary categories: outcome as perceived customer satisfaction and others. Perceived satisfaction represents a widely adopted conceptualization of customer value in service marketing research [21]. It can be either determined directly or equated to meeting and exceeding customers' expectations of service quality [21]. As [31] does not support the specification of any kind of control flow, six approaches are identified allowing for the direct determination and analysis of customer satisfaction on a service process level. Different mechanisms are provided by them for that purpose: (a) assigning customer requirements to process activities and measuring their fulfillment [9, 23, 33]; (b) giving scores of customer satisfaction per service transaction [59, 63]; (c) measuring customer satisfaction based on predefined methods [45] and identifying relevant and irrelevant process activities [49]. Since these mechanisms draw on well accepted theories on consumer behavior (e.g. the expectation-confirmation theory), related approaches should be considered mature regarding the outcome dimension.

More than half of the identified approaches (25) allow for the assessment of service quality. However, 19 of them primarily focus on the speed of service delivery and thus rely on an insufficient conceptualization of service quality. Therefore, they do not offer a high level of maturity with regard to service outcome. Nonetheless, six of these 25 approaches allow for a more holistic measurement of service quality, drawing upon methods such as the SERVQUAL instrument [60], Critical Incident Technique [52], Walk-Through Audit [62] and others [29, 48, 57]. They therefore can be considered as highly mature when it comes to service outcome.

Other conceptualizations of value, like net benefit, means-end, emotional value and private value, are not so much considered by the identified approaches. While the concepts of emotional value and private value are less known to service researchers, net benefit and means-end can be seen as well accepted conceptualizations of customer value [21]. However, only two of the identified approaches adopt net benefit and means-end for value determination on a process level [18, 34]. Other service blueprinting tools making use of these two conceptualizations, do not support the specification of process flows [42, 58]. As both concepts enable service providers to obtain important insights, including information about service process attributes to establish, issues that hinder the realization of customer benefits and desired customer objectives, we propose the following research question to encourage future research on their exploration: *How to determine customer value conceptualized in form of net benefit and means-end on a service process level? (FRQ-1).*

Provider Perspective. Of the total set of identified approaches, only 26 provide support for the assessment of service outcome from a service provider's perspective.

The majority (15) of these approaches focuses on simple measures of profitability, that is the difference between revenues and costs [5, 47, 54]. Following this conceptualization, one of the determining factors of provider value is value-in-exchange. However, the concept of value-in-exchange is considered as one of the foundations of Goods-Dominant Logic and has been strongly criticized in recent research on service [6, 64]. It is argued that a conceptualization of value in nominal exchange is not able to capture the 'real' value perceived through service. In a consequence, alternative conceptualizations of value have been proposed by service researchers, such as value-in-use and value-in-context [6, 15].

Other approaches that are identified, adopt the notion of value as internal efficiency [32, 46, 53]. It describes the relation between output and input from the service provider's point of view [8], and thus is similar to the concept of service productivity [4]. "In an efficiency-driven inside-out perspective the key performance indicators are time and costs. The main goal is to find a single time- and cost-efficient process design [...]" [8, p. 742]. This more concrete definition is also supported by [42]. The conceptualization of value as internal efficiency represents a manufacturing-based measure in service [4]. Its application in the context of service, therefore, is associated with limitations (see Sect. 2). As both concepts of value, profitability as well as internal efficiency, seem to be outdated according to recent research in the areas of SSME, the maturity regarding service outcome of the approaches relying on them, has to be considered low. Hence, future research should strive towards answering the following research question: *How to assess more recently proposed conceptualizations of value, such as value-in-use and value-in-context, from the service provider perspective on a service process level? (FRQ-2).*

Two of the identified approaches, allow for the assessment of value as net benefit [42] and means-end [58] from the service provider's viewpoint, however, not on a service process level. The two value types are determined by the service provider based on an actual or potential service experience. Therefore, they can be seen as concretizations of Service-Dominant Logic's notion of value-in-use. In order to contribute to the proposed future research question FRQ-2, the following research question might be addressed by future research: *How to determine provider value conceptualized in form of net benefit and means-end on a service process level? (FRQ-3).*

The conceptualizations of provider value (service process results, internal service quality and public value) presented and/or introduced in [24, 27, 28, 44] are not widely adopted in SSME research. It is therefore difficult to conclude on whether the maturity of the related approaches has to be considered high or low regarding the outcome dimension of service process.

4.3   Discussion on the Interrelations Between Service Input and Outcome

Service providers and customers, together, provide input resources to the service process in order to transform them (through resource integration) into outcome (value) [4, 15]. Realizing a desired outcome, both actors have to provide a required set of input resources to a specific level. Hence, it is important to know which input resources are required for the achievement of the outcome and to what extent the involved actors possess these resources and thus are able to provide them. Especially operant resources are considered as an important input, as they are considered as the fundamental source of competitive advantage [6, 15]. Nonetheless, as depicted in Table 1, the identified approaches fall short in considering these aspects.

With regard to provider input resources, 14 identified approaches allow for the specification of operand and operant resources. However, only five of these approaches consider the specified input resources when it comes to outcome determination. While two approaches draw upon well-known value conceptualizations [42, 58], the other three approaches define outcome as public value [28], internal quality [24] and service process results [27]. Especially the methods comprised by the approaches presented in [27, 58] can be considered as highly mature, since they allow for the specification of complex job descriptions and value compositions integrated into outcome assessment. Future research should aim for transferring these concepts to other service process specification and analysis approaches to allow their application to more well-known value conceptualizations on a service process level. Such future research would contribute to the research question of: *How to integrate provided input resources into the determination of well-known value conceptualizations on a service process level? (FRQ-4).*

This research question is also of interest with respect to customer input resources. Only six approaches allow for the specification of operand and operant resources from a customer perspective. Two of them however, do not allow for the analysis of customer outcome. The rest of these approaches have already been discussed from the provider perspective.

5   Conclusions, Limitations and Future Research

Our research findings indicate a high level of maturity for most of the identified approaches with regard to the control flow and resource integration dimensions. Having a look at the dimensions of service input and output, however, several shortcomings can be identified which should be addressed by future research. In this regard, the study at hand proposes a set of research questions.

As highlighted by the discussion in Sect. 5, the presented framework for maturity assessment has some limitations. Future research might enhance the conceptual framework regarding the input and output dimensions. This is required to also take into consideration different conceptualizations of outcome and the interrelations between service input and outcome. As a result, this would allow for a more adequate assessment of the maturity of existing service process specification and analysis approaches.

References

1. Sampson, S.E.: Visualizing service operations. J. Serv. Res. **15**, 182–198 (2012)
2. Scott, E.: Sampson: customer-supplier duality and bidirectional supply chains in service organizations. Int. J. Serv. Ind. Manage. **11**, 348–364 (2000)

3. Bitner, M.J., Ostrom, A.L., Morgan, F.N.: Service blueprinting: a practical technique for service innovation. Calif. Manage. Rev. **50**, 66–94 (2008)
4. Yalley, A.A., Sekhon, H.S.: Service production process: implications for service productivity. Int. J. Product. Perform. Manage. **63**, 1012–1030 (2014)
5. Shostack, G.L.: How to design a service. Eur. J. Mark. **16**, 49–63 (1982)
6. Vargo, S.L., Lusch, R.F.: Evolving to a new dominant logic for marketing. J. Mark. **68**, 1–17 (2004)
7. Sampson, S.E., Froehle, C.M.: Foundations and implications of a proposed unified services theory. Prod. Oper. Manage. **15**, 329–343 (2006)
8. Gersch, M., Hewing, M., Schöler, B.: Business process blueprinting – an enhanced view on process performance. Bus. Process Manage. J. **17**, 732–747 (2011)
9. Hara, T., Arai, T., Shimomura, Y., Sakao, T.: Service CAD system to integrate product and human activity for total value. CIRP J. Manuf. Sci. Technol. **1**, 262–271 (2009)
10. Kazemzadeh, Y., Milton, S.K., Johnson, L.W.: Service blueprinting and process-chain-network: an ontological comparison. Int. J. Qual. Res. Serv. **2**, 1–12 (2015)
11. Milton, Simon K., Johnson, Lester W.: Service blueprinting and BPMN: a comparison. Manage. Serv. Qual. Int. J. **22**, 606–621 (2012)
12. Žemguliené, J.: Productivity in the service sector: A service classification scheme for productivity measurement. Ekonomika 86 (2009)
13. Kitchenham, B.: Procedures for performing systematic reviews. Keele UK Keele Univ. **33**, 1–26 (2004)
14. Sampson, S.E.: Understanding Service Businesses: Applying Principles of Unified Services Theory. Wiley, New York (2001)
15. Vargo, S.L., Lusch, R.F.: Service-dominant logic: continuing the evolution. J. Acad. Mark. Sci. **36**, 1–10 (2007)
16. Mills, P.K., Chase, R.B., Margulies, N.: Motivating the client/employee system as a service production strategy. Acad. Manage. Rev. **8**, 301 (1983)
17. Tax, S.S., McCutcheon, D., Wilkinson, I.F.: The service delivery network (SDN) a customer-centric perspective of the customer journey. J. Serv. Res. **16**, 454–470 (2013)
18. Sampson, S.E.: Visualizing service operations. J. Serv. Res. **15**, 182–198 (2012)
19. Leyer, M.: A framework for systemising service operations management. Int. J. Serv. Oper. Manage. (forthcoming)
20. Netten, A., Forder, J.: Measuring productivity: an approach to measuring quality weighted outputs in social care. Public Money Manage. **30**, 159–166 (2010)
21. Ng, I.C.L., Smith, L.A.: An integrative framework of value. In: Vargo, S.L., Lusch, R.F. (eds.) Special Issue – Toward a Better Understanding of the Role of Value in Markets and Marketing, pp. 207–243. Emerald Group Publishing Limited, Bingley (2012)
22. Becker, J., Beverungen, D., Knackstedt, R., Matzner, M., Muller, O., Poppelbuss, J.: Bridging the gap between manufacturing and service through IT-Based boundary objects. IEEE Trans. Eng. Manage. **60**, 468–482 (2013)
23. Patrício, L., Fisk, R.P., e Cunha, J.F.: Designing multi-interface service experiences the service experience blueprint. J. Serv. Res. **10**, 318–334 (2008)
24. Lings, I.N.: Managing service quality with internal marketing schematics. Long Range Plann. **32**, 452–463 (1999)
25. Stender, M., Ritz, T.: Modeling of B2B mobile commerce processes. Int. J. Prod. Econ. **101**, 128–139 (2006)
26. Webster, J., Watson, R.: Analyzing the past to prepare for the future: writing a literature review. Manage. Inf. Syst. Q. **26**, 13–23 (2002)
27. Congram, C., Epelman, M.: How to describe your service: an invitation to the structured analysis and design technique. Int. J. Serv. Ind. Manage. **6**, 6–23 (1995)

28. Alford, J., Yates, S.: Mapping public value processes. Int. J. Public Sect. Manage. **27**, 334–352 (2014)
29. Johnston, R.: A framework for developing a quality strategy in a customer processing operation. Int. J. Qual. Reliab. Manage. **4**, 37–46 (1987)
30. Berkley, B.J.: Analyzing service blueprints using phase distributions. Eur. J. Oper. Res. **88**, 152–164 (1996)
31. Verboom, M., van Iwaarden, J., van der Wiele, T.: A transparent role of information systems within business processes: a case study. Manage. Serv. Qual. Int. J. **14**, 496–505 (2004)
32. Fließ, S., Kleinaltenkamp, M.: Blueprinting the service company: managing service processes efficiently. J. Bus. Res. **57**, 392–404 (2004)
33. Teixeira, J., Patrício, L., Nunes, N.J., Nóbrega, L., Fisk, R.P., Constantine, L.: Customer experience modeling: from customer experience to service design. J. Serv. Manage. **23**, 362–376 (2012)
34. Jüttner, U., Schaffner, D., Windler, K., Maklan, S.: Customer service experiences: Developing and applying a sequentialincident laddering technique. Eur. J. Mark. **47**, 738–769 (2013)
35. Holmström, J., Ala-Risku, T., Auramo, J., Collin, J., Eloranta, E., Salminen, A.: Demand-supply chain representation: a tool for economic organizing of industrial services. J. Manuf. Technol. Manage. **21**, 376–387 (2010)
36. Morelli, N.: Designing product/service systems: a methodological exploration1. Des. Issues **18**, 3–17 (2002)
37. Geum, Y., Park, Y.: Designing the sustainable product-service integration: a product-service blueprint approach. J. Clean. Prod. **19**, 1601–1614 (2011)
38. Lai, J., Lui, S.S., Hon, A.H.Y.: Does standardized service fit all?: novel service encounter in frontline employee-customer interface. Int. J. Contemp. Hosp. Manage. **26**, 1341–1363 (2014)
39. Mills, J., Purchase, V.C., Parry, G.: Enterprise imaging: representing complex multi-organizational service enterprises. Int. J. Oper. Prod. Manage. **33**, 159–180 (2013)
40. Polyvyanyy, A., Weske, M.: Flexible service systems. In: Demirkan, H., Spohrer, J.C., Krishna, V. (eds.) The Science of Service Systems, pp. 73–90. Springer, US (2011)
41. Trkman, P., Mertens, W., Viaene, S., Gemmel, P.: From business process management to customer process management. Bus. Process Manage. J. **21**, 250–266 (2015)
42. Campbell, C.S., Maglio, P.P., Davis, M.M.: From self-service to super-service: a resource mapping framework for co-creating value by shifting the boundary between provider and customer. Inf. Syst. E-Bus. Manage. **9**, 173–191 (2010)
43. Beverungen, D., Knackstedt, R., Winkelman, A.: Identifying e-Service potential from business process models: a theory nexus approach. E-Serv. J. **8**, 45–83 (2011)
44. Lings, I.N., Brooks, R.F.: Implementing and measuring the effectiveness of internal marketing. J. Mark. Manage. **14**, 325–351 (1998)
45. Nicklas, J.-P., Schlüter, N., Winzer, P.: Integrating customers' voice inside network environments. Total Qual. Manage. Bus. Excell. **24**, 980–990 (2013)
46. Womack, J.P., Jones, D.T.: Lean consumption. Harv. Bus. Rev. **83**, 58–68 (2005)
47. Coenen, C., von Felten, D., Schmid, M.: Managing effectiveness and efficiency through FM blueprinting. Facilities **29**, 422–436 (2011)
48. Field, J.M., Heim, G.R., Sinha, K.K.: Managing quality in the E-Service system: development and application of a process model. Prod. Oper. Manage. **13**, 291–306 (2004)
49. Tseng, M.M., Qinhai, M., Su, C.: Mapping customers' service experience for operations improvement. Bus. Process Manage. J. **5**, 50–64 (1999)
50. Biege, S., Lay, G., Buschak, D.: Mapping service processes in manufacturing companies: industrial service blueprinting. Int. J. Oper. Prod. Manage. **32**, 932–957 (2012)

51. Patrício, L., Fisk, R.P., Cunha, J.F., Constantine, L.: Multilevel service design: from customer value constellation to service experience blueprinting. J. Serv. Res. **14**, 180–200 (2011)
52. Stauss, B., Weinlich, B.: Process-oriented measurement of service quality: applying the sequential incident technique. Eur. J. Mark. **31**, 33–55 (1997)
53. Kim, H., Kim, Y.: Rationalizing the customer service process. Bus. Process Manage. J. **7**, 139–156 (2001)
54. Armistead, C.G., Clark, G.: Resource activity mapping: the value chain in service operations strategy. Serv. Ind. J. **13**, 221–239 (1993)
55. Szende, P., Dalton, A.: Service blueprinting: shifting from a storyboard to a scorecard. J. Foodserv. Bus. Res. **18**, 207–225 (2015)
56. Spraragen, S.L., Chan, C.: Service blueprinting: when customer satisfaction numbers are not enough. In: International DMI Education Conference. Design Thinking: New Challenges for Designers, Managers and Organizations (2008)
57. Brown, S.W.: Service Quality: Multidisciplinary and Multinational Perspectives. Lexington Books, Lanham (1991)
58. Lessard, L., Yu, E.: Service systems design: an intentional agent perspective. Hum. Factors Ergon. Manuf. Serv. Ind. **23**, 68–75 (2013)
59. Johnston, R.: Service transaction analysis: assessing and improving the customer's experience. Manage. Serv. Qual. Int. J. **9**, 102–109 (1999)
60. Kingman-Brundage, J.: Technology, design and service quality. Int. J. Serv. Ind. Manage. **2**, 47–59 (1991)
61. Becker, J., Beverungen, D.F., Knackstedt, R.: The challenge of conceptual modeling for product service systems: status-quo and perspectives for reference models and modeling languages. Inf. Syst. E-Bus. Manage. **8**, 33–66 (2009)
62. Koljonen, E.L.L., Reid, R.A.: Walk-through audit provides focus for service improvements for Hong Kong law firm. Manage. Serv. Qual. Int. J. **10**, 32–46 (2000)
63. Rasila, H.M., Rothe, P., Nenonen, S.: Workplace experience – a journey through a business park. Facilities **27**, 486–496 (2009)
64. Vargo, S.L., Maglio, P.P., Akaka, M.A.: On value and value co-creation: a service systems and service logic perspective. Eur. Manage. J. **26**, 145–152 (2008)

Social Media

Enterprise Social Networks: Status Quo of Current Research and Future Research Directions

Gerald Stei[1(✉)], Sebastian Sprenger[2], and Alexander Rossmann[1]

[1] Research Lab for Digital Business, Reutlingen University, Alteburg Str. 150,
72762 Reutlingen, Germany
{gerald.stei,alexander.rossmann}@reutlingen-university.de
[2] Friedrich-Alexander-University Erlangen-Nuremberg, Chair of IT-Management,
Lange Gasse 20, 90403 Nuremberg, Germany
sebastian.sprenger@fau.de

Abstract. This paper provides an introduction to the topic of Enterprise Social Networks (ESN) and illustrates possible applications, potentials, and challenges for future research. It outlines an analysis of research papers containing a literature overview in the field of ESN. Subsequently, single relevant research papers are analysed and further research potentials derived therefrom. This yields seven promising areas for further research: (1) user behaviour; (2) effects of ESN usage; (3) management, leadership, and governance; (4) value assessment and success measurement; (5) cultural effects, (6) architecture and design of ESN; and (7) theories, research designs and methods. This paper characterises these areas and articulates further research directions.

Keywords: Enterprise social networks · Enterprise 2.0 · Enterprise social software · Literature review · Research agenda

1 Introduction[1]

The rapid spread of Web 2.0 technologies allows users to play an important role in content creation on Internet platforms. A fundamental characteristic of these technologies is that they allow information to be created and exchanged via Internet-based platforms [1]. The advantages associated with this are not only of interest for private users but also for companies [2]. Organisations introduce Web 2.0 tools for internal applications too, e.g. to enable employees to share information and to support cooperation with customers and partners [3]. A group of these tools that has gained particular attention in an organisational context is called Enterprise Social Networks (ESN). These ESN are applied with the objective of improving both, organisational effectiveness and efficiency [4].

Following Li et al., we define ESN as a "set of technologies that creates business value by connecting the members of an organization through profiles, updates and notifications" ([5], p. 3).

---

[1] This paper is part of the dissertation project of Gerald Stei.

© Springer International Publishing Switzerland 2016
W. Abramowicz et al. (Eds.): BIS 2016, LNBIP 255, pp. 371–382, 2016.
DOI: 10.1007/978-3-319-39426-8_29

ESN research is part of the research field of Computer Supported Cooperative Work (CSCW). In general, CSCW examines how information technology can be used to gather the knowledge of employees and to solve work-related issues [6]. In the span of more than 25 years, research on CSCW has become a well-established area in information systems research [7]. Already in 1996, Teufel created a framework to classify CSCW applications in terms of their ability to support group *communication, coordination* and *cooperation* [8]. These three support functions can be drawn on to describe the aims of ESN usage.

The first aim refers to the improvement of *communication*. ESN support users in communicating with each other beyond spatial and temporal distances and in having conversations via the company network [3, 9]. Hence, ESN facilitate communication on an individual level, in private or public groups, and finally across an entire enterprise [4, 10]. This fosters communication between different organisational units and levels, but it may also challenge the formal hierarchy [11].

The second aim of implementation of ESN is the improvement of *coordination* within the enterprise. ESN have been found to support coordination in respect of work shared between employees within organisations by for example providing functions that help employees to fulfil subtasks [12, 13].

The third aim refers to *cooperation* and to the development of new forms of work arrangements [9]. ESN support collaborative cooperation in organisations by way of functions that range from mutual document processing and sharing [4, 14] to the creation of extensive collaborative work environments [7]. In this context, it has also been found that the cooperation is improved even beyond hierarchical structures [15].

The purpose of this paper is a systematic presentation of research potentials discussed in literature. The elaboration of such focal points can serve as a starting point for future research while contributing towards a better understanding of ESN and their effects on organisational work processes. For this reason, this paper answers the following questions:

Which main research areas can be derived from analysing existing research papers on Enterprise Social Networks?
Which specific research questions remain unanswered in current research on Enterprise Social Networks, and how might these questions contribute to a comprehensive research agenda?

In order to answer these questions, the paper is structured as follows: The next chapter presents and summarises existing literature reviews on ESN. On this basis, an extensive literature review focusing on current ESN research publications is performed. Afterwards we describe the main results of this review and outline crucial research areas for ESN research as well as specific research questions thus far unanswered. Finally, we set out a research agenda that addresses the previously identified gaps in ESN research.

2   ESN Literature

A raw analysis of current literature indicates a large amount of related research on ESN. In this paper, we first analyse ESN literature reviews in order to present the current state in the field of ESN research. Subsequently, we present the methodical approach used in our own supplementary literature review.

2.1 Related Work

As a first step this paper considers three exemplary scholarly publications featuring literature reviews on the topic of ESN to illustrate the status of current research. In doing so, the question focussed on is which core questions for future research are identified in each paper. For this reason, the research work by Altamimi [3], Williams et al. [16], and Berger et al. [11] is outlined in the following.

Altamimi [3] analysed the scientific literature on Social Software in a lexical study published in 2013. His objective was to systematically present the evolution of the literature since the development of social software and Web 2.0 technologies. As for identifying directions for future research, his claims are threefold: to have considered the advantages and disadvantages of Social Software usage in organisational scenarios; to have determined the effects and successes of Enterprise Social Software; and to have analysed the problems arising along the way. Altamimi further states that continuative literature reviews with innovative approaches, such as content analysis, are required [3].

Also in 2013, Williams et al. examined the extent to which scholarly publications on ESN meet the challenges and impacts currently faced by organizations [16]. The authors identify five ESN-specific topics in their analysis of the relevant scholarly literature: "overview", "adoption", "use", "impact" and "other". Williams et al. found that organisations are more interested in topics relating to information management and compliance challenges; the identification and measurement of the advantages of ESN; and the integration of ESN into business processes and business software. For this reason, the authors call for a new research track mainly focusing on the requirements of practice [16].

The review by Richter et al. [17] is a comprehensive account of research in Internet Social Networking. It gives an overview of Social Networking Sites (SNSs) and makes the body of research accessible to researchers as well as practitioners in the Enterprise 2.0 context. The authors analyse the research done in papers published from 2003–2009 and highlight areas where research is missing. Richter et al. specify three potentials for further research concerning SNSs in an enterprise context: Recruiting and Professional Career Development, Relationship Facilitation in Distributed Work Contexts, and SNSs as medium to engage with consumers [17].

In the papers mentioned, the focus of the reviews lies on different aspects of ESN. The paper at hand therefore is meant to supplement these papers by systematically presenting the research potentials as stated in the specific ESN literature from a wide number of scholarly publications.

2.2 The Review Process

This paper differs from the above reviews in various aspects. First of all, our main goal is to identify and work through in detail the ESN research gaps mentioned in literature, but without focussing on a certain cultural area. Furthermore, in order to give particular attention to currently relevant questions, only papers published no earlier than from 2013 on are considered for our analysis. To identify the relevant research papers for this review and to ensure a broad and suitable literature background [18], three different scientific databases were accessed (Business Source Complete, IEEE Xplore, and Science Direct). In addition, only papers published in scientific journals [18] and conference proceedings [19] were taken into consideration. The search for relevant research was based on the following keywords: "Enterprise Social Networks", "Enterprise Social Software", "Enterprise 2.0", "Enterprise Social Networking", "Enterprise Social Platform" and "Enterprise Social Networking Site". In total, the search found 86 relevant papers all of which were included in this review.

3   Research Areas on Enterprise Social Networks

As part of the analysis a plethora of research questions relating to ESN was identified. The implications for theory were collected from the papers and condensed through several iteration steps: The single papers were grouped based on the proximity of the content. As a result, seven fields of research were identified, which will be closely described in the following section: (1) user behaviour; (2) effects of ESN usage; (3) management, leadership, and governance for ESN; (4) value assessment and success measurement; (5) cultural effects; (6) architecture and design of ESN; and (7) theories, research designs and methods. The key elements of these research fields are characterised in the following.

3.1   User Behaviour

A consensus prevails that a better understanding of user behaviour is a key success factor for the operation of ESN. For example, Berger et al. pointed out the significance of examining user behaviour in relation to social networks in a company context [11].

The core question to be considered in this area refers to the motives driving the staff of an organisation to use or not use ESN [1]. In this context, the motivational factors that lead to using or opposing ESN could be studied [20]. Moreover, it seems promising to observe user behaviour in relation to the sexes or various age groups, since this allows us to evaluate differences in the attitude towards ESN among different user groups [21].

Furthermore, the relevant context factors impacting on ESN usage can be analysed. There is a great variety of relevant factors whose influence has not yet been conclusively determined [22]. Chin et al. call for the identification and validation of technological, organisational, social, individual and task-related factors that influence ESN usage [23]. One possible factor of great relevance is the interpersonal trust prevailing between the (potential) users of a network. This and other context factors might be claimed as subject matter for further research [24].

A gain in insight might also result from differentiating between user groups [25]. With a deeper analysis of the needs of heterogeneous user groups, success factors for the introductory phase of ESN could be identified and addressed [26].

Observation of ESN usage by teams is also to be seen as a promising line of inquiry since the way teamwork influences adoption of ESN can then be examined [27].

A further question of interest is whether (and if so, how) the users' requirements change over time. The performance of longitudinal studies could shed light on how the adoption and usage of ESN develop in the course of time [28].

Fulk and Yuan call for the deeper examination of the differences between conventional Information and Communication Technology and ESN [29].

One further aspect of ESN usage relates to the sharing of knowledge in organizations. It is still debatable which factors are of relevance for knowledge sharing in ESN. In future research, ESN usage can be evaluated in relation to task-related vs. social purposes [30]. In general, it makes sense to analyse the reasons why employees share their knowledge [31, 32]. After all, a key area of research lies in the identification of the (de-)motivating factors that have a significant influence on whether or not knowledge is shared in ESN [33]. Considering the specific vitality of ESN, the diffusion of the information and knowledge in companies can be examined more closely [34]. To conclude, it is desirable to understand the potentials of the development and flow of organisational knowledge and the role of ESN [29].

3.2   Effects of ESN Usage

Another research area of highly rated relevance is the impact of ESN on corporate performance. To this end, further case study research is necessary to describe the effects of ESN in a more holistic way [21]. In order to generate causal models of great practical relevance, the effects of ESN usage need to be described in empirical practice [3]. Empirical studies can determine which causal effects are actually in play and under which conditions [35].

ESN usage promises a large number of positive effects on organisational processes, despite these not being fully recorded and empirically verified. A deeper examination of the advantages of ESN usage is called for by Kuegler and Smolnik [36].

One advantage of ESN lies in the possibility of open communication and the exchange of knowledge. In this context, further evaluation of the contextual factors stimulating these effects can offer valuable insights [37].

In addition, the findings of qualitative preliminary studies indicate that ESN usage can support the creative potential of organisations. It seems promising to research the correlation between the functions of communication platforms and the creativity of users [38].

Guy et al. propose as a further research question to investigate how the use of social media contributes to an increasing level of expertise within organisations [39].

ESN enable employees to network with each other and to exchange information, whether this be work-related or of a social kind. A fruitful research approach could lie in extensive analysis of the relationships between the key users plus evaluation of this topic with respect to further relevant questions [40].

Apart from the positive effects of ESN, negative effects may also arise. These have not yet been researched in depth. In order to evaluate negative effects of ESN, Subramaniam et al. call for a holistic analysis of undesirable effects of ESN usage [41]. For example, it seems possible that ESN usage may promote hidden behaviour or cause dialectic tensions between the employees [37].

Worthy of mention too are aspects like the environmental friendliness of ESN by reducing travel costs. By pointing towards the term "Green IT", Agarwal and Nath outline the fact that further intensive studies relating to energy efficiency and environmental compatibility of ESN are needed [42].

3.3 Management, Leadership and Governance

A relevant question for theory and practice is the extent to which the introduction and operation of ESN can be influenced with respect to corporate goals. Experiences in companies show that ESN applications significantly differ from other IT systems by being more flexible and useable for a plethora of purposes. The development of suitable theories and management concepts would be helpful for such "malleable end-user software" [43]. Other authors also focus on the relevance of developing custom-made management and governance approaches for ESN [4, 44].

The idea of leadership also features in research within this context. In this regard, it is relevant to explore the role played by different forms of leadership in relation to ESN [45].

In the introductory phase of ESN, it is highly important to stimulate adoption of the system by users. The focus will be on which guiding principles are reasonable when considering an adoption strategy [46] and which elements are critical for that strategy's success [47].

Closely connected to the concepts of management and leadership is that of governance. Han et al. describe the implementation of a governance structure for the introduction of ESN, and call for further research into the development of general governance models within this context [48]. Alqahtani et al. also call for further studies on how organisations can effectively govern ESN [49]. Further in need of clarification is whether ESN usage can be stimulated by a range of training measures. In this case, a longitudinal study with time series data may prove helpful [33] and consideration of various environmental factors is certainly desirable [37].

Effective governance mechanisms also lead to the consideration and analysis of data integrity. A core question for organisations is how to address these additional risks [49]. Another challenge consists in the securing of data generated in ESN. Especially in view of the risk of unauthorised access to ESN and the risk of hacking, extensive data protection must be ensured. Berger et al. also call for further research on data security in ESN [11].

3.4 Value Assessment and Success Measurement

Another key research area concerns the value proposition with respect to ESN. Development of a compelling and holistic value assessment for ESN still has potentials for further research activities. In this context, Altamimi calls for a conceptualisation and evaluation of the "performance" of ESN [3].

Also, this being the basis for the decision to invest in ESN, it is essential for companies to be able to clearly determine their value. With regard to the business value, specific measurement models need to be defined and evaluated [50].

A measurement approach to how ESN contribute to the quality of collaboration and general performance is also called by Merz et al. [51]. ESN generate extensive data that can be correlated with the productivity of employees [52, 53].

Finally, for the purpose of measuring success in ESN, there are several barriers whose examination may prove rewarding. Herzog et al., for example, call for further research into possible barriers to measuring success [54].

3.5 Cultural Effects

Introducing ESN to companies means creating a new kind of communication channel. A change in the communication structure can radiate into various areas of the organisation [3]. Further examination of the interplay between cultural factors and ESN therefore is desirable [12].

Another area of interest relates to how ESN impact the interplay between single organisational sub-cultures [55]. Kuegler et al. call for the evaluation of organisational climate and further determinants of the ESN implementation process [56]. The role of organisational hierarchy is also relevant for ESN adoption [57]. It would also be appropriate to evaluate the influence of individual hierarchy levels on the setup of organisations upon adoption of ESN [58].

Considering the effects of country- and region-specific cultures on the usage of ESN, it is evident that a significant influence by such factors is insinuated in the literature. Hence, Trimi and Galanxhi point out the necessity of determining the effects of regional cultures on ESN [26]. This allows to evaluate the influence of various cultural conditions on communication behaviour in ESN [59]. Thus, conclusions can be drawn about the adoption of ESN in different countries [28, 48].

3.6 Architecture and Design of ESN

Another relevant research area concerns the technological architecture of ESN and its software design. Exemplarily, Baghdadi and Maamar present an interaction reference model and an interaction architecture for ESN [60]. This preliminary work can be developed and extended by the use of service-oriented architectures and web services [60]. Another promising approach to the architecture of ESN lies in the implementation of such systems in a scalable distributed computing environment [61].

The usage of the communication platform can be influenced by the design of ESN. Evidence about the needs and preferences of (potential) users can be taken into account. This in turn might have positive effects on the adoption of ESN. For this reason, Berger et al. call for the acquisition of additional knowledge concerning the software design [11].

3.7 Theories, Research Designs and Methods

In literature, several implications are stated for future research: they refer to theories, methods and research designs. The suggestions are difficult to generalise and need to be seen in the specific research context.

First to be mentioned is the call for elaboration of specific and relevant theoretical perspectives in relation to Enterprise Social Media, as outlined by Osch and Coursaris [62]. In addition, empirical tests for relevant multi-level constructs should be devised.

Regarding future research designs, examination of the development of ESN-related phenomena is called for in various papers. By evaluating time series data in longitudinal studies, conclusions can be drawn as to their developments [28, 33, 63].

Numerous proposals can also be found in the literature on selected research methods. Content analysis [59], semantic analysis [64] and ethnographic research methods [13] are identified as additional useful methods to be applied in future.

4   Conclusion

This paper analyses the existing literature on ESN research and, at the same time, provides an overview of exemplary papers dedicated to ESN-specific literature. On the basis of the insights gained therefrom, current research papers are analysed with a view to derive statements for future research. As a result, seven promising research fields can be identified: (1) user behaviour; (2) effects of ESN usage; (3) management, leadership and governance; (4) value assessment and success measurement; (5) cultural effects; (6) architecture and design of ESN; and (7) theories, research designs and methods. These fields are outlined and characterised on the basis of exemplary questions. It is evident that extensive research potentials in the field of ESN research does exist and that these lend themselves to elaboration in future research designs and scientific examination. Table 1 shows an overview of the papers in the individual research categories.

Table 1. Allocation of papers to the categories

User behaviour	[1, 11, 20–34]
Effects of ESN usage	[3, 21, 35–42]
Management, leadership and governance	[4, 11, 33, 37, 43–49]
Value assessment and success measurement	[3, 50–54]
Cultural effects	[3, 12, 26, 28, 48, 55–59]
Architecture and design of ESN	[11, 60, 61]
Theories, research designs and methods	[13, 28, 33, 59, 62–64]

In the analysed scholarly publications most papers belong to the "user behavior" category (17 papers), meaning that the bulk of the analysed papers anticipate research potential in this category. Second most papers (11) are assigned to "management, leadership and governance", indicating that extensive research is called for in this category, too. "Effects of ESN usage" and "cultural effects" each comprise ten papers, showing that different topics in these categories are expected to be fruitful. Rather less prominent in the analysed papers are research suggestions belonging to the following categories: "theories, research designs and methods" (7), "value assessment and success measurement" (6), and "architecture and design of ESN" (3).

This paper is of limited scope. In order to depict the status quo of research, only professional papers from 2013 to 2015 are considered. As part of a more comprehensive research agenda, the focus of the study could be extended to other years. In addition, the online libraries of other specialist subjects might be drawn on to gain a wider database and to stimulate interdisciplinary research work on ESN.

This paper gives an overview of the research potentials of ESN. A deeper analysis of the individual research fields would exceed the scope of this paper. Interested researchers might wish to scrutinize individual cited papers in detail and derive specific questions. However, the paper at hand makes a valuable contribution to further development in the area of ESN research.

References

1. Antonius, N., Xu, J., Gao, X.: Factors influencing the adoption of enterprise social software in Australia. Knowl.-Based Syst. **73**, 1–12 (2014)
2. Mukkamala, A.M., Razmerita, L.: Which factors influence the adoption of social software? An exploratory study of Indian Information Technology consultancy firms. J. Glob. Inf. Technol. Manag. **17**, 188–212 (2014)
3. Altumimi, L.: A lexical analysis of social software literature. Informatica Economica **17**, 14–26 (2013)
4. Hatzi, O., Meletakis, G., Nikolaidou, M., Anagnostopoulos, D.: Collaborative management of applications in enterprise social networks. In: IEEE 8th International Conference on Research Challenges in Information Science, pp. 1–9 (2014)
5. Li, C., Webber, A., Cifuentes, J.: Making the Business Case for Enterprise Social Networking: Focus on Relationships to Drive Value, pp. 1–23. Altimeter (2012)
6. Ackerman, M.S., Dachtera, J., Pipek, V., Wulf, V.: Sharing knowledge and expertise: the CSCW view of knowledge management. CSCW **22**, 531–573 (2013)
7. Schmidt, K., Bannon, L.: Constructing CSCW: the first quarter century. CSCW **22**, 345–372 (2013)
8. Teufel, S.: Computerunterstutzte Gruppenarbeit - Eine Einführung. Praxis des Workflow-Managements, pp. 35–63 (1996)
9. Jarle Gressgård, L.: Virtual team collaboration and innovation in organizations. Team Perform. Manag. **17**, 102–119 (2011)
10. Snell, J.M., Prodromou, E.: Activity Streams 2.0. w3.org/TR/activitystreams-core/
11. Berger, K., Klier, J., Klier, M., Probst, F.: A review of information systems research on online social networks. Commun. Assoc. Inf. Syst. **35**, 145–172 (2014)
12. Kuegler, M., Dittes, S., Smolnik, S., Richter, A.: Connect Me! Antecedents and impact of social connectedness in enterprise social software. Bus. Inf. Syst. Eng. **57**, 1–16 (2015)
13. Richter, A., Riemer, K.: The contextual nature of enterprise social networking: a multi case study comparison. In: 21st ECIS, pp. 1–12 (2013)
14. Williams, S.P., Schubert, P.: An empirical study of Enterprise 2.0 in context. In: 24th Bled eConference, pp. 42–55 (2011)
15. Awolusi, F.: The impacts of social networking sites on workplace productivity. J. Technol. Manag. Appl. Eng. **28**, 2–6 (2012)

16. Williams, S.P., Hausmann, V., Hardy, C.A., Schubert, P.: Enterprise 2.0 research: meeting the challenges of practice. In: 26th Bled eConference, pp. 251–263 (2013)

17. Richter, D., Riemer, K., vom Brocke, J.: Internet social networking. J. Bus. Inf. Syst. Eng. **3**, 89–101 (2011)

18. Levy, Y., Ellis, T.J.: A systems approach to conduct an effective literature review in support of information systems research. Inf. Sci. J. **9**, 181–212 (2006)

19. Webster, J., Watson, R.T.: Analyzing the past to prepare for the future: writing a literature review. MIS Q. **26**, 13–23 (2002)

20. Xiong, M., Chen, Q., Zhao, A.: The comparison study on the motivations of staffs' behaviors on public and enterprise social network: evidence from China. In: International Conference on Advances in Social Networks Analysis and Mining, pp. 802–807 (2014)

21. Cardon, P.W., Marshall, B.: The hype and reality of social media use for work collaboration and team communication. Int. J. Bus. Commun. **52**, 273–293 (2015)

22. Kuegler, M., Smolnik, S., Raeth, P.: Determining the factors influencing enterprise social software usage: development of a measurement instrument for empirical assessment. In: 46th HICSS, pp. 3635–3644 (2013)

23. Chin, C.P.-Y., Evans, N., Choo, K.-K.R., Tan, F.B.: What influences employees to use enterprise social networks? A socio-technical perspective. In: PACIS, pp. 1–11 (2015)

24. Buettner, R.: Analyzing the problem of employee internal social network site avoidance: are users resistant due to their privacy concerns? In: 48th HICSS, pp. 1819–1828 (2015)

25. Behrendt, S., Richter, A., Riemer, K.: Conceptualisation of digital traces for the identification of informal networks in enterprise social networks. In: 25th ACIS, pp. 1–10 (2014)

26. Trimi, S., Galanxhi, H.: The impact of Enterprise 2.0 in organizations. Serv. Bus. **8**, 405–424 (2014)

27. Leonardi, P.M.: When does technology use enable network change in organizations? A comparative study of feature use and shared affordances. MIS Q. **37**, 749–775 (2013)

28. Wang, T., Jung, C.-H., Kang, M.-H., Chung, Y.-S.: Exploring determinants of adoption intentions towards Enterprise 2.0 applications: an empirical study. Behav. Inf. Technol. **33**, 1048–1064 (2014)

29. Fulk, J., Yuan, Y.C.: Location, motivation, and social capitalization via enterprise social networking. J. Comput.-Mediat. Commun. **19**, 20–37 (2013)

30. Ellison, N.B., Gibbs, J.L., Weber, M.S.: The use of enterprise social network sites for knowledge sharing in distributed organizations: the role of organizational affordances. Am. Behav. Sci. **59**, 103–123 (2015)

31. Leroy, P., Defert, C., Hocquet, A., Goethals, F., Maes, J.: Antecedents of willingness to share information on enterprise social networks. Organ. Change Inf. Syst. **2**, 109–117 (2013)

32. Singh, J.B., Chandwani, R.: Adoption of web 2.0 technologies among knowledge workers: a theoretical integration of knowledge sharing and seeking factors. In: 22nd ECIS, pp. 1–11 (2014)

33. Engler, T.H., Alpar, P., Fayzimurodova, U.: Initial and continued knowledge contribution on enterprise social media platforms. In: 23rd ECIS, pp. 1–11 (2015)

34. Viol, J., Durst, C.: A framework to investigate the relationship between employee embeddedness in enterprise social networks and knowledge transfer. In: Pedrycz, W., Chen, S.-M. (eds.) Social Networks: A Framework of Computational Intelligence. Studies in Computational Intelligence, pp. 259–285. Springer, Heidelberg (2014)

35. Suh, A., Bock, G.-W.: The impact of enterprise social media on task performance in dispersed teams. In: 48th HICSS, pp. 1909–1918 (2015)

36. Kuegler, M., Smolnik, S.: Just for the fun of it? Towards a model for assessing the individual benefits of employees' enterprise social software usage. In: 46th HICSS, pp. 3614–3623 (2013)

37. Gibbs, J.L., Rozaidi, N.A., Eisenberg, J.: Overcoming the "Ideology of Openness": probing the affordances of social media for organizational knowledge sharing. J. Comput.-Mediat. Commun. **19**, 102–120 (2013)

38. Kuegler, M., Smolnik, S., Kane, G.C.: What's in IT for employees? Understanding the relationship between use and performance in enterprise social software. J. Strateg. Inf. Syst. **24**, 90–112 (2015)

39. Guy, I., Avraham, U., Carmel, D., Ur, S., Jacovi, M., Ronen, I.: Mining expertise and interests from social media. In: 22nd International Conference on World Wide Web, pp. 1–11 (2013)

40. Berger, K., Klier, J., Klier, M., Richter, A.: "Who is key…?" Value adding users in enterprise social networks. In: 22nd ECIS, pp. 1–16 (2014)

41. Subramaniam, N., Nandhakumar, J., Babtista, J.: Exploring social network interactions in enterprise systems: the role of virtual co-presence. Inf. Syst. J. **23**, 475–499 (2013)

42. Agarwal, S., Nath, A.: A study on implementing green IT in Enterprise 2.0. Int. J. Adv. Comput. Res. **3**, 43–49 (2013)

43. Richter, A., Riemer, K.: Malleable end-user software. Bus. Inf. Syst. Eng. **5**, 1–7 (2013)

44. Nedbal, D., Stieninger, M., Auinger, A.: A systematic approach for analysis workshops in Enterprise 2.0 projects. Procedia Technol. **16**, 897–905 (2014)

45. Richter, A., Wagner, D.: Leadership 2.0: engaging and supporting leaders in the transition towards a networked organization. In: 47th HICSS, pp. 574–583 (2014)

46. Louw, R., Mtsweni, J.: Guiding principles for adopting and promoting Enterprise 2.0 collaboration technologies. In: International Conference on Adaptive Science and Technology, pp. 1–6 (2013)

47. Louw, R., Mtsweni, J.: The quest towards a winning Enterprise 2.0 collaboration technology adoption strategy. Int. J. Adv. Comput. Sci. Appl. **4**, 34–39 (2013)

48. Han, S., Sörås, S., Schjødt-Osmo, O.: Governance of an enterprise social intranet implementation: the statkraft case. In: 23rd ECIS, pp. 1–17 (2015)

49. Alqahtani, S.M., Alanazi, S., McAuley, D.: The role of enterprise social networking (ESN) on business: five effective recommendations for ESN. In: 9th International Conference on Advances in Intelligent Systems and Computing, vol. 286, pp. 23–36 (2014)

50. Richter, A., Heidemann, J., Klier, M., Behrendt, S.: Success measurement of enterprise social networks. In: 11th International Conference on Wirtschaftsinformatik, pp. 1–15 (2013)

51. Merz, A.B., Seeber, I., Maier, R.: Social meets structure: revealing team collaboration activitites and effects in enterprise social networks. In: 23rd ECIS, pp. 1–16 (2015)

52. Matthews, T., Whittaker, S., Badenes, H., Smith, B.A., Muller, M.J., Ehrlich, K., Zhou, M.X., Lau, T.: Community insights: helping community leaders enhance the value of enterprise online communities. In: SIGCHI Conference, pp. 1–10 (2013)

53. Leftheriotis, I., Giannakos, M.N.: Using social media for work: losing your time or improving your work? Comput. Hum. Behav. **31**, 134–142 (2014)

54. Herzog, C., Richter, A., Steinhueser, M., Hoppe, U., Koch, M.: Methods and metrics for measuring the success of enterprise social software – what we can learn from practice and vice versa. In: 21st ECIS, pp. 1–12 (2013)

55. Koch, H., Leidner, D.E., Gonzalez, E.S.: Digitally enabling social networks: resolving IT-culture conflict. Inf. Syst. J. **23**, 501–523 (2013)

56. Kuegler, M., Luebbert, C., Smolnik, S.: Organizational climate's role in enterprise social software usage an empirical assessment. In: 12th International Conference on Wirtschaftsinformatik, pp. 1–16 (2015)

57. Chelmis, C., Srivastava, A., Prasanna, V.K.: Computational models of technology adoption at the workplace. Soc. Netw. Anal. Min. **4**, 1–18 (2014)
58. Stieglitz, S., Riemer, K., Meske, C.: Hierarchy or activity? The role of formal and informal influence in eliciting responses from enterprise social networks. In: 22nd ECIS, pp. 1–14 (2014)
59. Riemer, K., Stieglitz, S., Meske, C.: From top to bottom: investigating the changing role of hierarchy in enterprise social networks. Bus. Inf. Syst. Eng. **57**, 197–212 (2015)
60. Baghdadi, Y., Maamar, Z.: A framework for enterprise social computing: towards the realization of Enterprise 2.0. In: International Conference on Networking, Sensing and Control, pp. 672–677 (2013)
61. Liu, D., Wang, L., Zheng, J., Ning, K., Zhang, L.-J.: Influence analysis based expert finding model and its applications in enterprise social network. In: International Conference on Services Computing, pp. 368–375 (2013)
62. van Osch, W., Coursaris, C.K.: Organizational social media: a comprehensive framework and research agenda. In: 46th HICSS, pp. 700–708 (2013)
63. Beck, R., Pahlke, I., Seebach, C.: Knowledge exchange and symbolic action in social media-enabled electronic networks of practice: a multilevel perspective on knowledge seekers and contributors. MIS Q. **38**, 1245–1270 (2014)
64. Friedman, B.D., Burns, M.J., Cao, J.: Enterprise social networking data analytics within alcatel-lucent. Bell Labs Tech. J. **18**, 89–109 (2014)

Influencing Factors Increasing Popularity on Facebook – Empirical Insights from European Users

Rainer Schmidt[2], Michael Möhring[2(✉)], Ralf-Christian Härting[1],
Christopher Reichstein[1], and Barbara Keller[1]

[1] Aalen University of Applied Sciences, Business Information Systems, Aalen, Germany
{ralf.haerting,christopher.reichstein,
barbara.keller}@hs-aalen.de
[2] Munich University of Applied Sciences, Business Information Systems, Munich, Germany
{rainer.schmidt,michael.moehring}@hm.edu

Abstract. Popularity in social networks is a significant indicator of social success in western societies. The social capital within social networks has become an important element of social status. Therefore, the paper investigates why some users on Facebook receive more likes than others. The authors created eight hypotheses to test the influence of determinants on the popularity on the social media platform Facebook. The results show that especially gender, age, written posts and uploaded pictures or videos as well as adding new friends influences the popularity on Facebook.

Keywords: Social media behavior · Facebook · Likes · Popularity · Empirical study · Empirical · Popularity

1 Introduction

The popularity on Facebook and other social network sites has become the most significant indicator of social success in western societies. Facebook offers individuals a platform to present themselves and their social network [1]. In turn, Facebook can collect data from its users. To foster Facebook as a social network site and to embrace new user groups, Facebook continuously extends its offerings. According to Facebook Report 2014 [2] 1.39 billion people used Facebook actively during one month meaning they are logged into the network once or more per month. The number of users on average day aggregates to 890 million worldwide (301 million in Europe) and has increased by 18 % compared to 2013 [2]. While recent literature focuses on the maximization of brand post popularity [43], factors that might maximize post popularity for each are neglected so far.

Facebook exposes a huge amount of personal information. Likes express personal preferences and by analyzing them, some private traits and attributes can be predicted [3]. Therefore, it does not surprise that concerns about the protection of privacy arise. In Europe, deep concerns about the protection of privacy in social networks exist [4]. On the other hand, it was found that users consider self-disclosure of basic and sensitive information as helpful for receiving benefits from Facebook [5]. By establishing and maintaining connections in social network sites the user can increase his social capital [6]. Social

© Springer International Publishing Switzerland 2016
W. Abramowicz et al. (Eds.): BIS 2016, LNBIP 255, pp. 383–394, 2016.
DOI: 10.1007/978-3-319-39426-8_30

capital [6] is expressed both as static connections (friends) and dynamically as e.g. the number of Likes. Users with a lot of friends are e.g. rated as popular people, consequently to this, people with many generated Likes could be rated as interesting and artful. Accordingly, the status of a person within a group can increase because of this social capital [6] (e.g. number of Likes). Often, a high number of Likes is an indicator for the future creation of a Facebook friendship. In general, Facebook Like behavior shows aspects of user behavior as well as different relations in social networks. It is very interesting to discover different aspects of social media user's behavior to understand which factors influence their behavior and therefore how their behavior can be stimulated or antagonize. In particular, the understanding is very meaningful, also for research, because the behavior in social media environments can completely differ from the common manner of peoples' behave. There is a sparse research of the question how user behavior can be affected in terms of receiving Facebook Likes. Therefore, this paper investigates potential drivers of "Likes" on Facebook. Moreover, the research presented is in relation to other works. For instance, Quercia et al. describe factors for preferring either offline or online friends [7]. Recently, there are 350 friends per user on average [8]. Furthermore, gender differences in young Facebook users' assumptions on privacy are explored by [9]. In this context, Likes can evoke positive feelings suggesting that Likes received for a post might be seen as acts of acceptance from other users within a social network [42]. Finally, this study shows how personality and social capital is associated with friendship patterns, and how the individual can increase it by Likes.

The paper starts with a description of the research context and the research design based on a systematic literature review. Then the research methods and the data collection are discussed. The results are presented in the following section. Finally, a conclusion is given.

2   Background and Research Design

In order to investigate the reason why some users on Facebook receive a high number of Likes (e.g. for own posts, pictures/videos, etc.) and others do not, we design a research model based on the recent literature [10] as well as core elements of the social network Facebook as shown in Fig. 1.

Regarding to our study design (Fig. 1), the authors found eight determinants based on a systematic literature review of the last decade according to [10] with keywords as "Facebook", "like", etc. in databases such as SpringerLink, EbscoHost, IeeeXplore, AISeL, ACM Digital Library, and ScienceDirect that might influence the received "Likes" on the social media platform Facebook. First, we assume that the number of written posts have a positive influence on the number of Likes a user gets in average. Users with a high degree of extraversion tend to interact more with social media sites [11]. According to Forest and Wood, users who frequently update their Facebook status are getting more "Likes" from their friends [12]. General aspects of this topic were explored in a recent study taken place in the US by Choi et al. [13]. In particular, positive emotions posted on Facebook lead to more feedback on "Likes" [41]. This leads us to H1:

H1: The number of written posts is positively associated with received Facebook Likes.

Second, once users became friends on Facebook, they are much more likely to comment or to like the status of each other [14, 15]. The level of relationship with the "poster" is much more relevant to like user generated content than the content itself [14, 16]. As a result, user with a higher than average number of added friends tend to receive more Likes on Facebook [11, 15], which leads us to the following hypothesis:

H2: The number of friends in Facebook positively influences the received Facebook Likes.

Third, we hypothesize that there is a positive link between the amount of shared posts and the received average "Likes" a user gets for it. Oeldorf-Hirsch and Sundar [17] detected that people get more involved in news-stories if they comment and like shared news-stories on Facebook. Once involved, they tend to comment and like more often shared posts about these news-stories [17]. Choi et al. [13] found a positive link between the number of overall posts of an individual and the "Comments" and "Likes" received from friends [13] as reciprocity is an important factor in defining relationship maintenance. This leads us to the following hypothesis:

H3: The more posts an individual shares, the more Likes received on Facebook.

In the fourth place, there are several references describing user behavior in social media and the relationship among friends on Facebook. For instance, Moore and McElroy [18] argue that there are some dimensions of personality such as extraversion to explain and better understand Facebook usage. Hence, an extraverted user on Facebook tend to have more friends and more social interactions than the average [11, 19]. In accordance to Amichai-Hamburger and Vinitzky [20] as well as Bachrach et al. [21], who found a positive relation between extraversion and the number of Facebook friends, we assume:

H4: The more often a person adds friends, the higher the received Facebook Likes.

Fifth, existing studies show a relationship between the age of a Facebook user and its activity [20, 22]. Both Hampton et al. [20] as well as McAndrew and Jeong [22] reveal negative correlations between age and Facebook activity. Thus, younger people are said to be more active on Facebook [9]. As we assume, a high activity also implies a high number of Likes allocated by users, we formulated:

H5: The age of Facebook user does have a negative impact on the number of received Facebook Likes.

As a sixth point, we assume that the number of "Fan-Pages" liked by a user positively influences the number of "Likes" a user gets in average for a post. Arteaga Sanchez et al. examined the relation between the usage of Facebook and shared interests in an educational context [23]. The results show that individuals with common interests tend to interact more, so we hypothesize:

H6: The number of pages liked by a user is positively associated with the number of Facebook Likes received.

Seventh, female users statistically contribute to more user generated content than men – while women averagely make 21 status updates a month, men only make six on their Facebook profile [20]. Comparing the mean of the questionnaire by Hampton et al. [20]

with respect to the frequency of Facebook activities by gender in 2011, one can see huge differences in the number of allocated Likes by gender. Overall, female users are said to be more active regarding the time spent on Facebook [9] resulting in the hypothesis:

H7: Women receive more Facebook Likes than men do.

Authors found out that user-generated content, like uploaded videos or photos in social media, is perceived as a mere self-presentation to impress others [16]. As a basis for social comparison, these self-presentations might influence users' life satisfaction by evoking negative feelings such as envy [16]. In this context, we assume that positive feelings like being delighted for somebody are overlaid by negative emotions, like disfavor and envy, if a user uploads a high number of videos or photos, for instance impressions from vacation, to show its social standing. Hence, we conclude that a higher number of videos and photos posted are negatively associated with the number of "Likes" received:

H8: The amount of photos and videos posted on Facebook has a negative influence on the received Likes.

The literature review is summarized in the following table (Table 1):

Table 1. Summary of literature review

Determinant	Author
Written posts	[11, 12, 13, 42]
Amount of Facebook friends	[11, 14, 15, 16]
Shared posts	[13, 17]
Frequency of adding new friends	[7, 11, 18, 19, 20, 21]
Age	[9, 20, 22]
Page likes	[23]
Gender	[9, 20, 22]
Post of pictures and videos	[16]

To examine the impact of the specified determinants above on the received number of Likes on Facebook, the online-based responses of the participants ranged on a Likert scale [24] of one to five (1: never; 2: rarely; 3: sometimes; 4: often; 5: very often). To prove hypothesis six (H6: The number of pages liked by a user is positively associated with the number of Facebook Likes received) we introduced a scale from one to six to answer the question "How many publicly accessible fan pages did you like on Facebook?" (1: <5; 2: 5–10; 3: 11–15; 4: 16–20; 5: 21–25; 6: >25). For questions concerning the age and the number of friends on Facebook, we implemented a text field for entering numbers in our online-survey. Gender was finally set as a dichotomous variable (1 = female; 0 = male). All questions were designed according to general quantitative study guidelines [25].

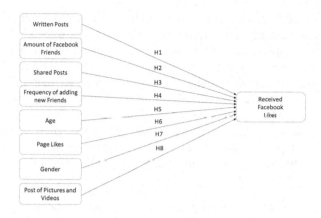

Fig. 1. Research model

3   Research Methods and Data Collection

According to our research model, we investigate our hypotheses based on a quantitative research approach via a web-based online survey [26] in the German language. Additional to qualitative research methods like interviews, quantitative research methods like questionnaires are applied to evaluate the fit of a theoretical model with empirical data [26]. Furthermore, questionnaires are often used to extract Facebook user behavior (e.g. [27–29]). The survey was implemented via the open source software LimeSurvey [30] in Germany. First, a Pre-test of the study was implemented in November 2014 to ensure a high quality of research. Based on the results the questionnaire was improved. The main study started in December 2014 and ended in January 2015. We collected a sample of n = 772 answers of different participants. We asked only young people (e.g. students) because of the typical use of Facebook in this socio group [22, 31]. Therefore, we use the age as a strong selection question as well as a question if the person uses Facebook.

After data cleaning, we got a final sample of n = 485 German Facebook users. The mean age was 23.19 years. 60 % of the asked persons were female and 40 % male. The amount of Facebook friends of every asked participant within this study are 322 persons on average. Furthermore, each asked person receives averagely 19 Likes for a written

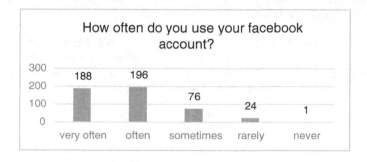

Fig. 2. Facebook usage

post, uploaded video or picture, status change, etc. The majority (approx. 79 %) of the asked persons uses their Facebook account quite a few times as seen in the Fig. 2.

Regarding the sample of our survey, most of the people (n = 196) answered that they often use their Facebook account; 188 respondents even stated to use Facebook very often. In fact, only one person never uses its own Facebook account. In addition, it has to be noted that the income per month of most people is lower than 1.000 Euros as well as the expenses are, which might be caused by the fact that this study predominantly asked young people about their Facebook activity. Moreover, most of the interviewed persons (n = 306) stated to post something with respect to their leisure time. For further themes regarding the posts on Facebook, please refer to Fig. 3 below.

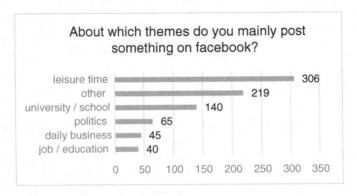

Fig. 3. Posting themes on Facebook (several choices possible)

To analyze our theoretical model with empirical data, we used a multivariate analysis through multiple, linear regression. In general, linear regression can be used to explore functional relationships between independent and dependent variables [32–34]. Linear regression is very common to analyzing theoretical models with empirical data in information systems research (e.g. [36, 37]). There are two main types of regression: a simple (bivariate) and multivariate regression. In our research model, we explore the effect of different independent variables such as written or shared posts, ages, etc. on the dependent variable of received Facebook likes. Therefore, we used a multivariate regression to examine the relationship between this eight quantitative independent, and the one quantitative depended variable [35]. The regression equation [39] of our research model is formulated as follows:

$$Received\_Facebook\_Likes = \beta_0 + \beta_1 * Written\_Posts + \beta_2 *$$
$$Amount\_of\_Facebook\_Friends + \beta_3 * Shared\_Posts + \beta_4 *$$
$$Frequency\_of\_adding\_new\_Friends + \beta_5 * Age + \beta_6 * Page\_Likes + \beta_7 * Gender + \beta_8$$
$$* Post\_of\_pictures\_and\_videos + \varepsilon$$

Equation 1. Regression model

There are a number of different software tools like IBM SPSS, R project or Microsoft Excel to implement a regression analysis. We used IBM SPSS 22 because of the good implementation and statistical validation.

4 Results

After analyzing the empirical sample with a multiple linear regression model, we got the following results (Fig. 4):

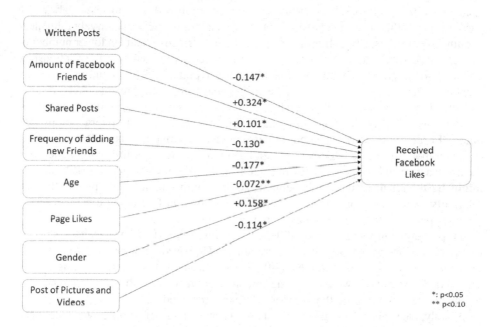

Fig. 4. Linear regression model with coefficients

Seven of eight (Beta) values are significant ($p < 0.05$) according to Table 2. Furthermore, the coefficient of the determination (R^2) is also in a satisfying range ($0.307 > 0.19$) [38]. All standardized regression coefficients (Beta values) are defined in Table 2.

The first hypothesis explores the influence of the written posts on the received Facebook Likes. Based on our study, we cannot confirm H1 because of a negative value (-0.147) of the regression coefficient of our model. Therefore, people cannot increase their Facebook Likes by generating more posts on their own Facebook profile. Users with a high amount of written posts tend not to interact more proportional with other users in this social network.

According to different current research [16], H2 investigates the influence of the amount of Facebook friends on the received Facebook Likes. There are significant influences ($+0.324$) regarding this connection. Hypothesis 2 can also be confirmed. In fact, more friends indicate a higher likelihood of getting back more Facebook Likes.

Shared posts are explored via hypothesis 3. Regarding the results of the regression model, H3 can be confirmed, because of a positive regression coefficient value (+0.101). However, the influence is very weak. With regard to the hypotheses 1 and 3, we can assume that users get more "Likes" for self-penned posts than for shared posts. Personal information disclosed by users seems to be more interesting for interactions between users on Facebook.

The influence of the frequency of adding new friends on Facebook was discovered in hypothesis 4. The main question here is, if adding new friends can increase the likelihood of getting more Facebook Likes. The results show that there is a negative influence (−0.130). Therefore, hypothesis 4 cannot be confirmed. As our results above (in particular hypothesis 2) already showed that the number of friends positively influences the number of Likes, a high frequency of adding new friends results in lower number of Likes. This might be the case, because of a lower intensity or quality of the relationship.

Socio-demographic factors are examined via hypothesis 5 (age) and hypothesis 7 (gender). The analysis of both hypotheses generates interesting results. According to hypothesis 5, a lower age indicates a higher likelihood of getting more Facebook Likes. Younger users of Facebook are more active and, therefore, tend to interact more with others. Furthermore, female persons get also more Likes (hypothesis 7). Based on these results hypothesis 7 and hypothesis 5 can be confirmed. It is not just that female users are more active and do spend more time on Facebook [9], they also get more Likes which might be the result of their engagement. As it is with female users, people that are relatively young are more active and spend more time on Facebook, too. As a result, younger people receive more Likes than higher aged users.

Hypothesis 6 investigates the influence of adding Likes (to other Facebook sites) to the received Facebook Likes. We cannot confirm H6 based on our results because of a weak negative regression coefficient (−0,072). Furthermore, this factor is not significant ($p = 0.084$). Therefore, we have no significant support for the influence of the number of pages liked by a user on the number of "Likes" received.

Finally, the analysis of hypothesis 8 shows a negative impact on the posted videos and pictures to the received Facebook Likes (−0.114). Therefore, H8 can be confirmed. Hence, posting more photos and videos leads to a lower number of "Likes" received on average.

Table 2. Regression coefficients

Variable	Standardized coefficients of regression (Beta)	Significance (P values)
Age	−0.177	0.000
Amount of friends	+0.324	0.000
Posts of pictures and videos	−0.114	0.030
Written posts	−0.147	0.004
Gender	+0.158	0.000
Frequency of adding new friends	−0.130	0.003
Page likes	−0.072	0.084
Shared posts	+0.101	0.025

Signal theory according to Spence [44] can be used to explain some facts of our results. Regarding to this theory of consumer psychology, people like Facebook users send different signals (like written or shared posts, pictures, etc.) to their Facebook environment. They receive the signal and process it (e.g. interpretation of a shared post). Finally, they can give a feedback back. This feedback can be for instance a Like on Facebook, if the evaluation of the signal through the signal processing is positive.

Important values of the multiple linear regression are shown in Table 2.

5 Conclusion

Based on a multivariate analysis of empirical data through linear regression, we investigated how Facebook Likes are influenced by different factors. The empirical study shows various interesting influence factors according to our quantitative web-based survey in Germany. Especially the gender, age as well as written posts and uploaded pictures or videos and adding new friends are important influencing factors for the Likes someone gets on Facebook.

Our results contribute to the current research. Research on social media behavior can benefit from new knowledge about Likes on Facebook. Therefore, current approaches can adopt this knowledge and improve theories of Facebook and other social media user behavior. Furthermore, different social media platforms can be evaluated with our model and research results according to the basic feature of "Likes". This paves the way to concrete optimization steps on these platforms. We also used signal theory to explain some aspects of the user behavior.

There are also practical implications. First, users can evaluate their social media behavior in relation to getting more or less Facebook Likes. This helps them to improve their social capital and possibly improve their social status. Second, enterprises can use these results to calculate different personal attributes based on the received Likes of individual Facebook users. This approach can help to get more detailed (private) personal data. For instance, it might be possible for recruiters to monitor applicants on Facebook analyzing user status updates and the amount of Likes [40].

There are some limitations to discuss. First, our sample consists only of young Facebook users. Second, we only asked persons in Germany. Future research should cover these limitations. Furthermore, our quantitative research approach could not investigate all possible influence factors. Most of the explored relationships are significant, but one factor ("Page Likes") is only on the significance level of 10 percent. Further, it could be a good opportunity to try to get some information via direct Facebook access and not only through a survey.

Therefore, future research should enlarge the sample of the study to people of all ages as well as other countries. A qualitative research approach (e.g. user interviews) can also be a good opportunity for future research as we need more insights about the processes behind the interactions on Social Networking Sites. Consequently, more Facebook attributes and information (e.g. tagged persons, time differences) can be observed in the future, e.g. via the use of Facebook API Graph.

Acknowledgment. We thank Johannes Schlunsky and Kristina Martyn for supporting our research.

References

1. Ellison, N.B., Steinfield, C., Lampe, C.: The benefits of facebook 'friends': social capital and college students' use of online social network sites. J. Comput. Mediat. Commun. **12**(4), 1143–1168 (2007)
2. Zuckerberg, M.: Facebook Q4 2014 Earnings Call Transcript (2014)
3. Kosinski, M., Stillwell, D., Graepel, T.: Private traits and attributes are predictable from digital records of human behavior. Proc. Natl. Acad. Sci. **110**(15), 5802–5805 (2013)
4. Lusoli, W., Bacigalupo, M., Lupiáñez Villanueva, F., Monteleone, S., Maghiros, I.: Pan-European Survey of Practices, Attitudes and Policy Preferences as regards Personal Identity Data Management. Publications Office of the European Union (2012)
5. Chang, C.-W., Heo, J.: Visiting theories that predict college students' self-disclosure on facebook. Comput. Hum. Behav. **30**, 79–86 (2014)
6. Coleman, J.S.: Social capital in the creation of human capital. Am. J. Sociol. **94**, 95–120 (1988)
7. Quercia, D., Lambiotte, R., Stillwell, D., Kosinski, M., Crowcroft, J.: The personality of popular facebook users. In: Proceedings of the ACM 2012 Conference on Computer Supported Cooperative Work, New York, NY, USA, pp. 955–964 (2012)
8. Statista, Facebook - Durchschnittliche Anzahl an Freunden in den USA nach Altersgruppe 2014 | Umfrage, Statista (2014). http://de.statista.com/statistik/daten/studie/325772/umfrage/durchschnittliche-anzahl-von-facebook-freunden-in-den-usa-nach-altersgruppe/. Accessed 19 Feb 2015
9. Hoy, M.G., Milne, G.: Gender differences in privacy-related measures for young adult facebook users. J. Interact. Advertising **10**(2), 28–45 (2010)
10. Cooper, D.R., Schindler, P.S., Sun, J.: Business Research Methods, vol. 9. McGraw-hill, New York (2006)
11. Correa, T., Hinsley, A.W., De Zuniga, H.G.: Who interacts on the web?: The intersection of users' personality and social media use. Comput. Hum. Behav. **26**(2), 247–253 (2010)
12. Forest, A.L., Wood, J.V.: When social networking is not working individuals with low self-esteem recognize but do not reap the benefits of self-disclosure on facebook. Psychol. Sci. **23**(3), 295–302 (2012)
13. Choi, M., Panek, E.T., Nardis, Y., Toma, C.L.: When social media isn't social: friends' responsiveness to narcissists on facebook. Pers. Individ. Differ. **77**, 209–214 (2015)
14. Koroleva, K., Stimac, V., Krasnova, H., Kunze, D.: I like it because I ('m) like you–measuring user attitudes towards information on facebook. In: ICIS 2011 (2011)
15. Ong, E.Y., Ang, R.P., Ho, J.C., Lim, J.C., Goh, D.H., Lee, C.S., Chua, A.Y.: Narcissism, extraversion and adolescents' self-presentation on facebook. Pers. Individ. Differ. **50**(2), 180–185 (2011)
16. Krasnova, H., Wenninger, H., Widjaja, T., Buxmann, P.: Envy on facebook: a hidden threat to users' life satisfaction? Wirtschaftsinformatik **92**, 1477–1492 (2013)
17. Oeldorf-Hirsch, A., Sundar, S.S.: Posting, commenting, and tagging: effects of sharing news stories on facebook. Comput. Hum. Behav. **44**, 240–249 (2015)
18. Moore, K., McElroy, J.C.: The influence of personality on facebook usage, wall postings, and regret. Comput. Hum. Behav. **28**(1), 267–274 (2012)

19. Wehrli, S.: Personality on social network sites: an application of the five factor model. Working Paper No. 7, Zurich: ETH Sociology (2008)
20. Hampton, K., Goulet, L.S., Marlow, C., Rainie, L.: Why most Facebook users get more than they give. Pew Research Center, Washington, DC, USA (2012)
21. Bachrach, Y., Kosinski, M., Graepel, T., Kohli, P., Stillwell, D.: Personality and patterns of facebook usage. In: Proceedings of the 3rd Annual ACM Web Science Conference, pp. 24–32 (2012)
22. McAndrew, F.T., Jeong, H.S.: Who does what on facebook? Age, sex, and relationship status as predictors of facebook use. Comput. Hum. Behav. 28(6), 2359–2365 (2012)
23. Sánchez, R.A., Cortijo, V., Javed, U.: Students' perceptions of facebook for academic purposes. Comput. Educ. 70, 138–149 (2014)
24. Kothari, C.R.: Research Methodology: Methods and Techniques. New Age International, Darya Ganj (2011)
25. Hewson, C.: Internet Research Methods: A Practical Guide for the Social and Behavioural Sciences. Sage, Beverly Hills (2003)
26. Rea, L.M., Parker, R.A.: Designing and Conducting Survey Research: A Comprehensive Guide. Wiley, London (2012)
27. Ryan, T., Xenos, S.: Who uses facebook? An investigation into the relationship between the big five, shyness, narcissism, loneliness, and facebook usage. Comput. Hum. Behav. 27(5), 1658–1664 (2011)
28. Maier, C., Laumer, S., Eckhardt, A., Weitzel, T.: Giving too much social support: social overload on social networking sites. Eur. J. Inf. Syst. 24, 447–464 (2014)
29. Sagioglou, C., Greitemeyer, T.: Facebook's emotional consequences: why facebook causes a decrease in mood and why people still use it. Comput. Hum. Behav. 35, 359–363 (2014)
30. T. L. Project Team: LimeSurvey - the free and open source survey software tool! 24 Apr 2011. https://www.limesurvey.org/de/. Accessed 15 Feb 2015
31. Social networking fact sheet, pew research center's internet and American life project. http://www.pewinternet.org/fact-sheets/social-networking-fact-sheet/. Accessed 10 Feb 2015
32. Stigler, S.M.: The Story of Statistics: The Measurement of Uncertainty Before 1900. Harvard University Press, Cambridge (1986)
33. Boscovich, R.J.: De litteraria expeditione per pontificiam ditionem, et synopsis amplioris operis, ac habentur plura ejus ex exemplaria etiam sensorum impressa. Bononiensi Sci. Artum Inst. Atque Acad. Comment. 4, 353–396 (1757)
34. Boscovich, R.J.: De recentissimis graduum dimensionibus, et figura, ac magnitudine terrae inde derivanda. Philos. Recentioris Benedicto Stay Romano Arch. Publico Eloquentare Profr. Versibus Traditae Libri X Adnot. Suppl. P Rogerii Joseph Boscovich SJ Tomus II, 406–426 (1757)
35. Mason, C.H., Perreault, W.D.: Collinearity, power, and interpretation of multiple regression analysis. J. Mark. Res. XXVIII, 268–280 (1991)
36. Gefen, D., Straub, D., Boudreau, M.C.: Structural equation modeling and regression: guidelines for research practice. Commun. Assoc. Inf. Syst. 4(1), 2–77 (2000)
37. Schmidt, R., Möhring, M., Maier, S., Pietsch, J., Härting, R.-C.: Big data as strategic enabler - insights from central European enterprises. In: Abramowicz, W., Kokkinaki, A. (eds.) BIS 2014. LNBIP, vol. 176, pp. 50–60. Springer, Heidelberg (2014)
38. Chin, W.W.: The partial least squares approach to structural equation modeling. Mod. Meth. Bus. Res. 295, 295–336 (1998)
39. Keller, G.: Statistics for Management and Economics, pp. 1–674. Cengage, Boston (2014)
40. Marjanovic, O., Rothenhoefer, M.: Improving knowledge-intensive business processes through social media. In: Proceedings of the AMCIS 2014. AISNET, Savannah (2014)

41. Schoendienst, V., Dang-Xuan, L.: Investigating the relationship between number of friends, posting frequency and received feedback on facebook. In: ECIS 2012 Proceedings, p. 234 (2012)
42. Kietzmann, J.H., Hermkens, K., McCarthy, I.P., Silvestre, B.S.: Social media? Get serious! Understanding the functional building blocks of social media. Bus. Horiz. **54**, 241–251 (2011)
43. De Vries, L., Gensler, S., Leeflang, P.S.: Popularity of brand posts on brand fan pages: an investigation of the effects of social media marketing. J. Interact. Mark. **26**(2), 83–91 (2012)
44. Spence, M.: Job market signaling. Q. J. Econ. **87**(3), 355–374 (1972)

Applications

Persuasive Design Principles of Car Apps

Chao Zhang[1(✉)], Lili Wan[2(✉)], and Daihwan Min[3]

[1] Linton School of Global Business, Hannam University, Daejeon, Korea
simplegeminizc@gmail.com
[2] College of Business Administration,
Hankuk University of Foreign Studies, Seoul, Korea
wanlili@hufs.ac.kr
[3] Department of Digital Management, Korea University, Seoul, Korea
mismdh@korea.ac.kr

Abstract. This study attempts to identify the persuasive design principles of car-related smartphone apps that assist users in driving or managing their vehicles. We developed a guideline by experts for evaluating persuasive design principles of car apps and recruited four evaluators who were trained to apply the guideline and given 35 car apps for evaluation. The value of Fleiss' Kappa was 0.782, over the excellent criterion of 0.75, which means the inter-rater reliability of persuasive design guideline was reliable. We collected 697 car apps from Apple iTunes Store and Google Play and examined which design principles were implemented by these car apps. The result shows that nine persuasive design principles are found, such as reduction, trustworthiness, real-world feel, self-monitoring, personalization, reminder, suggestion, expertise, and verifiability. The results from this study would suggest some implications for car app developers and automakers to develop better car apps in the future.

Keywords: Mobile application · Car-related mobile apps · Persuasive technology · Persuasion · Persuasive system design principles

1 Introduction

Mobile applications, thanks to the explosive growth of smart phones, have been widely used in our daily lives. A mobile application (hereafter app) is a computer program designed to run on mobile devices such as smartphones or tablets. Nowadays, Apps are available to download from different distribution platforms, such as Apple iTunes App Store, Google Play, BlackBerry App World, Windows Phone Store, Samsung Apps Store, Nokia Store, etc. Some apps are free to download, while others require direct payment or in-app purchase for full functions. As of March 2016, over 3.82 million apps have been published on the Apple iTunes Store and Google Play [9].

This research was supported by Basic Science Research Program through the National Research Foundation of Korea (NRF) funded by the Ministry of Education (NRF-2014R1A1A2059510).

W. Abramowicz et al. (Eds.): BIS 2016, LNBIP 255, pp. 397–410, 2016.
DOI: 10.1007/978-3-319-39426-8_31

Apps on smartphones allow users to do practically anything, such as online payment, playing games, sending emails, banking and portfolio investment, downloading music and video, online shopping, and so on.

There are about five million app developers in the world who are competing for users' time and/or money [7] and that number is growing quite fast [8]. For app developers, one of the most urgent issues is what kinds of design principles are the most effective ones in order to satisfy the needs of mobile users. In other words, app developers want to know how to design an effective and persuasive app that may attract user's interests, persuade users to use, and increase app's using life cycle. This involves psychological and behavioral issues.

This study is an attempt to evaluate empirically the applicability of the persuasive design principles. The objects of the evaluation are mobile apps used for a car, i.e., car-related mobile apps (hereafter car apps). App developers and car manufacturers have great interests in car apps, as "IoT" (Internet of Things) draws attention and investment. For example, "TPMS" (tire pressure monitoring system) is a kind of in-car application, which has built-in sensors to alert driver when tire pressure is abnormal. "Blue Link" system enables Hyundai car users to control car by using a smartphone application. At the intersection of the smartphone and the automobile lies a vast opportunity for increasing in speed, efficiency, and entertainment.

As the first step, an evaluation guideline was developed to investigate persuasive design principles of car apps. About seven hundred car apps were identified from the two most popular app distribution platforms – 273 apps from Apple iTunes App Store and 424 apps from Google Play Store. One of the researcher groups which consists with two researchers of this paper and two master program students together evaluated all 697 car apps according to the guideline. As the second step, 35 car apps were randomly selected and given to an evaluation group with four other evaluators who were trained to apply the guideline. The results from four evaluators were compared in order to calculate the inter-rater reliability. The next two sections describe the evaluation results and the inter-rater reliability.

2   Related Work

Several prior studies have performed some experimental methods to check the persuasive design of systems. The review of the prior studies highlights the following issues.

First, there is no clear description about the theoretical interpretation of behavior change after using mobile apps. Moreover, until now there is no clear research about the theoretical interpretation of persuasive design for mobile apps that caused behavior change. Most of the prior studies focus on the behavior change after some experiments in medical research field, such as physical exercises method designed with persuasive technology to help participants to control bodyweight. Most of them did not focus on mobile app designing process. Second, until now, it is hard to find any empirical research about the relationship between persuasive design and behavior change. Some scholars and researchers in the field of "persuasive technology" have investigated issues of human behavioral change and habit-forming in the use of information technology.

Fogg [3] has provided a widely utilized framework to help developers to understand the persuasive technology which is defined as any interactive technical system for the purpose of changing people's attitudes or behaviors. However, it cannot be used directly as a guide for developers of technical systems to follow [5]. Oinas-Kukkonen [4] has identified distinct software features in order to confirm and evaluate the significance of persuasive systems. Oinas-Kukkonen and Harjumaa [6] have suggested 28 persuasive design principles in four categories such as primary task support, dialogue support, system credibility support, and social support. However, those design principles have not been evaluated by empirical research.

3 Persuasive Design Principles

3.1 Persuasive Technology

Persuasive technology is defined as any interactive technical system designed for the purpose of changing people's attitudes or behaviors [3]. In other words, persuasive technology is developed and designed to change user's attitudes or behaviors not by coercion, but through persuasion and social influence. If persuasion is possible or available, a person or a group can receive an intervention from other people or group in a particular setting [3]. Persuasive technology can be found in mobile apps or websites with behavior-oriented designs like Amazon and Facebook, which can persuades users to buy more often or stay logged in. Many mobile apps, such as some health-oriented apps that incentivize weight loss and help to manage addictions and other mental health issues. According to B.J. Fogg's Behavior Change Wizard, "Persuade users to use or change their attitudes or behaviors" means after using a persuasive designed app, user's behavior change can be found one time, more than one time, or even habit formation. However, not any exist research has focused on the causal relationship between persuasive design features and behavior change.

Persuasive technology, as a kind of interactive information technology, is a fast-growing research topic, especially for app design. It is proposed for changing users' behaviors and attitudes [3]. Both researchers and app developers are focusing on increasing the app's persuasive characteristics in order to motivate and influence users. These kinds of interactive technologies and persuasive technologies can absolutely change user's attitudes and behaviors. Persuasive theory is the main representatives of B.J. Fogg who was the first scientist to describe the overlap between persuasion and computers. He established the Behavior Model and found that behavior (B) occurs only when three elements converge at the same moment, that is motivation (M), ability (A), and a trigger (T), which leads to B = MAT [3].

3.2 Persuasive System Design Model (PSD)

The "Persuasive System Design Model" (PSD Model) focuses on the detailed analysis of the persuasion context, the event, and the strategy [5, 6]. The persuasion context means the intended change in behaviors and attitudes. The event refers to the context of use and users of technologies. The strategy means to develop the content of message and delivery route tightly integrated with target users. The PSD Model also provides 28 detailed design principles in four categories, "Primary Task Support" category includes the principles of reduction, tunneling, tailoring, personalization, self-monitoring, simulation, and rehearsal. "Dialogue Support" category consists with the principles of praise, rewards, reminders, suggestion, similarity, liking, and social role. System Credibility Support" category includes the principles of trustworthiness, expertise, surface credibility, real-world feel, authority, third-party endorsement, and verifiability. "Social Support" category includes social learning, social comparison, normative influence, social facilitation, cooperation, competition, and recognition. Until now, most persuasive system design principles are not evaluated by using empirical method.

4 Inter-Rater Reliability

Before performing the evaluation process, an evaluation guideline for car app's persuasive design was developed. This evaluation guideline was new developed based on literature review. Therefore, the evaluation guideline may involve ambiguity in characteristics descriptions or subjective judgments. In order to ensure the empirical evaluation method of design principles in car apps, inter-rater reliability is needed. It can help to address the degree of agreement among raters and to determine this evaluation guideline is appropriate for measuring persuasive design principles of car apps.

Actually, before this evaluation process, over 1000 mobile apps were collected and analyzed from both Google Play and Apple iTunes Store by searching with keywords, such as "car", "driving", "maintenance", "automobile", and etc. from October 2013 till March 2015. After screened out some game apps and cartoons, based on app's functionality, utility, and features, a systematic categorizing method was developed. The purpose of our research focuses on car app's functionality and utility, car app's persuasive design features, and user's behavior change, therefore, these car apps were only coded the app name, price, and category code number. After two rounds of reliability analysis by experts, such as app developers and vehicle system researchers, the results of Cohen's Kappa (kappa1 = 0.886, kappa2 = 0.828) showed evaluators' agreements with this categorizing method. A total of 697 available car apps left and were classified into eight categories: news and basic information about car, buying and selling, driver's communication, location service, safe driving service, A/S maintenance management, renting service, and car expenses monitoring. This categorizing method was developed and evaluated to be reliable. It will not be discussed in detail in this paper because it has been discussed and it is under reviewing process for publishing.

Four evaluators (three assistant professors and one student in a master program) were recruited and trained with the evaluation guideline. Two assistant professors came from English-speaking countries with deep knowledge of business management. The third assistant professor came from Malaysia who was interested in app design. One student came from China whose major is in mobile business. The evaluators were asked to follow three rules to evaluate the car apps. First, read and understand the implications of every persuasive design principles. Second, download or use simulators to try out apps. Third, make judgments and evaluate apps according to the guideline.

35 car apps were randomly selected from 697 car apps which we have investigated their functionalities and utilities. This 35 car apps were given to all four evaluators. Appendix 2 shows an example about the evaluation results of three car apps' persuasive design features.

Generally, Cohen's kappa is a statistical measure for assessing the degree of agreement. Cohen's Kappa refers to a measurement of concordance or agreement between two raters or two methods of measurement. However, in this research, Fleiss' kappa should be used with nominal-scale ratings to assess the agreement among four evaluators. Fleiss's Kappa is an extension of Cohen's Kappa for evaluating concordance or agreements among multiple raters (generally more than 2 raters), but no weighting is applied. Four evaluators ($m = 4$) independently analyzed 26 persuasive design features for 35 car apps ("subject" $n = 910$) by giving agree (Y) to persuasive design principle exist or giving disagree (N) to persuasive design principle not exist ("two evaluation categories" $k = 2$). Based on the equation of Fleiss' kappa, the inter-rater agreement can be thought of as follows: if four evaluators ($n = 4$) assign 3640 ratings or decisions ("decisions" $= n*m$, each subject is evaluated m times) to 910 subjects, the value of kappa will show how consistent the ratings are. The p-value of z-test is significant which means the value of kappa is not equal to zero. The population kappa's confidence interval is between 0.755 (Lower) and 0.808 (Upper). The interpreting Fleiss' Kappa value is 0.782. Fleiss' guideline characterizes values below 0.40 as poor, 0.40 to 0.75 as fair to good, and over 0.75 as excellent [2]. Altman characterizes values <0.20 as poor agreement, 0.21–0.40 as fair, 0.41–0.60 as moderate, 0.61–0.80 as good, and 0.81–1 as very good agreement [1]. Therefore, the evaluation method for car app's persuasive design principles is considered as reliable (Table 1).

Table 1. The statistic result of Fleiss' Kappa measurement for inter-rater reliability

Kappa	0.782
Standard error	0.014
Z-statistics	57.725
P-value	0.000
Lower	0.755
Upper	0.808
alpha = 0.05; two tails	

5 Persuasive Design Principles in Car Apps

28 persuasive design principles from "PSD Model" were used for checking which persuasive features current car apps possess. A guideline for evaluating car apps was developed. Due to limitations on space, the detail information about persuasive design guideline will not be further mentioned in this paper. Only a brief description will be shown (See Appendix 1).

Although this data mining process took quite a long time, a complete investigation can help researchers and app developers to know well about the current status of car app market. Another reason for this long-term investigation is that few branded car apps can be found with high download amounts, ratings, and reviews. The current situation for applying persuasive technology into car app design process is that only a few persuasive design principles are used and it is hard to find a represented car app or branded car app. Sample survey can hardly reflect the status of the population. Moreover, current car apps and vehicles are still independent and lack of connection. The functions and content of current car apps do not satisfy all needs of car users. Most car apps have a common shortcoming: lack of integration and singleness of functions. It is hard for a car app to satisfy the full demands of users. For example, some persuasive design features such as praise or rewards have gained widely attention because of the habit-forming effectiveness. However, it is hard to find these features in current car apps.

Table 2. Persuasive design principles of car apps

Persuasive design principle		Evaluation Result (Total 697 Car Apps)		
		YES	**NO**	**Y/T (%)**
Primary Task Support	**Self-Monitoring**	**689**	**8**	**98.85%**
	Reduction	**697**	**0**	**100.00%**
	Personalization	**304**	**393**	**43.62%**
	Rehearsal	0	697	0.00%
	Tunneling	83	614	11.91%
	Simulation	0	697	0.00%
	Tailoring	9	688	1.29%
Dialogue Support	**Reminder**	**507**	**190**	**72.74%**
	Praise	7	690	1.00%
	Suggestion	**636**	**61**	**91.25%**
	Rewards	7	690	1.00%
	Similarity	3	694	0.43%
	Social Role	5	692	0.72%
	Liking	Not Evaluated		
System Credibility Support	**Trustworthiness**	**697**	**0**	**100.00%**
	Real-World Feel	**693**	**4**	**99.43%**
	Expertise	**689**	**8**	**98.85%**
	Verifiability	**267**	**430**	**38.31%**
	Authority	7	690	1.00%
	Third Party Endorsements	3	694	0.43%
	Surface Credibility	Not Evaluated		
Social Support	Social Comparison Sharing	5	692	0.72%
	Cooperation	0	697	0.00%
	Normative Influence	0	697	0.00%
	Social Facilitation	0	697	0.00%
	Competition	0	697	0.00%
	Recognition	3	694	0.43%
	Social Learning	0	697	0.00%
Note: Shaded cell= Principle was selected; Clear cell= Principle was not selected				

The evaluation process for persuasive design principles started from June 2014 and ended in March 2015. A pilot test was performed and two participants were asked to read the guideline carefully and evaluate a few car apps. One participant is a mobile app developer and another participant is a professor majored in e-business. Some descriptions for persuasive design principles were modified. After that, except the members of first evaluation group who have participated the evaluation process for persuasive design guideline, another evaluation group which consisted with two researchers of this study and two students in a Master program were fully involved in the whole evaluation process for a total of 697 car apps. In the investigation period, from Apple iTunes Store, we investigated and evaluated 218 free apps and 55 paid apps. From Google Play, we investigated and evaluated 395 free apps and 29 paid apps. Because the price of app was fluctuated, the price of app depended on the situation of the investigation day.

Car apps were evaluated in two ways. One is to download and try out a car app directly. The other way is to use some simulators such as BlueStacks (a popular Android simulator used for running applications on PC) or Xcode (an iOS simulator used for running applications on Mac). There is no difference between the two ways. In the evaluation process, only a few car apps were evaluated by using simulators, most of evaluated car apps were directly downloaded and tried out in order to address the persuasive characteristics.

As shown in Table 2, the result shows that current car apps have implemented at most nine design principles: self-monitoring, reduction, personalization, reminder, suggestion, trustworthiness, real-world feel, expertise, and verifiability. Seven persuasive design principles are universal in car apps and two persuasive design principles are present in less than half of the apps.

6   Discussion

In our first research stage, an investigation was used to find out the current status of car apps and to check the persuasive design characteristics of current car apps. Total six hundred and ninety-seven apps were checked and nine persuasive design principles were found to be the common features.

Over 90 % of car apps implemented the principles of self-monitoring, reduction, suggestion, trustworthiness, real-world feel, and expertise, while 73 % implemented the reminder principle and less than 50 % implemented the principles of personalization and verifiability. However, it is hard to find some persuasive design features such as praise or rewards in current car apps, even though these features have gained wide attention due to the habit-forming effectiveness. Among them, about 38.31 % car apps had "verifiability" feature, and about 43.62 % car apps had "Personalization" feature. Although the number of car apps for "Verifiability" and "Personalization" was not over 50 %, it still could be considered as common features because most of the raters had no argument with their conceptions and evaluation rules. Raters can easily confirm the two features by using their evaluation rules. During the evaluation process, some persuasive design principles caused disagreement among raters. For instance, the variable of "Tunneling", about 11.91 % car apps were considered to have this persuasive design feature. The principle of tunneling was discussed a lot by evaluators during the evaluation process. Although it has been well used in a lot of game apps or healthy care apps, this design feature was still not well used in current car app design which caused a little more arguments among raters. Furthermore, during the investigation process, the variable of "surface credibility" and "liking" are not investigated for the reason that everyone has a subjective evaluating point of view.

7 Future Research

This paper shows the result of our first research stage from an empirical evaluation of persuasive design principles in car apps. The following research is proceeded as follows:

First, scale development for persuasive design features is a critical problem that should be solved as quickly as possible. Most of the concepts of persuasive design features are not normalized or formal concepts, and lack standardization. They cannot be easily used to perform empirical studies. Therefore, scale development process for these nine or even more variables will be compiled in a future study.

Second, the purpose of this long-term research is to evaluate user's perceiving level for the persuasive design characteristics and check out the causal relationship between car user's behavior change and car app's persuasive design features. Therefore, a quasi-experiment was developed to estimate user's behavior change, to find out user's perception level for the persuasive design characteristics of car apps in different behavior change group, and to confirm the causal relationship between persuasive design characteristics and behavior change. Experimental data has been collected and we now focusing on data analysis and the evaluation process. The result will be reported in the near future.

8 Conclusion

This paper reports the result from an empirical evaluation of persuasive design principles in car apps. First, the evaluation revealed that six design principles of self-monitoring, reduction, suggestion, trustworthiness, real-world feel, and expertise are universal, while reminder is implemented in the majority of car apps and two principles of personalization and verifiability are found in less than half of car apps. Second, the inter-rater reliability showed that Fleiss' kappa = 0.782, which means the strength of the agreement is "good" or "substantial."

Current car apps do not have some persuasive design features, such as praise and rewards, which are highly recommended by famous app designers or researchers. For example, popular game apps or some branded apps including super apps (such as Wechat) have these two features of rewards and praise. Unfortunately, these two persuasive design features are not found in current car apps. Car manufactures and app developers can take the results from this study as a reference to design more persuasive car apps.

Appendix 1

A brief description of Evaluation Guideline for Persuasive Design Principles

Guideline	Implementation	Case: Modu parking
Self-monitoring	Mobile app can eliminate the tedium of tracking performance or status and help users track their behavior, status, task schedule, or performance	Tracks the parking place and routines
Reduction	Mobile app can increase user's perceived motivation and be more persuasive by making a behavior easier to perform or reducing complex behavior to simple tasks or a few simple steps	Reduces the complexity of finding parking place and makes it easier to find out the parking price
Personalization	Mobile app can offer personalized or tailored content and services so that users can perceive more credible and trustworthy	Inputs phone number, car number, credit card number, email; Uses environment setting, reminding setting
Virtual rehearsal	Mobile app can offer a simulated environment in which users can practice a target behavior or rehearse an actual behavior to change their attitudes or behavior in real world	N/A
Tunneling	Mobile app can set up a "tunnel" to provide a process or a step-by-step systematic approach for users to proceed a behavior along the way. It can weaken user's self-determination level and provide opportunities for persuasion	A searching results filter is designed step by step. Users should select every condition to screen parking lot results along a way; Introduces a new parking lot and users should follow a process to input the information
Simulation	Mobile app can offer simulated cause-and-effect scenarios, to allow users to observe the causality of their behavior and persuade them to change their behaviors or attitudes	N/A

(*Continued*)

(Continued)

Guideline	Implementation	Case: Modu parking
Tailoring	Mobile app can offer tailored information to match individual's needs, personality, interests, or usage context and it can be more persuasive to help users find appropriate information	A searching results filter can be selected by user's preference
Reminder	Mobile app can provides automatized, event-triggered, or customizable reminders via e-mail, SMS, screen prompt, and other methods to remind users of their target behavior and help them achieve their goals	Notices for events or drive information; Uses push messages to remind users about latest news and latest update
Praise	Mobile app can provides positive feedback based on user's behavior by using praise via words, images, symbols, cartoons, animations, or sounds	Not found
Suggestion	Mobile app offers suggestions, coaching message, help tips, using guide, or recommended options during the using process	Uses guide, question emails, or Kakao talk to help users
Rewards	Mobile app provides virtual rewards to increase the frequency and activeness of user's target behavior performing via sounds, game items, bonus points, or other virtual prize	Coupon can be downloaded; Bonus points can be used as money
Similarity	Mobile app imitates users in personality, preferences, language habits, interests, or affiliation	Not found
Social role	Mobile app offers an avatar playing the role of a specialist	Not found
Liking	Not evaluated	
Trustworthiness	Mobile app provides truthful, fair, and unbiased information and complies with laws, design regulations, privacy policies, or social expectations	Declares privacy policies and user agreements, users can resign membership freely

(Continued)

(Continued)

Guideline	Implementation	Case: Modu parking
Real-world feel	Mobile app provides a direct experience of daily routines and making the impact on everyday life clearly via decrease imagination or suspension of disbelief	Uses virtual map of Korea and timeliness parking information
Expertise	Mobile app provides expert knowledge, specialized experience, or professional competence and offers timeliness, efficient, and frequently updated sources	Uses push message to remind users about latest news and latest update. Provides professional information about parking lots and charging standards
Verifiability	Mobile app can be more persuasive and its credibility perceptions can be enhanced by providing means to verify the accuracy of its content via Tel, email, website or other outside sources	Uses email, telephone, Kakao Talk to confirm the content
Authority	Mobile app leverages roles of authority or some official, authoritative, and reputable information	Not found
Third party endorsements	Mobile app shows certified, well-known, or respected sources as its endorsements via Certification logos, high usability, or other industry standards	Not found
Surface credibility	Not evaluated	
Social comparison sharing	Mobile app provides comparison information of other users to determine personal attitudes, behaviors, or a possible way forward	Not found
Cooperation (intrinsic motivation)	Mobile app provides a cooperation platform and leverages human nature to cooperate via group working, pressing need, or win-win cognition	N/A

(Continued)

(*Continued*)

Guideline	Implementation	Case: Modu parking
Normative influence	Mobile app leverages conformity or belongingness to match personal attitudes, behaviors, and expectations with groups via adopting or avoiding a behavior, or gathering users with common goal	N/A
Social facilitation	Mobile app uses connected technology to provide a virtual social group in which users can realize and observe other user's participating, virtual presence, or even aware of being observed by other users	N/A
Competition	Mobile app provides a competition platform and leverages human nature to compete via racing game, outcome comparison, and prize or non-prize activities	N/A
Recognition	Mobile app leverages the motivating power of recognition via awards, reputation logos, donate stickers, citations, ranking list, or other incentive programs	Not found
Social learning	Mobile app offers an observing platform especially a rewarded observing platform to learn new attitudes and behaviors from others	N/A

Appendix 2

An example of evaluation results about car app's persuasive design principles

An example of evaluation results about car app's persuasive design principles

Car App Name	Mudu Parking				Car Butler				Autoist Diary			
Evaluator	Raters				Raters				Raters			
Evaluator Id	1	2	3	4	1	2	3	4	1	2	3	4
Self Monitoring	Y	Y	Y	Y	Y	Y	Y	Y	Y	Y	Y	Y
Reduction	Y	Y	Y	Y	Y	Y	Y	Y	Y	Y	Y	Y
Personalization	Y	Y	Y	Y	Y	Y	Y	Y	Y	Y	Y	Y
Rehearsal	N	N	N	N	N	N	N	N	N	N	N	N
Tunneling	Y	N	N	Y	N	Y	N	Y	Y	N	Y	N
Simulation	N	N	N	N	N	N	N	N	N	N	N	N
Tailoring	Y	N	Y	N	N	Y	N	N	N	Y	N	Y
Reminder	Y	Y	Y	Y	N	N	Y	N	Y	Y	Y	Y
Praise	N	N	N	N	N	Y	Y	N	N	N	N	N
Suggestion	Y	Y	Y	Y	Y	Y	N	Y	Y	Y	Y	Y
Rewards	Y	Y	Y	Y	N	N	N	N	N	N	N	N
Similarity	N	N	N	N	N	N	N	N	N	N	N	N
Social Role	N	N	N	N	N	N	N	N	N	N	N	N
Liking	Not Evaluated											
Trustworthiness	Y	Y	Y	Y	Y	Y	Y	Y	Y	Y	Y	Y
Real-World Feel	Y	Y	Y	Y	Y	Y	Y	Y	Y	Y	Y	Y
Expertise	Y	Y	Y	Y	Y	Y	N	Y	Y	Y	Y	Y
Verifiability	Y	Y	Y	Y	N	Y	N	Y	Y	Y	Y	Y
Authority	N	N	Y	N	N	Y	N	N	N	N	N	N
Third Party Endorsements	N	N	N	N	N	Y	N	N	N	N	N	N
Surface Credibility	Not Evaluated											
Social Comparison Sharing	N	N	N	N	N	N	N	N	N	N	N	N
Cooperation	N	N	N	N	N	N	N	N	N	N	N	N
Normative Influence	N	N	N	N	N	N	N	N	N	N	N	N
Social Facilitation	N	N	N	N	N	N	N	N	N	N	N	N
Competition	N	N	N	N	N	N	N	N	N	N	N	N
Recognition	N	N	N	N	N	N	N	N	N	N	N	N
Social Learning	N	N	N	N	N	N	N	N	N	N	N	N
Self-Monitoring	N	N	N	N	N	N	N	N	N	N	N	N

References

1. Altman, D.G.: Practical Statistics for Medical Research. Chapman and Hall, London (1991)
2. Fleiss, J.L.: Statistical Methods for Rates and Proportions, 2nd edn. Wiley, New York (1981). ISBN: 0-471-26370-2 (1981)
3. Fogg, B.J.: Persuasive Technology: Using Computers to Change What We Think and Do. Morgan Kaufmann Publishers, San Francisco (2003)

4. Oinas-Kukkonen, H.: Requirements for measuring the success of persuasive technology applications. In: MB 2010, Eindhoven, the Netherlands, 24–27 August 2010
5. Oinas-Kukkonen, H., Harjumaa, M.: Towards deeper understanding of persuasion in software and information systems. In: Proceedings of the First International Conference on Advances in Human-Computer Interaction (ACHI 2008), pp. 200–205. Electronic Publication (2008). ISBN: 978-0-7695-3086-4
6. Oinas-Kukkonen, H., Harjumaa, M.: Persuasive systems design: key issues, process model, and system features. Commun. Assoc. Inf. Syst. **24**(1), Article 28 (2009)
7. Salz, P.A.: The changing economics of app development. Harvard Bus. Rev. (2015)
8. Schuermans, S., Vakulenko, M., Voscoglou, C.: Developer Megatrends H1 2015: Five Key Trends in the Developer Economy, Vision Mobile Report (2015). http://www.developer economics.com/reports/
9. Statistica: Number of apps available in leading app stores as of July 2015 (2015). http://www. statista.com/statistics/276623/number-of-apps-available-in-leading-app-stores/

TRaining AssigNment Service (TRANS) to Meet Organization Level Skill Need

Atul Singh[(✉)], Rajasubramaniam T., Gurulingesh Raravi, Koyel Mukherjee, Partha Dutta, and Koustuv Dasgupta

Xerox Research Centre India, Bengaluru, India
atul.singh@xerox.com

Abstract. The need for training employees in new skills in an organization generally arises due to the changing skill requirements coming from the introduction of new products, technology and customers. Efficient assignment of employees to trainings so that the overall training cost is minimized while considering the career goals of employees is a challenging problem and to the best of our knowledge there is no existing work in literature that solves this problem. This paper presents TRaining AssigNment Service (TRANS) that minimizes an organization's overall training costs while assigning employees to trainings that match their learning ability and career goals. TRANS uses an ORGanization and Skills ontology (ORGS) to calculate the cost for training each available employee for a potential role taking into account constructivist learning theory. TRANS uses TRaining assIgnMent algorithm (TRIM), based on Hungarian method for bipartite matching, for assigning employees to trainings. In our experiments with real-world data, proposed allocation algorithm performs better than the existing strategy of the organization.

Keywords: Applications · Training management system · Learning theory · Organization and skills ontology · Training cost optimization

1 Introduction

Organization level training need arise due to reasons such as a change in the organization's corporate plan, and the introduction of new products, process, technology etc. Unfulfillment of organization level training needs may lead to problems such as underutilization of resources, delay and escalation in costs of new projects due to unavailability of resources and a decrease in the employee morale. Training management systems are 'software packages that organize, deliver and track training through a central interface over an Intranet or the Internet' and are typically bundled as a part of talent management suites [1]. In the existing training management systems, the assignment of employees to trainings is manual and does not take into account an employee's learning abilities, career goals, and the organization level training needs. This paper presents TRAining AssigNment Service (TRANS) that can be integrated with an organization's

© Springer International Publishing Switzerland 2016
W. Abramowicz et al. (Eds.): BIS 2016, LNBIP 255, pp. 411–423, 2016.
DOI: 10.1007/978-3-319-39426-8_32

training management system to recommend training to employees. The training recommendations minimize the overall training costs to meet organization level training needs while allowing employees to acquire skills that allow them to meet their career goals. The key component of the solution is a TRaning assIgnMent (TRIM) algorithm based on Hungarian method [2], a well-known algorithm for bipartite matching. TRIM uses an ORGanization and Skills ontology (ORGS), introduced in this work, to determine the relevant trainings and to calculate the total training cost to move an available employee to a potential role.

Learning theories provide a set of principles that can be used to understand the learning process. There are multiple learning theories [3,4] and TRIM uses a novel approach based on the theoretical foundation of the constructivist learning theory to determine the relevant trainings and to calculate the total training cost to move an available employee to a potential role. TRIM uses employee's existing skills to determine a sequence of relevant trainings so that the employee can incrementally build the skills required for a potential role. The total training cost for transitioning an available employee to a potential role is adjusted to take into account trainings that help an employee to reach their career goals.

The rest of the paper is organized as follows: Sect. 2 presents an overview of TRANS. Section 3 presents the ORGS ontology and the approach used for estimating the cost of training an employee. Section 4 presents TRIM that optimizes organization's cost of assigning employees to trainings. Section 5 presents the evaluations of the training assignment algorithm on both synthetic and real-world organization data. Section 6 presents the related work and concludes.

2   TRaining AssigNment Service (TRANS)

TRANS service takes as its input an employee id and returns the trainings recommended for the employee. TRANS system architecture (shown in Fig. 1) includes the following key components: Instantiate Ontology, Skill Gap Matrix, Cost Matrix, and TRIM. The steps below describe the interaction of these components for allocating employees to trainings to fulfil organization level training needs as well as to help employees to move closer to their potential target roles.

Step 1. Instantiate ORGanization and Skills (ORGS) ontology. Instantiate Ontology component uses the information specific to the organization (such as employees, their skills, roles, etc.), to create an instantiation of the ORGS ontology (presented in Sect. 3.1).

Step 2. Determine the skill gap matrix. A cell of the ith row and the jth column in the skill gap matrix TM indicates the trainings that an employee e_i needs to complete to take up role r_j. Section 3.2 presents the algorithm used by Skill Gap Matrix component for computing TM.

Step 3. Compute the cost matrix. With the help of the skill gap matrix, we then compute the cost matrix TC. A cell of the ith row and the jth column in the cost matrix TC indicates the cost incurred by the organization if an employee e_i is assigned to role r_j which require trainings of certain skills that are listed in

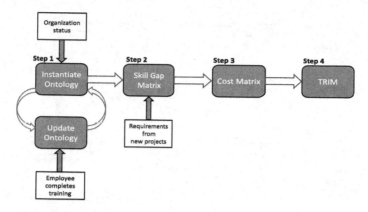

Fig. 1. TRANS architecture uses TRIM and ORGS for assigning employees to trainings to meet organization level training needs

the corresponding cell of the skill gap matrix. Cost Matrix component uses the approach presented in Sect. 3.3 for computing TC.

Step 4. Assigning trainings to employees. Using the cost matrix computed in Step 3, we optimally allocate employees to trainings with the objective of minimizing the overall cost for the organization. For doing this, we propose a polynomial time-complexity algorithm, namely TRIM, based on the Hungarian method [2]. With the way we compute the cost, minimizing the overall cost also ensures that the target roles of employees (i.e., their career goals) are considered as well while performing the allocation. The algorithm is discussed in Sect. 4.

3   Training Cost Estimation

In this section we present an approach for estimating the overall cost by an organization for training each employee in the skills required for each potential role. Section 3.1 presents the ORGS ontology that is used to estimate employee's skill gap for a required role as well as for a target role. Section 3.2 presents an algorithm for generating a training sequence for each employee using ORGS ontology. Section 3.3 presents the approach used to estimate the overall cost for training.

3.1   ORGanization and Skills (ORGS) Ontology

Figure 2 presents a visual representation of key classes and relationships in ORGS. ORGS reuses the concepts of organization, employee, and roles from the World Wide Web Consortium (W3C) recommended organization ontology [5]. In the W3C recommended organization ontology, an employee is represented using a `foaf:Person` class from the Friend of a Friend (`foaf`) ontology [6]. A foaf:Person is a type of `foaf:Agent` and the employment relationship between an employee and an organization is represented using the class

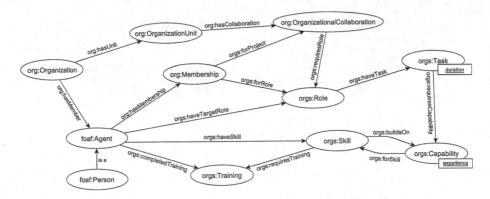

Fig. 2. Key classes of the ORGS ontology used by the TRIMS algorithm

`org:Membership` which captures the n-ary relationship between a `foaf:Agent` and an `org:Role`. The org:Membership class includes a `time:Interval` during which an employee will be associated with the org:Role. A department is represented by class `org:OrganizationalCollaboration` and a project is represented by class `org:OrganizationalUnit`.

ORGS extends the org:role class by associating it with a collection of tasks which an employee (represented by foaf:agent class) assigned to an org:role is expected to perform. The `Task` class includes a `Duration` data property which captures the time for which an employee is asked to perform the task. To perform a task an employee requires skills represented by the `Skill` class. ORGS captures the n-ary relationship between a task, the skill and the skill level required to perform the task, and the required experience in the skill through the `Capability` class. The `Experience` in a skill is a data property of the Capability class. An employee represented by a foaf:agent class has capabilities. TRIM recommends trainings on the basis of an employee's aspiration of a target role and ORGS introduces an object property `hasTargetRole` that captures the relationship between an employee and her target role.

3.2 Training Sequence Generation

This section describes the approach used for populating skill gap matrix TM introduced in Sect. 2. Let $S(r_j)$ be the set of required skills for a role $r_j \in Role$ and $S(e_i)$ be the set of skills of an employee $e \in Agent$ then:

$$S(r_j) \equiv \{s \mid \forall_{c \in \text{Capability}, t \in \text{Task}} \exists \text{haveTasks}(r_j, t) \cap \text{requireCapability}(t, c) \cap \text{haveSkill}(c, s)\}$$
$$S(e_i) = \{s \mid \forall_{c \in \text{Capability}} \exists \text{haveCapabilities}(e_i, c) \cap \text{haveSkill}(c, s)\}$$

Let r_c be an employee's existing role and r_j be a potential role to which an employee e_i can be assigned. Then $\Delta S(r_j, e_i)$ be the skill gap that the employee needs to fulfil for moving from current role r_c to a potential role r_j is defined as:

$$\forall r_j \in \text{Role}, e_i \in \text{Agent}, \ \Delta S(r_j, e_i) \equiv S(r_j) - S(e_i)$$

Algorithm 1. Algorithm findIntermediateSkills to determine intermediate skills required by an employee to learn a new skill

---

   Input   : S(E): set of existing skills of an employee E.
   Input   : NS: new skill to be learned.
   Output: LP: Set of intermediate skills to learn new skill NS
1 LP = NULL ;
2 **foreach** *Skill S in S(E)* **do**
3   find SP the set of nodes on the shortest path between S and NS ;
4   **if** *SP == NULL* **then** continue; ;
5   **if** *LP == NULL or size(SP) less than size(LP)* **then** LP = SP ;
6 **end**
7 return LP

As shown in the ORGS ontology the skills have dependencies, which means that an employee who wants to learn a new skill for a potential role has to learn intermediate skills on which the new skills depend. Let us look at an example. Alice is a UNIX systems programmer in an Information Technology (IT) services company and aspires to pick up the role of a JAVA web application programmer. For her target role Alice requires new skills in JAVA programming and J2EE framework. However, to learn JAVA she needs to pick up Object Oriented Programming a skill on which JAVA programming depends. Similarly to pick up J2EE framework she needs basic networking skills which she has due to her UNIX systems programming background. Algorithm 1 presents the pseudo-code findIntermediateSkills that is used to determine the set of intermediate skills required by an employee e_i to learn a new skill. The algorithm takes as its input the set $S(e_i)$ of employee's existing skills and NS the new skill that the employee has to learn for a potential role. A shortest path algorithm is applied on the ORGS ontology (containing the skill information) to determine the set of intermediate skills that an employee is required to obtain a new skill NS.

Let $S_{intermediate}$ be the set of skills that an employee e_i is required to learn to obtain the skills from the skill gap $\Delta S(r_j, e_i)$ to move from the current role r_c to a potential role r_j:

$$S_{intermediate}(r_j, e_i) \equiv \bigcup_{s \in \Delta S(r_j, e)} \{\acute{s} \mid \text{findIntermediateSkills}(S(e_i), s)\}$$

$T(r_j, e_i)$ be the set of trainings that an employee e has to undergo to takeup a potential role r_j and is used to populate the skill gap matrix $TM(i, j)$:

$$T(r_j, e_i) \equiv \{t \mid \forall s \in \Delta S(r_j, e_i) \cup S_{intermediate}(r_j, e_i), \text{learnFrom(s, t)}\}$$

3.3 Training Cost Calculation

Cost matrix represents total cost incurred by the organization in assigning each employee $e_i \in Employee$ to each of the required role $r_j \in Role$. The total cost contains two components, cost of training and cost of convenience, and is calculated as follows:

$$TC_{ij} = \alpha \times CT_{ij} - \beta \times CC_{ij}$$

Cost of training (CT_{ij}) represents the cost to train an employee e_i to acquire the skills required for a required role r_j. CT_{ij} is the total expenses incurred by an organization in arranging and making the employee undergo the trainings from the set $T(e_i, r_j)$ described in the previous section. This cost is derived from the constructivist theory of learning according to which an employee with shorter skill gap for a required role can easily build the required knowledge by building on prior knowledge.

According to constructivist theory, motivation has a positive influence on learning. Hence, assigning employees to trainings that are required to reach their target role will not only motivate them to learn better but will also reduce the long-term training expenses of an organization. Let $r_t \in R$ be the target role of an employee $e_i \in E$ and $r_j \in R$ be a required role to which the employee can be assigned. Cost of convenience (CC_ij) is the total expenses incurred by an organization in arranging and making the employee undergo the trainings from the set $T(e_i, r_j) \cap T(e_i, r_t)$. Target roles have a time line and hence an employee who has shorter buffer time to learn a skill has to be preferred over an employee with longer buffer time. Cost of convenience captures this by multiplying each training from the set $T(e_i, r_j) \cap T(e_i, r_t)$ with a scaling factor $\frac{needExperience(r_t, t)}{\text{time line for the target role}}$, where the term in the numerator returns the experience required by the target role in the skill imparted by the training.

In Eq. 3.3, α and β are tunable weights that can be adjusted based on the organization preferences. For example, an organization that employs mostly contract workers may want to minimize its immediate cost of training and doesn't worry much about its long term cost-benefit can set $\alpha = 1$ and $\beta_C = 0$. As another example, an organization that has many full term employees may want to give equal priority to immediate cost as well as to long term cost can set $\alpha = 0.5$ and $\beta = 0.5$.

Determining this cost for each required role for each employee TC_{ij} will give the cost matrix TC.

4   TRaining AssIgnMent Algorithm (TRIM)

In this section, we present the algorithms for allocating employees to trainings given the cost matrix TC computed in Sect. 3.3. Observe that it is trivial to determine the trainings that each employee needs to go through using the skill gap matrix TM created in Sect. 3.2.

First, we formulate the problem as an Integer Linear Program (ILP) which upon solving gives the desired solution. Such a solution obtained after solving the ILP formulation is optimal in the sense that it has the least overall cost. However, solving ILP formulation may incur a very high run-time complexity, as no polynomial algorithm is known to exist for general ILP. Hence, we propose an algorithm for our problem based on the Hungarian method [2] for bipartite matching. The algorithm is optimal and has polynomial time complexity. Though ILP would suffice most real-life scenarios where managers can afford the needed computation time to get the optimal solution, polynomial time heuristic algorithm will be of great help to perform what-if analysis on multiple strategies of the organization in real-time.

4.1   ILP Formulation of the Problem

In this section, we show how the problem of assigning employees to roles (which in turn can be mapped to required trainings) can be formulated as an Integer Linear Program (ILP). The formulation for allocating employees to roles is shown in Fig. 3. The formulation is an ILP on x_{ij} variables where $i \in \{1, 2, \ldots, n\}$ is an index of employees and $j \in \{1, 2, \ldots, m\}$ is an index of potential roles. Each variable x_{ij} indicates whether an employee $e_i \in \mathcal{E}$ is assigned to a role $r_j \in \mathcal{R}$ ($x_{ij} = 1$) or not ($x_{ij} = 0$). The objective of the ILP formulation is to minimize the overall cost of the training — the first component is the cost of training considering the two types of costs described earlier in the paper (i.e., training cost and convenience cost) and the second component is the cost/penalty in failing to meet demand for roles, if any; γ indicates the penalty for not meeting one role. The first set of constraints specifies that each employee can be assigned to at most one role. The second set of constraints specifies that the number of employees assigned to a role should not exceed the demand for that role (denoted by D_j). The third set of constraints specifies that each employee must be integrally allocated to a role, if at all.

Trainings to employees can be assigned using the solution obtained by solving the ILP formulation as follows. If $x_{ij} = 1$ then it implies that employee e_i needs to be allocated role r_j. From this information, we can extract the set of trainings that employee e_i needs to complete to take-up the assigned role r_j from the training matrix — specifically, trainings listed in the cell of the ith row and the jth column of the training matrix. The overall cost of the solution is obtained by $\left(\sum x_{ij} \times TC_{ij} \right) + \left(\sum_{j \in \mathcal{R}} \left(D_j - \sum_{i \in \mathcal{E}} x_{ij} \right) \right) \times \gamma$.

Minimize $\left(\sum x_{ij} \times TC_{ij} \right) + \left(\sum_{j \in \mathcal{R}} \left(D_j - \sum_{i \in \mathcal{E}} x_{ij} \right) \right) \times \gamma$ subject to:

I1.	$\sum_{j \in \mathcal{R}} x_{ij} \leq 1$	$(i = 1, 2, \ldots, n)$
I2.	$\sum_{i \in \mathcal{E}} x_{ij} \leq D_j$	$(j = 1, 2, \ldots, m)$
I3.	$x_{ij} \in \{0, 1\}$	$(i = 1, 2, \ldots, n)$
		$(j = 1, 2, \ldots, m)$

Fig. 3. ILP formulation*— for allocating employees in E to skills in S.

4.2   TRIM Algorithm

In this section, we present a polynomial time-complexity optimal algorithm for allocating employees to roles. The algorithm works as follows. First, it creates a bi-partite graph with employees as one set of nodes and roles as another set of nodes and a set of edges between these nodes are added with their weights set to the corresponding cost involved in training. Then, it determines the optimal assignment of employees to roles in polynomial time by determining the maximum weight matching in this bipartite graph using Hungarian algorithm, a well known algorithm from the literature. Now, we describe these two steps in detail.

First, algorithm creates a bipartite graph as follows. It creates $|E|$ nodes in one partition where E is the set of employees who are available to be assigned to the roles, and $|R'|$ nodes in the other partition where R is the set of roles to be filled; since there may be multiple openings for a role $j \in R$ (i.e., $D_j \geq 1$ number of openings for role j), algorithm adds D_j nodes in the second partition for each role j. So, it holds that $|R'| = \sum_{j \in R} D_j$. For each employee node i in the first partition, it then adds edges between node i and all the role nodes j in the second partition. The weight of an edge between a node i in the first partition and a node j in the second partition is set equal to $\gamma - TC_{ij}$ where γ is the penalty factor for failing to meet the demand for a role and c_{ij} is the overall cost (which includes both training cost and convenience cost) incurred by the organization in training employee i for the role j obtained by cost matrix TC.

Every employee node in the first partition can only be matched to at most one role node in the second partition (ensuring that each employee can be assigned at most one role) and every role node can be matched to at most one employee node (ensuring that every role is assigned at most D_j employees). The size of the graph created is polynomial in the input size as shown in Eq. (1). Here we assume that the maximum demand for any role does not exceed the available number of employees.

$$\text{The number of nodes in the graph: } |E| + |R'| = |E| + \sum_{j \in R} D_j$$
$$\leq |E| + |E| \times |R|$$
$$\leq |E| \times (1 + |R|)$$
$$\text{The number of edges in the graph: } |E| \times |R'| = |E| \times |E| \times |R|$$
$$= |E|^2 \times |R|$$

Then, the algorithm then uses the Hungarian method to find the maximum weight matching. By optimality of the Hungarian algorithm, the maximum weighted matching in this graph can be obtained in polynomial time. In terms of the ILP notation (of Fig. 3), the resultant weight of the employee assignments to roles maximizes $\sum_{i \in E, j \in R} TC_{ij}$. Since $\gamma \times \sum_{j \in R} D_j$ is a system parameter, the resultant assignment of employee to roles minimizes $\sum_{i \in E, j \in R} x_{ij} \times c_{ij} + \sum_{j \in R} \left(D_j - \sum_{i \in E} x_{ij} \right) \times \gamma$, while respecting the ILP constraints. Hence, this gives an optimal solution to the ILP in polynomial time.

Finally, using the assignment of employees to roles, TRIM algorithm outputs the trainings that each employee needs to go through by looking into the training matrix TM. For example, the trainings that employee i need to undergo to take up the assigned role j are listed in the ith row and the jth column of the training matrix TM — the algorithm outputs this training list.

5   Simulations

In this section, we describe the experimental evaluations of TRIM algorithm. The first set of evaluations are done using synthetic data and performance of TRIM algorithm is compared against the ILP formulation (referred to as *ILP approach* from here on) presented in Sect. 4.1. The second set of simulations are done using real-world data from an organization. The performance of TRIM algorithm is compared with a competitive approach referred as BAsic Training Assignment (BATA) algorithm currently used by that organization.

ILP is run using IBM ILOG CPLEX Optimization Studio[1]. TRIM and BATA algorithms are implemented in JAVA. The evaluations were run of a virtual machine with 64 bit OS, 2.0 GHz processor, and 16 GB RAM. In the evaluations, we set equal weights to training cost as well as convenience cost (i.e., $\alpha = \beta = 0.5$) which indicates that the organization gives equal importance to both its short-term goals as well to the long-term career goals of its employees.

5.1   Comparison of TRIM with ILP Approach Using Synthetic Data

In this set of evaluations, we compare the performance of TRIM with ILP approach in terms of (i) the running time, and (ii) the cost of the solution output by the algorithms; specifically, Cost of Training incurred per Employee (CTE). It is trivial to see that the total cost of the solution can be obtained by multiplying CTE with the number of employees allocated for training.

The synthetic data is generated as follows. Training cost matrix TC is generated using a Zipf distribution of 0.7. In this generated TC matrix, the number of required roles is 0.4 times the number of available employees.

TRIM and ILP are applied on the synthetically generated TC matrix. The number of employees is varied from 50 to 2500 and the time taken by both the approaches is recorded in seconds as shown in Fig. 4a. Observe that the time taken by the ILP approach exponentially increases as the number of employees increase, while the time taken by TRIM increases polynomially. Also, as can be seen in Fig. 4b, the cost of training per employee in the solution output by TRIM is exactly same as that of the ILP approach. This re-confirms the optimality of TRIM algorithm with respect to cost of the solution. To summarize, in our evaluations with synthetic data, TRIM exhibited same performance as ILP in terms of the cost of the solution, however, it significantly outperformed ILP approach in terms of running time.

---

[1] http://www-01.ibm.com/software/commerce/optimization/cplex-optimizer/.

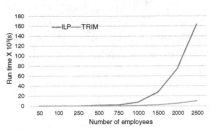
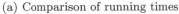
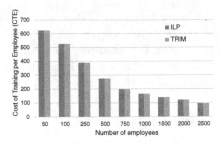

(a) Comparison of running times   (b) Comparison of cost per employee

Fig. 4. Performance of TRIM algorithm and ILP approach for synthetic data

5.2 Comparison of TRIM with BATA Using Real-World Data

In this set of evaluations, we compare the performance of TRIM with BATA algorithm currently being used by an organization using the real-world data from that organization. The comparison is based on the cost of the solution output by the algorithms; specifically, based on the Cost of Training incurred per Employee (CTE) in each solution. BATA is a two step training assignment algorithm: a) First step assigns employees to trainings required for potential roles so to minimize the cost of training alone b) Second step assigns employees to trainings required for their target roles so as to minimize the corresponding cost. As can be seen, the main difference between TRIM and BATA is that, TRIM considers both training cost and convenience cost together while performing the allocation whereas BATA does the allocation in two phases and hence fails to take into account the cost of convenience while fulfilling the required roles. This leads to incurring higher cost as shown by evaluations in this section.

The real-world data of an organization that is used in this set of evaluations contains information about 2000 available employees, 22 roles and 25 projects. Each project has roles and has multiple vacancies for each role to be filed which is referred in this paper as potential role. For these evaluations, it is assumed that each employee has a target role which is their supervisor's role. The evaluations consider that all the 2000 employees are available. Since the organization data does not have cost for each skill training, we assume cost for each training is constant, i.e., 2000 units (could be any currency).

CTE for both the solutions are measured by varying $AEtoPR$ ratio. $AEtoPR$ is the ratio of the number of available employees to the number of potential roles to be filled. $AEtoPR$ is varied from 1.0 to 3.0 and its effect on CTE is observed (see Fig. 5a). To vary $AEtoPR$, projects are selected from the data based on their projected start date till the desired $AEtoPR$ ratio is reached. To meet the desired ratio a subset of potential roles may be chosen from the last selected project. The project selection process takes care that as $AEtoPR$ increases the number of employees whose target roles aligns with the required roles also increases. As seen in Fig. 5a, as $AEtoPR$ increases CTE decreases. As mentioned earlier, this happens because as the number of available employee's increase it becomes more

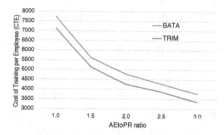
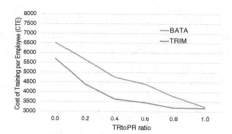
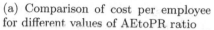

(a) Comparison of cost per employee for different values of AEtoPR ratio

(b) Comparison of cost per employee for different values of TRtoPR ratio

Fig. 5. Performance of TRIM algorithm and existing algorithm of the organization for real-world data

and more possible to find employees whose current skill set and target role aligns with the available potential roles. As can be seen from Fig. 5a, CTE of $TRIM$ is on an average 1000 units lesser than $BATA$, which signifies that the overall cost when all the employees assigned to trainings are considered, $TRIM$ outperforms $BATA$ in terms of overall cost of the solution.

Further, CTE for both the solutions are measured by varying $TRtoPR$ ratio. $TRtoPR$ is the ratio of number of target roles present in the potential roles to the number of potential roles. $TRtoPR$ is varied from 0.0 to 1.0 and its effect on CTE is observed. To vary $TRtoPR$ ratio potential roles are selected from the projects and the required vacancies are varied till the required ratio is reached. Figure 5b plots the variation of CTE with respect to $TRtoPR$ for both the algorithms. As can be seen, for most of the $TRtoPR$ values, $TRIM$ performs better than $BATA$. Also, it can be seen that, as the number of target roles present in the potential role increase, the gap in CTE of $TRIM$ and $BATA$ reduces. As $TRtoPR$ ratio approaches 1.0, there is a high chance that every employee gets a potential role that is actually the target role, hence the gap between $TRIM$ and $BATA$ reduces to almost 0. To summarize, evaluations with real-world data show that $TRIM$ always provides a cost saving while assigning trainings when compared to $BATA$.

6   Related Work

The literature contains multiple works that focus on using ontologies for automating skill management activities in an organization. Kunzman et al. present Professional Learning Ontology [7] and a reference model for integrating human resource processes with ontology catalogues in [8]. Fazel-Zarandi and Fox present an ontology for capturing skill and competency [9] and use it to determine an individual's skills and the skill gap for a role [10]. Authors in [11] present an ontology for competency and use it to determine skill gap of employees. An ontology for competency and a functional architecture of a web-based system that can be used for skill management activities is presented in [12]. The existing ontologies do not integrate skills, organization structure and project requirements thereby making it unusable for recommending trainings. Furthermore, ORGS enriches the concept of skill with experience that has not been covered in earlier ontologies and is a crucial aspect of skill.

The existing work on skill management in an organization using ontologies discussed in the previous paragraph focus on identifying an individual's skill gap. An extensive review of systems that recommend learning resources to students in the education domain is presented in [13]. Shen and Shen [14] present a semantic recommender that use an ontology of concepts and a student's competency gaps to recommend learning resources. However, they do not focus on using the ontology in an organization to suggest learning paths that can be used by employees to reach their target goals. A recent work [15] suggests using ontologies to guide employees through their career goals in an organization. The work suggests that ontologies can be used to recommend competencies to employees but does not present any algorithms for the same. This work minimizes an organization's costs while assigning training to employees while taking into account their career goals and learning abilities. To the best of our knowledge there are no algorithms in literature that does such a training assignment for an organization.

7   Conclusion

Cost efficient allocation of employees to trainings to meet the organization level skill requirements while considering the career goals of employees is a challenging problem. For this problem, we presented a service TRANS that can be integrated with an existing training management system. TRANS uses an optimal polynomial time-complexity algorithm, TRIM, based on Hungarian method. Further, in the process of designing the algorithm, we also presented ORGS ontology that is used to calculate the cost for training each available employee for a potential role taking into account constructivist learning theory. To the best of our knowledge, no such solution exists in the literature and hence this is the first one. In our experiments with real-world data from an organization, we observed that our algorithm performs better than the existing strategy of the organization.

References

1. Piskurich, G.M., Beckschi, P., Hall, B.: The ASTD Handbook of Training Design and Delivery: A Comprehensive Guide to Creating and Delivering Training Programs, Instructor-led, computer-Based, or Self-Directed. McGraw-Hill, New York (2000)
2. Kuhn, H.W.: The hungarian method for the assignment problem. Nav. Res. Logistics Q. **2**(1–2), 83–97 (1955)
3. Olson, M.H., Hergenhahn, B.R.: An Introduction to Theories of Learning. Pearson/Prentice Hall, Upper Saddle River (2009)
4. Schunk, D.H.: Learning theories. Printice Hall Inc., New Jersey (1996)
5. W3C: The organization ontology (2014)

6. Brickley, D., Miller, L.: Foaf vocabulary specification 0.98. Namespace Document 9 (2012)
7. Schmidt, A., Kunzmann, C.: Towards a human resource development ontology for combining competence management and technology-enhanced workplace learning. In: On the Move to Meaningful Internet Systems: OTM Workshops (2006)
8. Schmidt, A., Kunzmann, C.: Sustainable competency-oriented human resource development with ontology-based competency catalogs. In: eChallenges (2007)
9. Fazel-Zarandi, M., Fox, M.S.: An ontology for skill and competency management. In: FOIS (2012)
10. Fazel-Zarandi, M., Fox, M.S.: Reasoning about skills and competencies. In: Camarinha-Matos, L.M., Boucher, X., Afsarmanesh, H. (eds.) PRO-VE 2010. IFIP AICT, vol. 336, pp. 372–379. Springer, Heidelberg (2010)
11. Sicilia, M.A.: Ontology-based competency management: infrastructures for the knowledge intensive learning organization. Europe 17 (2014)
12. Draganidis, F., Chamopoulou, P., Mentzas, G.: An ontology based tool for competency management and learning paths. In: 6th International Conference on Knowledge Management (2006)
13. Manouselis, N., Drachsler, H., Vuorikari, R., Hummel, H., Koper, R.: Recommender systems in technology enhanced learning. In: Ricci, F., Rokach, L., Shapira, B., Kantor, P.B. (eds.) Recommender Systems Handbook, pp. 387–415. Springer, Heidelberg (2011)
14. Shen, L., Shen, R.-M.: Learning content recommendation service based-on simple sequencing specification. In: Liu, W., Shi, Y., Li, Q. (eds.) ICWL 2004. LNCS, vol. 3143, pp. 363–370. Springer, Heidelberg (2004)
15. Malzahn, N., Ziebarth, S., Hoppe, H.U.: Semi-automatic creation and exploitation of competence ontologies for trend aware profiling, matching and planning. Knowl. Manage. E-Learn. Int. J. 5(1), 84–103 (2013)

Portfolio of Global Futures Algorithmic Trading Strategies for Best Out-of-Sample Performance

Aistis Raudys[✉]

Faculty of Mathematics and Informatics,
Vilnius University, Didlaukio 47, 08303 Vilnius, Lithuania
aistis.raudys@mif.vu.lt

Abstract. We investigate two different portfolio construction methods for two different sets of algorithmic trading strategies that trade global futures. The problem becomes complex if we consider the out-of-sample performance. The *Comgen* method blindly optimizes the Sharpe ratio, and *Comsha* does the same but gives priority to strategies that individually have the better Sharpe ratio. It has been shown in the past that high Sharpe ratio strategies tend to perform better in out-of-sample periods. As the benchmark method, we use an equally weighted (1/N, naïve) portfolio. The analysis is performed on two years of out-of-sample data using a walk forward approach in 24 independent periods. We use the mean reversion and trend following datasets consisting of 22,702 and 36,466 trading models (time series), respectively. We conclude that *Comsha* produces better results with trend-following methods, and *Comsha* performs the same as *Comgen* with other type of strategies.

Keywords: Portfolio construction · Algorithmic trading · Sharpe ratio · Optimization · *Comgen* · *Comsha*

1 Introduction

When a computer makes a decision to buy or sell an asset and executes the trade automatically, it is called algorithmic trading. This type of trading comes in many names and flavours and is known as algorithmic trading, automated trading, robot trading, bot trading, trading robots, program trading, mechanical trading, systematic trading, high frequency trading, low latency trading, ultra low latency trading, black-box trading, trading models and quant trading. Some of the names may refer to one specific aspect or type of trading, but they all are very similar – a human creates a computer program that later analyses the data automatically and makes a decision to buy or sell by itself [1]. The order is then sent to the market and executed with the anticipation of a profit. Automated trading firms have been known for a while now and are famous for trend following and more recently for high frequency trading. Portfolio construction for automated trading systems (ATS) is a very new research topic and typically involves a huge number of ATS. Very few research articles have been published in this field, most likely due to secrecy.

W. Abramowicz et al. (Eds.): BIS 2016, LNBIP 255, pp. 424–435, 2016.
DOI: 10.1007/978-3-319-39426-8_33

The important issue in financial engineering is to find the best portfolio optimisation method. These methods were first introduced more than half a century ago [2]. A number of surveys and research articles have subsequently appeared, based on a proposed portfolio optimisation idea [3–6].

In portfolio optimisation we construct a multidimensional weight vector $w = (w_1, w_2, ..., w_p)$, that establishes optimal proportions of assets $\mathbf{X} = (x_1, x_2, x_3, ... x_p)$ for the investment. Usually, profit or losses, x_{ij} ($i = 1...p, j = 1...n$) are used, where p is the number of assets in the asset universe and n is the number of days in each asset.

In [2], a standard method for achieving the best result for the investment is to calculate the ratio between the mean and standard deviation (SD). The mean and SD ratio have two important and useful features: mean represents the profit and SD represents the risk. The ratio of these two measures is generally known as the information ratio or Sharpe ratio [7]: $S(x_i) = $ mean $(x_i)/SD(x_i)$. In the original formula it contains risk free rate but for simplicity, we assume the risk free rate to be zero. Portfolio can be described as follows:

$$x_P = \sum_{j=1}^{p} w_j x_j = \mathbf{X} \times \mathbf{w}^T \tag{1}$$

Some authors have proposed supervised learning decision systems that modify portfolio weights to reflect various changeable factors [8, 9].

It can be concluded that, for successful portfolio creation, there are three main areas: data sample size, input dimensionality relationships and performance [10].

Another aspect is that few research papers deal with thousands of candidates in the portfolio construction as there are number of issues. First, covariance matrix can be invalid if number of assets is larger than number of days. Second, computational intensity grows exponentially and for thousands of asses it becomes virtually impossible to construct portfolio using typical Markowitz mean variance method. Authors in [14] proposed heuristic method that is computationally capable of construction portfolios for thousands of assets. Another attempt to speed up computations was presented in [13] where authors analyze and compare the runtime of these algorithms on a set of benchmark problems and demonstrate the most sophisticated version is several orders of magnitude faster than the standard implementation. In [11] authors use exceptionally fast portfolio optimizer that uses an active set method, which was enhanced by using penalty function methodology to gain dramatic increases in speed to reach a true global optimal solution for large, real world portfolio optimization problems with thousands of assets.

In addition to problems described above, numbers of researches confirm [16] that optimal portfolios do not hold if measured in out-of-sample basis. In [12] authors show that it is indeed difficult to avoid backtest overfitting. Any perseverant researcher will always be able to find a backtest with a desired Sharpe ratio regardless of the sample length requested. Thus, evaluation must be performed in out-of-sample and walk forward fashion.

2 Problem Formalisation

Systematic trading firm provided us daily series of profit and loss (PNL) generated by ATS. Each series correspond to simulated run of ATS. Simulations are very realistic and include adequate slippage and commissions.

2.1 Classic Portfolio vs. Algorithmic Portfolio

Portfolios constructed from ATS are different from classic portfolios constructed from assets. In classic case portfolio is constructed from set of assets such us stocks, bonds, currencies, commodities, etc. and for each asset weight is assigned. The first difference between ATS and classic portfolio is size. In classical approach number of assets is typically smaller i.e. less than 200. In AT case you can create thousands of different versions of ATS. This is not always correct as for example for stocks, as there are tens of thousands of stocks.

In the algorithmic trading it is quite often that you have hundred thousands of algorithms and their variations to choose from. Also quite often weights are just binary or discrete. Algorithm trades one lot or one futures contract. One cannot use Markowitz portfolio construction method as number of assets is typically much smaller than number of days (data points). There are some methods that can solve large scale portfolio optimisation problem. One may use some heuristic optimization methods such as *Comgen* [14, 15] that uses greedy optimization technique and adds algorithms to the portfolio one by one maximizing Sharpe ratio.

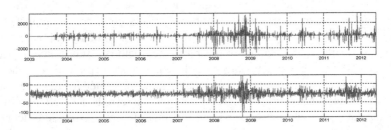

Fig. 1. Daily PNL of the typical ATS (top) time series in our $n = 3195$ sized dataset and asset (E-mini S&P 500) traded (bottom) by that ATS.

The second difference between the classic and ATS portfolios is in the daily PNL series characteristics. Not all trading models constantly hold position in the market. Some models trade once a week. This makes ATS time series quite sparse and even more difficult to construct correct covariance matrix and portfolio. Asset and ATP on the same asset example is illustrated in Fig. 1. In upper graph we can see periods of not trading as a flat line.

2.2   Overfitting and Out-of-Sample Performance

One big problem in algorithmic trading is overfitting. In financial literature and slogan this is also called curve fitting, bias, data mining, etc. This is a problem where your algorithm during optimisation adapts to random data fluctuations rather than general pattern in the data. It occurs due to numerous factors including: over-optimisation, small data set, small number of trades, complex models, number of model parameters. This phenomenon is known but not researched enough especially in algorithmic trading environments. The problem is that out-of-sample results are typically much worse than in-sample. See illustrative Figs. 2 and 3.

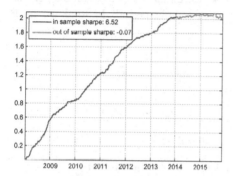

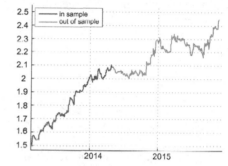

Fig. 2. Out-of-sample performance (from 2014-01-01) of the portfolio is noticeably worse than in-sample (till 2014-01-01).

Fig. 3. Out-of-sample portfolio performance from 2014-01-01 on the different set of trading methods. We note that out-of-sample performance is worse but positive.

Evaluation of the results on the out-of-sample dataset is a big problem in scientific literature. Quite often researches do not bother to test their portfolios/strategies out-of-sample. Results obtained are good in-sample but out-of-sample performance is poor. This is acute problem and some authors already reported it [17]. To avoid this problem we create a portfolio on one set of data and verified it on the unseen future data. We repeated this procedure several times to make sure winning method systematically produces better results.

2.3   Walk Forward Optimisation

We can distinguish 3 types of portfolio evaluation. First, we can create a model, calibrate model parameters and test performance and quality on the same data set. This type is highly criticised because more complex models can adapt to the training data and demonstrate superior result. Second, we calibrate the model on one set and validate it on the future unseen testing data. It is criticised as well as having one good out-of-sample result can be a matter of luck. This approach is very common but limits us to a small amount of unseen data. Good improvement is to use cross validation.

Take random subsets from the original data and repeat experiment multiple times. The last method is walk forward approach. Here we train the model on one set and test it on the small period of future data. Next, we shift the training period by period x and we shift the testing period by the same period x. We repeat this procedure until there is no data to shift our training and testing data periods. This is illustrated in Fig. 4. Walk forward analysis is gaining more and more popularity. It is time consuming process but allows one to view potential results in out-of-sample with longer time periods.

In our study we organised data into k month time intervals z_i of 21 working days (months). Initially we create a portfolio using $z_1 \ldots z_m$ intervals and test it on z_{m+1} interval. In the next step we create a portfolio using $z_1 \ldots z_{m+1}$ range and test on z_{m+2}. This process is repeated until we reach z_{m+k}. So we have totally k out-of-sample periods that we can concatenate and get one long out-of-sample period. In total we have $m + k$ periods.

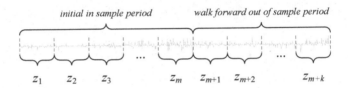

initial in sample period   *walk forward out of sample period*

z_1 z_2 z_3 z_m z_{m+1} z_{m+2} z_{m+k}

Fig. 4. Walk forward testing, z_{m+1} is first out-of-sample period.

There are several modifications of this approach. Some use all available history to calibrate parameters, some use fixed size recent history window. For example to test on z_{p+1} one uses $z_1 \ldots z_p$ data for training, but to test on z_{p+2} one uses $z_2 \ldots z_{p+1}$ data for training. We used all available history, so to test on z_{p+2} we used $z_1 \ldots z_{p+1}$ data for optimisation. Totally we used 2 years out-of-sample data, $(12 \times 2 = 24$ months$)$.

2.4   Data Description

All strategies under investigation are traded on 46 the most liquid futures from US and European exchanges and include: indexes, energy, metals, interest rates, currencies, agriculture and others. The data range varied from 2004 to 2016 depending on the instrument. Some futures started trading in 2006 or later and available data is shorter. We did not use proxy data for non existing history.

We used two types of trading models. From the systematic trading firm we received big series of daily PNL series of simulated model results. All models were optimised till 2014-01-01. Data from 2014 till 2016 was out-of-sample.

2.5   Short Term Trend Following (TF) Systems

Trend following strategies keep position as long as trend continues in the right direction. Systems measures trend strength or momentum of the market and tries to find good time to enter into a position. ATS were split into long and short sides. So strategy can either take long position and later exit and stay flat until the next long position. Short strategies will take short position on downtrend and will exit short position when tend finishes. We do not have detailed information on how ATS are implemented as this is sensitive information and was not shared by the systematic trading firm that provided the data. Typical returns are a series of small losses followed by one big profitable trade if model finds a good trend. To summarise the data we have: 36,466 trading systems, 3,094 days of history. In-sample period is from 2004-01-01 to 2014-01-01. Out-of-sample period from 2014-01-01 to 2016-01-01.

2.6   Short Term Mean Reversion (MR) Systems

This type of trading systems tries to make a profit on short term trend reversals. Positions are kept for a very short time (one or two days). The logic is opposite to trend following systems. If market moved in one direction too fast and/or too much there will be a correction and system tries to make a profit on this correction. Exact logic is not known as trading firm did not shared this information with us. Models are one directional long or short only. System can be short-flat or long-flat. Typical returns are a series of small wins followed by one big loss then market failed to reverse. Such types of models are less popular as they sometimes generate sharp losses, but at the same time they are less crowded. Logic is similar to market making but scale if much larger – days or sometimes weeks. Summary: 22,702 trading systems, 1,823 days of history. In-sample period from 2009-01-01 to 2014-01-01. Out-of-sample period from 2014-01-01 to 2016-01-01.

3   Portfolio Construction Methods

Unfortunately we cannot compare our method to classic mean variance optimiser as it is too slow. We used MATALB in our experiments and tried quadratic optimised (quadprog, frontcon, fminbnd) included in the MATLAB. However it becomes too slow if number of time series goes above 600. In our experiments we used 36,466 and 22,702 sized datasets. We know there are some fast optimisers but ones we found were commercial. We plan to include them in our future research.

3.1   Definitions

First we want to define some symbols used in our analysis.

O – pool of trading systems $O = \{o_1, o_2, \dots o_n\}$
P – portfolio of trading systems $P = \{p_1, p_2, \dots p_n\}$
Pi – portfolio ith element, $P_i = \sum_j^n p_{ij}$. Note that our portfolio gives equal weights to each system.
p_i – time series from the pool

p_{ij} – jth element of the time series
sharpe(P) – Sharpe of the portfolio P
sharpe(p_i) – Sharpe of the trading system p_i

3.2   Benchmark Method

Equally weighted portfolio also known as Naïve portfolio is a popular benchmark method. As the name suggests all assets/trading strategies are given equal weight in the portfolio. This method is sometimes called 1/N method as weights are all equal to 1/N, where N is the number of assets/trading strategies in the portfolio.

3.3   *Comgen* Algorithm

Comgen algorithm [14, 15] is based on greedy optimisation principle. We make series of locally optimal decisions and think that it will lead to a globally optimal (or near optimal) solution. In each step we examine all possible candidates to the portfolio and select one that increases portfolio Sharpe the most. This procedure is repeated until desired number of algorithms is selected or desired risk level is achieved. Typically we do not allow inclusion of the same model more than once but it can be changed. We present algorithm description below.

Algorithm 1. Comgen

```
select best Sharpe ratio strategy a from pool O
select a as a first portfolio P = {a}
for t=0 to Z
begin
     for all b strategies in the pool O
     begin
          add b to the portfolio P^new = {P^old b}
          measure Sharpe ratio s^new = sharpe(P^new)
     end
     select best new Sharpe ratio s^best
     add new b^best to portfolio P = {P^old b^best}
end
```

Here Z is the size of the portfolio. In all our experiments we used Z = 60.

3.4   *Comsha* Algorithm

In portfolio construction a problem of overfitting is as acute as in individual ATS development. There is an expectation that portfolio will be profitable in the future but if portfolio is overfitted, results will be poor if not loss making. The same problem is with the correlation. You try to add strategies into portfolio that are not correlated in-sample but correlation arises in out-of-sample period.

Comsha method is based on the idea that the highest Sharpe ratio strategies in-sample tend to perform better in out-of-sample. This phenomenon has been noted by several authors including [12, 18].

The essence of the algorithm is to give priority to the strategies that have the highest Sharpe ratio while also paying attention into correlation between strategies. We sort strategies by Sharpe ratio and add ones that has the highest Sharpe ratio first. We also aim to increase portfolio Sharpe. Two criteria are used in the selection process: individual Sharpe ratio and portfolio Sharpe ratio. I.e. add only if portfolio Sharpe increases by $s^{threshold}$. This number $s^{threshold}$ can vary but we selected 0.01 as optimal. It indicates minimum Sharpe ratio increase in order to include new strategy to the portfolio. Below we present algorithm pseudo code.

Algorithm 2. Comsha

```
make portfolio empty P = {}
sort strategies by Sharpe ratio in pool O
select best Sharpe strategy a and add P = {a}
for t=0 to Z
begin
      for all b strategies in descending Sharpe order in O
      begin
            try to add b to the portfolio P^new = {P, b}
            s^new = sharpe(P^new)
            if s^new - s > s^threshold
            begin
                  add b to the portfolio P = {P , b}
                  s = s^new
                  continue
            end
      end
      if no new strategies added
      begin
            break #not possible to add any new strategy
      end
end
```

---

4 Experiments

We performed all experiments using out-of-sample data. Sometimes we used walk forward approach; sometimes we used logic similar to cross validation. I.e. we take a subset O^{subset} of the original pool O, create a portfolio and measure out-of-sample performance. Next we take another subset of the original pool, create a portfolio and measure out-of-sample performance. We repeat this process 1000 times and averaged the results.

Both methods *Comsha* and *Comgen* have several parameters to select. We selected $z = 60$ strategies from the original pool. This number is sufficient to grasp majority of trading strategies coverage. We tried larger numbers but noticed no significant difference. Also we did not allow the same strategy to be included twice. Allowing this sometimes lead in multiple selection of one very good strategy. For *Comsha* algorithm we used $s^{\texttt{threshold}} = 0.01$.

4.1   Experiment 1, Walk Forward

In Figs. 5 and 6 we can see the results of walk forward experiments. *Comsha* method is showing better results for TF type of strategies and produced better results in 17 out of 23 cases. For MR type strategies the difference between *Comgen* and *Comsha* results is not so clear. Both methods perform similarly (11/24 cases). Equally weighted method was depicted in green and indicated "ew" label.

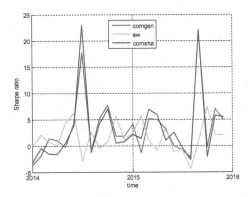
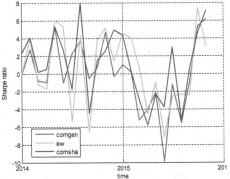

Fig. 5. Sharpe ratio in 23 out-of-sample periods on TF dataset. *Comsha* method outperforms *Comgen* in 17/23 cases (higher is better).

Fig. 6. Sharpe ratio in 24 out-of-sample periods on MR dataset. *Comsha* method outperforms *Comgen* in 11/24 cases (higher is better).

4.2   Experiment 2, *Comsha* vs. *Comgen*

In this experiment we randomly generated subsets from both datasets and tried to create 2 portfolios one with C*omgen* and one with *Comsha* algorithms. We measured Sharpe ratio in out-of-sample from 2014-01-01 till 2016-01-01 and in Fig. 7 plotted Sharpe ratio scatter diagram with *Comsha* on horizontal axis and *Comgen* on vertical axis. We repeated experiment 1000 times. In both experiments subset size was 10,000. We can see that *Comsha* algorithm is producing better portfolios in 60 % of the cases on mean reversion type strategies and 89 % of the cases on trend following strategies.

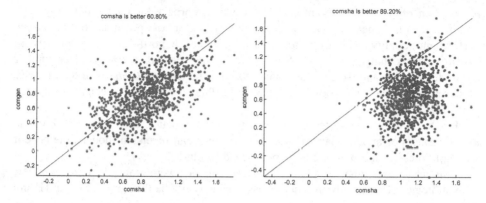

Fig. 7. Sharpe of *Comsha* method vs. Sharpe of *Comgen* method on 1000 random data subsets. On the left MR type strategies, on the right TF type strategies. The dot below the diagonal line indicates that *Comsha* produces better results. In-sample for both was from 2009-01-01 till 2014-01-01 and out-of-sample from 2014-01-01 to 2016-01-01.

4.3 Summary of the Experiments

From the Table 1 we can clearly see that newly proposed *Comsha* algorithm is better than *Comgen*. Results obtained are the same good or better. This depends on the trading systems type. It works best with trend following systems and produces results in some cases 90 % better. The statement that high Sharpe ratio strategies tend to perform better in out-of-sample is correct (especially on trend following type of strategies) and our experiments support this statement.

Table 1. Summary of the experiments. How many times *Comsha* was better than *Comgen* on 1000 random portfolio subset trials.

No.	Model	Size	Start	Comsha is better
1	TF	1000	2004	60.80 %
2	TF	10 000	2004	89.20 %
3	TF	20 000	2004	91.30 %
4	MR	10 000	2009	52.00 %
5	TF	10 000	2009	64.40 %

As we can see in all instances *Comsha* produces better results than *Comgen* algorithm.

5 Conclusions

In this paper we investigate several aspects of algorithmic trading portfolio construction: types of algorithms, portfolio construction methods and out-of-sample behaviour. We all know that by using too much optimisation very good in-sample results can be achieved. The problem arises when the final portfolio is tested with unseen data. The results can often be disappointing. We use the Sharpe ratio metric not only as a goodness/quality measure but also as an indicator of potential out-of-sample performance/quality. The *Comsha* algorithm creates a portfolio giving priority to trading models that have a high Sharpe ratio. This method is compared to another method, *Comgen,* and we can note that *Comsha* is better than *Comgen* in the majority of cases. Also, both methods are much better than a naïve, 1/N, equally weighted portfolio.

For trend following systems, *Comsha* was always better, in close to 90 % of the experiments. For another dataset that consisted of mean reversion models, the benefit was not so clear, and both methods performed similarly.

This leads us to other conclusion that TF strategies were more overfitted than MR type strategies. This is only true in our case and may not be the same for other TF and MR type strategies.

We also note that experimental results on the financial data can often be misleading due to the presence of large amounts of noise. A slight change in the configuration of the experiment can make results positive or negative. The influence of random change is high. It is very difficult to achieve consistently better results than other methods. Though difficult, it is possible for robust methods.

From the market perspective, we note that 2015 was less favourable for both algorithmic trading strategies. We performed a series of experiments, and it was difficult to achieve good results for 2015. Regardless of the portfolio construction method, the most important aspect is the profitability of the underlying models. If the trading system is not profitable in an out-of-sample period, no portfolio construction method can create good results.

Also, we note that the newly proposed *Comsha* method is very fast and produces results sometimes 20 times faster than *Comgen*. This makes the new algorithm suitable for large scale application where the number of portfolio candidates exceeds tens of thousands.

The *Comsha* and *Comgen* methods can be extended to include simple (one symbol not exceeding more than x % of the portfolio) constraints. Support of more complex constraints needs to be investigated in the future.

Acknowledgments. This work was supported by the Research Council of Lithuania under the grant MIP-100/2015. Authors also want to express their appreciation to Vilnius University.

References

1. Narang, R.K.: Inside the Black Box: A Simple Guide to Quantitative and High Frequency Trading. Wiley, New York (2013)
2. Markowitz, H.: Portfolio selection. J. Finan. **7**(1), 77–91 (1952)
3. Raudys, S., Raudys, A., Pabarskaite, Z.: Sustainable economy inspired large-scale feed-forward portfolio construction. Technol. Econ. Dev. Econ. **20**(1), 79–96 (2014)
4. McNelis, P.D.: Neural Networks in Finance: Gaining Predictive Edge in the Market. Academic Press, London (2005)

5. Ustun, O., Kasimbeyli, R.: Combined forecasts in portfolio optimization: a generalized approach. Comput. Oper. Res. **39**(4), 805–819 (2012)
6. Yamamoto, R., Ishibashi, T., Konno, H.: Portfolio optimization under transfer coefficient constraint. J. Asset Manag. **13**(1), 51–57 (2011)
7. Sharpe, W.F.: Mutual fund performance. J. Bus. **39**(1), 119–138 (1966)
8. Hung, K., Cheung, Y., Xu, L.: An extended ASLD trading system to enhance portfolio management. IEEE Trans. Neural Netw. **14**(2), 413–425 (2003)
9. Freitas, F.D., De Souza, A.F., de Almeida, A.R.: Prediction-based portfolio optimization model using neural networks. Neurocomputing **72**(10–12), 2155–2170 (2009)
10. Wang, J., Qiu, G., Cao, X.: Application of genetic algorithm based on dual mutation in the optimal portfolio selection. J. Nanchang Hangkong Univ. (Nat. Sci.) **4**, 006 (2009)
11. Jivendra, K.: Portfolio optimization using the quadratic optimization system and publicly available information on the WWW. Manag. Finan. **35**(5), 439–450 (2009)
12. Bailey, D.H., Borwein, J.M., de Prado, M.L., Zhu, Q.J.: Pseudomathematics and financial charlatanism: the effects of backtest over fitting on out-of-sample performance. Not. AMS **61**(5), 458–471 (2014)
13. Stein, M., Branke, J., Schmeck, H.: Efficient implementation of an active set algorithm for large-scale portfolio selection. Comput. Oper. Res. **35**(12), 3945–3961 (2008)
14. Raudys, A., Pabarskaite, Z.: Discrete portfolio optimisation for large scale systematic trading applications. In: 2012 5th International Conference on Biomedical Engineering and Informatics (BMEI). IEEE (2012)
15. Raudys, S., Raudys, A.: High frequency trading portfolio optimisation: integration of financial and human factors. In: 2011 11th International Conference on Intelligent Systems Design and Applications (ISDA). IEEE (2011)
16. Haley, M.R.: Shortfall minimization and the Naive (1/N) portfolio: an out-of-sample comparison. Appl. Econ. Lett. 1–4 (2015)
17. de Prado, M.L.: Recent trends in empirical finance. J. Portfolio Manag. **42**(1), 29–33 (2015)
18. Chan, E.: Algorithmic Trading: Winning Strategies and Their Rationale. Wiley, New York (2013)

Towards Federated, Semantics-Based Supply Chain Analytics

Niklas Petersen[1,2]([✉]), Christoph Lange[1,2], Sören Auer[1,2],
Marvin Frommhold[3,4], and Sebastian Tramp[4]

[1] Fraunhofer Institute for Intelligent Analysis and Information Systems (IAIS),
Sankt Augustin, Germany
[2] Enterprise Information Systems (EIS), University of Bonn, Bonn, Germany
niklas.petersen@iais.fraunhofer.de
[3] University of Leipzig, Leipzig, Germany
[4] eccenca GmbH, Leipzig, Germany

Abstract. Supply Chain Management aims at optimizing the flow of goods and services from the producer to the consumer. Closely interconnected enterprises that align their production, logistics and procurement with one another thus enjoy a competitive advantage in the market. To achieve a close alignment, an instant, robust and efficient information flow along the supply chain between and within enterprises is required. However, less efficient human communication is often used instead of automatic systems because of the great diversity of enterprise systems and models. This paper describes an approach and its implementation SCM Intelligence App, which enables the configuration of individual supply chains together with the execution of industry accepted performance metrics. Based on machine-processable supply chain data model (the SCORVoc RDF vocabulary implementing the SCOR standard) and W3C standardized protocols such as SPARQL, the approach represents an alternative to closed software systems, which lack support for inter-organizational supply chain analysis. Finally, we demonstrate the practicality of our approach using a prototypical implementation and a test scenario.

1 Introduction

In the past decades, internal enterprise information systems have experienced significant technical and scientific advancement. However, comparatively little progress has been made to improve the exchange of information *between* enterprises [6]. Until today, most of the communication between enterprises is done via informal channels, such as emails (including file attachments) or phone calls. Only tier-1 suppliers of major Original Equipment Manufacturers (OEM) are usually fully integrated into the information exchange and corresponding IT support (e.g. electronic data interchange connections) as they are expensive to deploy and maintain. Since they are the most dependent on well structured value chains, they are the ones who are urging their suppliers to share crucial information [2].

© Springer International Publishing Switzerland 2016
W. Abramowicz et al. (Eds.): BIS 2016, LNBIP 255, pp. 436–447, 2016.
DOI: 10.1007/978-3-319-39426-8_34

Informal communication is time-consuming, costly and can become inefficient when crucial information is passed on among many different people, each using their own vendor-specific business process model and Supply Chain Management software system. Automatic communication and analysis among various enterprises requires a common process model as a basis. The industry-agnostic *Supply Chain Operation Reference* (SCOR) [14], backed up by many global players (including IBM, HP and SAP), precisely aims at tackling this challenging task. By providing 201 different standardized processes and 286 metrics, it offers a well-defined basis that allows to describe supply chains within and between enterprises. The applicability of SCOR, however, is currently limited, since the standard stays on the conceptual and terminological level and major effort is required for implementing the standard in existing systems and processes.

Also, the systems implementing such a common process and data model need to provide secure interfaces to allow exchange of crucial supply chain information with their partners. However, proprietary software systems do not focus on offering flexibility by utilizing open standards [4]. While the market-dominating providers such as SAP and Oracle have recently introduced REST[1] interfaces, their main goal is not to share information with competing software systems. This led us to using standards and protocols that enable enterprises to project *entire* supply chains and to optimize them.

In this article, we present an approach for *representing, exchanging and analysing* supply chain data adhering to the SCORVoc vocabulary. SCORVoc is an RDF vocabulary implementing the SCOR standard, which bridges between the conceptual/terminological level of SCOR and the operational/executional level of IT systems. Based on the machine-processable supply chain data model and W3C standardized protocols such as SPARQL, supply chain information can be exchanged and analyzed in a distributed and flexible way. For example, the vocabulary can be easily extended with domain-specific classes or properties, thus tailoring the data mode for specialized domains. In pharmaceutical and food supply chains, for example, regulatory compliance information (such as temperature or expiration dates) plays a crucial role.

The remainder of this article is structured as follows: In Sect. 2, we present the requirements a system needs to fulfill to allow instant semantic-based supply chain analytics. Based on those requirements, we developed a system design, which is described in Sect. 3. Section 4 presents an implementation of this system including the data structure, and the client-server model application structure. Sect. 5 reports on an evaluation of the usefulness and usability of our approach by describing an application scenario and providing a demonstration of our implementation. Finally, we describe in Sect. 6 how open current market leading systems are w.r.t. accessing crucial supply chain data before concluding the paper with an outlook on future work in Sect. 7.

---

[1] https://scn.sap.com/community/developer-center/hana/blog/2014/12/10/sap-hana-sps-09-new-developer-features-rest-api,
https://blogs.oracle.com/stevenChan/entry/introducing_oracle_e_business_suite.

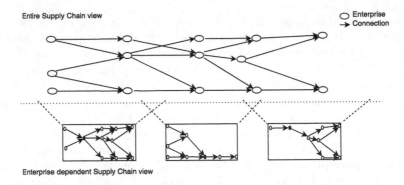

Fig. 1. Examples of individual supply chain views on specific parts of a supply network.

2   Requirements

Our requirements are driven by the *LUCID project*[2] which aims at improving the information flows between highly connected enterprises. Furthermore, a global operating manufacturer provided us with useful insights on day-to-day supply chain challenges and helped us to gain a better understanding of the SCOR. Instead of human intervention, the rationale is to automate data exchange of logistical information, order management or event propagation. Automating information exchange can, for example, help preventing the bullwhip effect, which causes supply chain inefficiencies due to excessive production line adaption as a response to business forecasts [9]. To attain the goal of an interactive, federated analysis tool for supply chain management, we first collect all requirements it needs to fulfill.

These are the requirements:

R1 Data sovereignty The major requirement is that each enterprise keeps the full sovereignty of its data. This requirement precludes systems based on external data hubs or cloud solutions.
R2 Secure data access Giving external enterprises access to one's own information system requires secure interfaces.
R3 Configurable supply chain networks Since every supply chain partner maintains a different view on the entire network (see Fig. 1, it is important that each partner can configure its own view in the system.
R4 Customizable data models: Since there is no single globally accepted process model, one needs to be able to customize and extend the data models.
R5 Customizable analysis techniques key performance indicators (KPIs) and analysis methods vary by industry, region and abstraction model.

Sharing information inter-organizationally is usually strictly governed by legal barriers. We assume in our approach that specific data access contracts have been established and thus ignore legal requirements in our technical implementation.

---

[2] Linked valUe ChaIn Data, http://www.lucid-project.org/.

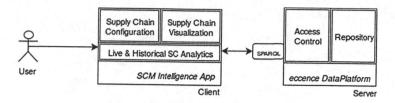

Fig. 2. SCM Intelligence App Architecture

3 Architecture

Figure 2 displays the overall architecture of our approach including the services that a server is going to provide and which the front-end takes advantage of. From the host's SPARQL endpoint, SCM Intelligence App is able to retrieve specific information relevant to the supply chain, or to compute KPI metrics, implemented as SPARQL queries, directly on the endpoint. The provider of the server thus provides access control (via standard HTTP authentication mechanisms) and a triple store. The SCM Intelligence App provides three services in total described in the sequel.

Configuration. First, the application allows each enterprise to configure their individual view on the supply chain. To achieve this, they are required to specify the endpoint and the credentials of each node in the supply chain. Furthermore, every connection within the supply chain needs to be described, that is: Which enterprise is the supplier/client of another enterprise.

Visualization. Second, once the information is specified in the configuration step, the application is able to generate a supply chain visualization. This visualization generally enables users to view supply chain information related to different aspects (e.g. supply chain throughput, health) and on different levels of granularity (e.g. tier 1 suppliers). In particular, it enables users to identify critical links and suppliers/clients.

Analytics. Third, the user can run pre- or self-defined analytic methods to compute KPIs on the entire supply chain. This main service helps to identify weak performing links which then can be either optimized or replaced. Thus, helping to keep the entire supply chain efficient and competitive.

4 Implementation

Our implementation follows the client-server model. We first introduce how the client is implemented and then describe the server including the services it provides.

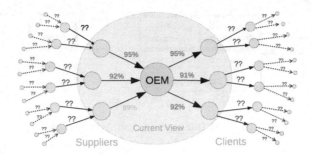

Fig. 3. Supply chain workflow example: Reliability between enterprises [12]

Data Structure. A major issue when it comes to the inter-organizational exchange of data is the vast *variety* of systems and data models among and within enterprises. Taking this into account, instead of aiming for a "one size fits all" data model, we provide a common structure enterprises can use as a basis and extend in a straightforward way.

Before defining a model, one needs to choose a meta language for expressing the model. In 1999, the World Wide Web Consortium (W3C) published RDF [3] to provide a "vendor-neutral and operating system-independent system of metadata"[3]. With the goal to facilitate knowledge sharing and exchange, it has matured over time and allows on one hand to express simple terminologies, taxonomies and vocabularies but also complex ontologies using the knowledge representation language OWL [11]. RDF and OWL thus provide us with an optimal basis such that an enterprise can adapt their supply chain models at any time to enable the exchange of specific use case descriptions among each other and to make sense of them.

Once the meta language has been chosen, the data model itself for the information planned to be exchanged needs to be specified. There are many reasons to distrust any system that claims to have a data model represents everything that multiple enterprises along a supply chain may want to exchange. Different interests in granularity, legal reasons, domain-specific requirements make finding an accepted standard hard; this is considered one of the major bottlenecks in the current Industry 4.0 initiative [2].

As a step towards a more standardized way to represent supply networks, each enterprise has to agree on how each connection is measured. Thus, one needs a process model with a clear definition for each process and how they are evaluated. The APICS Supply Chain Council[4] tackles this challenge, and elaborates a reference model named Supply Chain Operations Reference Model (SCOR) [14]. The industry-agnostic reference is backed up my many global players (including IBM, HP, and SAP) and contains 201 process definitions and 286 metrics. A metric represents a KPI that is used to measure processes.

---

[3] https://www.w3.org/Press/RDF.

[4] http://www.apics.org/sites/apics-supply-chain-council/frameworks/scor.

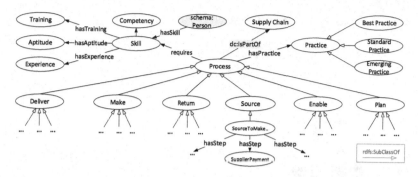

Fig. 4. Overview of the SCORVoc vocabulary (the namespace prefix schema refers to schema.org and dc to Dublin Core).

```
@prefix rdfs: <http://www.w3.org/2000/01/rdf-schema#> .
@prefix skos: <http://www.w3.org/2004/02/skos/core#> .
@prefix xsd:  <http://www.w3.org/2001/XMLSchema#> .
@prefix     : <http://purl.org/eis/vocab/scor#> .
:Enable rdfs:subClassOf :Process ;
        rdfs:comment  "Enable describes the ...";
        rdfs:label    "Enable"@en , "Permitir"@es ;
        skos:notation "E" ;
```

Listing 1. Concept definition example

Figure 3 represents a typical supply chain workflow. Each node represents an enterprise, each arrow a connection. The values besides the connection can have many dimensions: The reliability of a delivery, the costs involved, the time it takes to deliver from one place to another.

SCORVoc [12] is an RDFS vocabulary that fully formalizes the SCOR reference. It contains definitions for for the processes and KPIs ("metrics") in the form of SPARQL [7] queries. A process is a basic business activity. Listing 1 displays an example definition of a process and Fig. 4 gives on overview on the entire vocabulary.

As an example, there are multiple delivery processes: scor:SourceStock Product, scor:SourceMakeToOrderProduct and scor:SourceTo EngineerToProduct depending on whether a delivery is unloaded into the stock, used in the production lines or used for special engineered products. Each time such an activity takes place, all data that is needed to evaluate how well this process performed is captured. Such as whether the delivery was on time, it was delivered in full or if all documents were included. Each process contains at least one metric definition. Due to its depth, we chose SCORVoc as our common data basis. The vocabulary is available on GitHub[5] including definitions and localisation for each concept.

---

[5] https://github.com/vocol/scor/blob/master/scor.ttl.

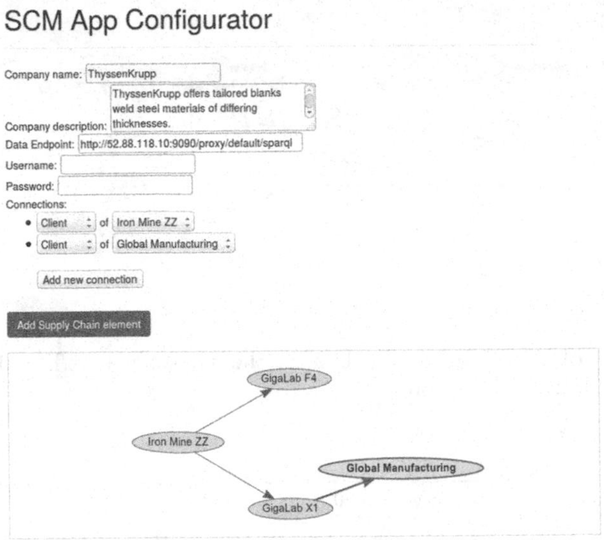

Fig. 5. SCM Intelligence App Configurator View

Client. The client is realized as a web application for two reasons: First, it prevents possible installation issues or required operating-system dependent preconfigurations. Thus, having a web browser is the only prerequisite for using our system. Second, due to the vast use of open web standards such as RDF or SPARQL, providing a web client is only consequential, considering that, without an active network connection, data exchange is impossible.

The user interface is developed by using the markup language HTML5[6] and the JavaScript libraries jQuery[7] and vis.js[8]. The entire source code is available at the repository hosting platform GitHub[9].

Figure 5 depicts the SCM configuration view for the user. Step by step the user can add each enterprise to the supply chain and thus model the entire view. For each enterprise, one needs to specify the SPARQL endpoint, the access credentials and its position in the supply chain. Once all this information has been defined, the entity is available for future supply chain analyses.

Figure 6 depicts the analysis view of the client. By specifying the timespan and analysis method (in the example: "Order Delivered in Full"), one can run the analysis on live or historical data and have strong and weak links highlighted. Thus, one can identify connections that require special attention and should therefore be acted upon or be replaced with alternative suppliers.

Predefined methods are described in [12]. Example dimensions include *Reliability*, *Responsiveness*, *Agility*, *Costs* and *Asset Management*.

---

[6] https://www.w3.org/TR/html5/.

[7] http://jquery.com/.

[8] http://visjs.org/.

[9] https://github.com/EIS-Bonn/SCMApp.

SCM App

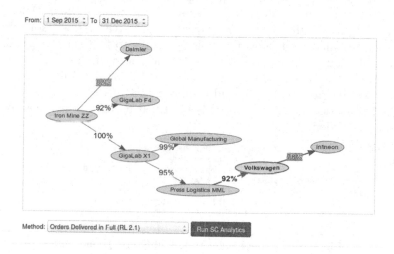

Fig. 6. SCM Intelligence App Analytics View

Server. As a server, we chose the eccenca DataPlatform[10] (basis described in [15]) because of its access control support and SPARQL endpoint.

To be more precise, it offers a triple store as a knowledge base and a SPARQL endpoint as an interface to access the data. The SPARQL endpoint is secured using the OAuth 2.0[11] framework for authentication. Authorization is based on the Authorization vocabulary[12] as described in [16]. By providing context-based access conditions (eccauth:AccessCondition), it allows providers (suppliers or clients) to decide which data to share with whom on a RDF named graph level.

However, the server may be exchanged with any other system which provides a secured data access interface. The knowledge base, a triple store, may also be exchanged as long as SPARQL queries can be executed. While in theory access control is not important for the analysis itself, in reality, for enterprises it is. Thus, if publishing supply chain information openly is an option, any open SPARQL endpoint may be sufficient.

5   Evaluation

We provide a proof-of-concept evaluation for the feasibility of our approach.

Figure 6 describes the following scenario: The raw material producer *Iron Mine ZZ* supplies the enterprises *Daimler, GigaLab F4* and *GigaLab X1*. Sub-

---

[10] https://www.eccenca.com/en/products/linked-data-suite.html.

[11] http://tools.ietf.org/html/rfc6749.

[12] https://vocab.eccenca.com/auth/.

sequently, *GigaLabX1* then supplies *Global Manufacturing* and *Press Logistics MML* and so on. Once this network is specified, one can choose the analytic method to measure the quality. In our example, we choose *Order Delivered in Full* (according to SCOR). Listing 2 represents the SPARQL query for this metric. In order to calculate this result, it is necessary to first select all deliveries executed and then to divide their count by the count of those that were not fully executed. Whether a delivery is considered full or not full is specified by the properties `hasMetricRL_33` and `hasMetricRL_50`.

In the SCOR standard, this metric is identified by *RL 2.1* and contains the following general definition:

```
@prefix  xsd: <http://www.w3.org/2001/XMLSchema#> .
@prefix     : <http://purl.org/eis/vocab/scor#> .
SELECT (xsd:decimal(?full) / (xsd:decimal(?notFull)) * 100) AS ?result
WHERE { { SELECT COUNT(?deliveredInFull) AS ?full
          WHERE { ?deliveredInFull  :hasMetricRL_33  100 .
                  ?deliveredInFull  :hasMetricRL_50  100 . } }
        { SELECT (COUNT(?allDeliveries)) AS ?notFull
          WHERE { ?allDeliveries  a  :Process . } }}
```

Listing 2. Orders Delivery in Full SPARQL metric

Percentage of orders which all of the items are received by customer in the quantities committed. The number of orders that are received by the customer in the quantities committed divided by the total orders [14]

Thus, the percentage *92 %* between *Iron Mine ZZ* and *GigaLab F4* means that *8 %* of all orders received by *GigaLab F4* were not received in the quantities originally ordered in the timespan of September 1 2015 up to December 31 2015. For the sake of better identifying problematic links, values below 90 % were highlighted with a red font color and a yellow background. We have deployed a demo[13] with multiple server instances including artificial test data.

6   Related Work

In 2014, according to a Gartner study[14], the Supply Chain Management software market reached a turnover total of \$9.9 billion (2015 data expected to be available in May 2016). The SCM software market offers a rich diversity of vendors providing systems and models to manage supply chains. However, most of the systems offer limited remote access or only recently introduced it.

We review some existing products based on their openness. In particular, we check if the products support connecting to data sources of another system without human interaction. For example, connectivity is provided by REST or SOAP [1] interfaces to build customizable applications, which make us of the underlying data.

---

[13] https://rawgit.com/EIS-Bonn/SCMApp/master/scm-app.html.
[14] http://www.gartner.com/newsroom/id/3050617.

Table 1. Openness of SCM products.

	Market share	Includes REST	Includes SOAP	Interface introduced
SAP SCM 7.0	25.8 %	+	−	2011
Oracle Business Suite	14.6 %	+	+	2014
JDA Software	4.4 %	−	−	−
Manhattan Associates	1.9 %	−	+	2010
Epicor ERP 10	1.6 %	+	+	2014

Based on a market share study by Gartner, SAP is the current market leader with 25.8 % market share, followed by Oracle and JDA Software with 14.6 % and 4.4 %. Using this information, we selected the strongest software vendors for our analysis.

Table 1 presents the results of our study. While the trend is slowly going towards more openness, even major vendors provide incomplete interface support. Usually, these systems are optimized for being connected to other products of the same vendor, as described in the technical documentation.[15]

Most of the existing research on formalizing SCOR into an ontology focuses on the benefits of having semantics clearly represented [5,13]. Some of these approaches further discuss use cases and an potential impact [10,17] of employing these ontologies or use it for simulation [5]. However, none of these works are based on the latest SCOR version (11.0), have defined executable KPIs or developed a SCOR-compliant application.

7 Conclusions and Future Work

The use of data-centric approaches in engineering, manufacturing and production are currently widely discussed topics (cf. Industry 4.0, smart manufacturing or cyber-physical systems initiatives). The complexity of supply chain management in general is considered to be one of the major bottlenecks of the field. Interoperability, decentralization, real-time capability and service orientation are four of the six design principles of the Industry 4.0 initiative [8]. Our approach tackles these challenges using a decentralized system, which provides direct access to crucial supply chain relevant information. While proprietary software systems are slowly introducing access services such as REST or SOAP, also a semantic description of the data is required to gain a common understanding and thus reduce the semantic barrier. Aligning data structures on industry standards increases interoperability, but in order to meet enterprise-specific demands, any data structure needs to allow customization without breaking the model. From a

---

[15] http://help.sap.com/saphelp/_scm700/_ehp03/helpdata/en/58/d4cc537cbf224 be10000000a174cb4/frameset.htm.

non-technical perspective, a challenge also lies in establishing trust among supply chain partners. Providing remote access to internal knowledge bases is often hindered by legal concerns and the fear of giving away one's business secrets.

For the future, we plan to support further standards in addition to SCOR, such as *GS1 EDI*[16] or *ISO/TC 184/SC 4*[17]. We plan to demonstrate the added value of integrating open data into supply chain management, such as spatial data from *LinkedGeoData*, company data from *OpenCorporates* or open product data. Furthermore, we plan to test our system in a production setting including real data.

Acknowledgments. This work has been supported by the German Ministry for Education and Research funded project LUCID, and by the European Commission under the Seventh Framework Program FP7 for grant 601043 (http://diachron-fp7.eu).

References

1. Box, D., et al.: Simple Object Access Protocol (SOAP) 1.1. W3C Note. World Wide Web Consortium (W3C), 8 May 2000. http://www.w3.org/TR/2000/NOTE-SOAP-20000508
2. Brettel, M., et al.: How virtualization, decentralization and network building change the manufacturing landscape: An Industry 4.0 Perspective. Int. J. Sci. Eng. Technol. **8**(1), 37–44 (2014)
3. Cyganiak, R., Wood, D., Lanthaler, M.: RDF 1.1 Concepts and Abstract Syntax. W3C Recommendation (W3C), 25 February (2014). http://www.w3.org/TR/2014/REC-rdf11-concepts-20140225/
4. Dalmolen, S., Moonen, H., Hillegersberg, J.: Industry-wide Inter- organizational systems and data quality: Exploratory findings of the use of GS1 standards in the Dutch retail market. In: 21st Americas Conference on Information Systems (AMCIS) (2015)
5. Fayez, M., Rabelo, L., Mollaghasemi, M.: Ontologies for supply chain simulation modeling. In: 37th Conference on Winter Simulation (2005)
6. Galasso, F., et al.: A method to select a successful interoperability solution through a simulation approach. J. Intell. Manuf. **27**(1), 217–229 (2014)
7. Harris, S., Seaborne, A.: SPARQL 1.1 Query Language. W3C Recommendation (W3C), 21 March 2013. http://www.w3.org/TR/2010/REC-sparql11-query-20130321/
8. Hermann, M., Pentek, T., Otto, B.: Design Principles for Industrie 4.0 Scenarios: A Literature Review. Working Paper 1. Dortmund, TU (2015)
9. Lee, H.L., Padmanabhan, V., Whang, S.: Information distortion in a supply chain: the bullwhip effect. Manage. Sci. **50**(12 supplement), 1875–1886 (2004)
10. Leukel, J., Kirn, S.: A supply chain management approach to logistics ontologies in information systems. In: Abramowicz, W., Fensel, D. (eds.) Business Information Systems. LNBIP, vol. 7, pp. 95–105. Springer, Heidelberg (2008)

---

[16] http://www.gs1.org/edi.
[17] http://www.iso.org/iso/standards_development/technical_committees/list_of_iso_technical_committees/iso_technical_committee.htm?commid=54158.

11. OWL Working Group. OWL 2Web Ontology Language: Document Overview(Second Edition). W3C Recommendation. (W3C), 11 December 2012. http://www.w3.org/TR/2009/REC-owl2-overview-20121211/

12. Petersen, N., et al.: SCORVoc: vocabulary-based information integration and exchange in supply networks. In: Tenth IEEE ICSC (2016)

13. Sakka, O., Millet, P.-A., Botta-Genoulaz, V.: An ontological approach for strategic alignment: a supply chain operations reference case study. Int. J. Comput. Integr. Manuf. **24**(11), 1022–1037 (2011)

14. Supply Chain Council. Supply chain operations reference model (SCOR). Version 11 (2012)

15. Tramp, S., Piris, R.N., Ermilov, T., Petersen, N., Frommhold, M., Auer, S.: Distributed linked data business communication networks: the LUCID endpoint. In: Gandon, F., Guéret, C., Villata, S., Breslin, J., Faron-Zucker, C., Zimmermann, A. (eds.) The Semantic Web: ESWC 2015 Satellite Events. LNCS, vol. 9341, pp. 154–158. Springer, Heidelberg (2015)

16. Unbehauen, J., Frommhold, M., Martin, M.: Enforcing authorization on SPARQL queries in enterprise scenarios. ESWC (submitted) (2016)

17. Zdravković, M., et al.: An approach for formalising the supply chain operations. Enterp. Inf. Syst. **5**(4), 401–421 (2011)

Author Index